Land Use Change and Mountain Biodiversity

Land Use Change and Mountain Biodiversity

EDITED BY **Eva M. Spehn,**
Maximo Liberman, and Christian Körner

Taylor & Francis
Taylor & Francis Group
Boca Raton London New York

A CRC title, part of the Taylor & Francis imprint, a member of the
Taylor & Francis Group, the academic division of T&F Informa plc.

Published in 2006 by
CRC Press
Taylor & Francis Group
6000 Broken Sound Parkway NW, Suite 300
Boca Raton, FL 33487-2742

International Standard Book Number-10: 0-8493-3523-X (Hardcover)
International Standard Book Number-13: 978-0-8493-3523-5 (Hardcover)
Library of Congress Card Number 2005047213

Library of Congress Cataloging-in-Publication Data

Land use change and mountain biodiversity / [edited by] Eva Spehn, Maximo Liberman, and Christian Körner.
   p. cm.
     Selected papers from 2 workshops, the first held in Moshi, Tanzania, Aug. 19-24, 2002 and the second held in La Paz, Bolivia, Aug. 20-23, 2003.
   Includes bibliographical references.
   ISBN 0-8493-3523-X (alk. paper)
   1. Mountain ecology--Congresses. 2. Land use--Environmental aspects--Congresses. 3. Biological diversity--Congresses. I. Körner, Christian, 1949- II. Spehn, E. M. Eva M.) III. Liberman, Máximo.

QH541.5.M65L36 2005
577.5'3--dc22
                                                          2005047213

Taylor & Francis Group
is the Academic Division of Informa plc.

Visit the Taylor & Francis Web site at
http://www.taylorandfrancis.com

and the CRC Press Web site at
http://www.crcpress.com

Preface

SUSTAINABLE USE AND BIODIVERSITY OF SUBTROPICAL AND TROPICAL HIGHLANDS

Within the worldwide biodiversity program of DIVERSITAS, the Global Mountain Biodiversity Assessment (GMBA) seeks to assess the biological richness of high-elevation biota around the world. Mountains provide an excellent opportunity for a global biodiversity research network, as they exist in every climatic zone. GMBA has a high-elevation focus, including the uppermost forest regions or their substitute rangeland vegetation, the treeline ecotone, and the alpine and the nival belts. Although acknowledging the significance of lower-montane biota, they fall outside the GMBA agenda. Beyond description, GMBA aims at explaining the causes of biological richness in mountains and its change over time. Given that changes in biodiversity most often result from human land use, one specific GMBA agenda is the assessment of land use impacts. Such assessments have priority in low-latitude regions, where land use pressure on upland biota is greatest. Upland grazing, often facilitated by fire management, is the most widespread utilization of mountain terrain, often followed by erosion and enhanced risk for valley and foreland environments. High-elevation forests have disappeared in most regions, and the few relicts are under intense use. Cultivation of formerly pristine areas and intensification of agriculture in montane areas are often associated with a loss of mountain biodiversity. Both problems are most severe in the tropics and subtropics.

This book is the second volume produced by the Global Mountain Biodiversity Assessment (GMBA) of DIVERSITAS, following *Mountain Biodiversity: A Global Assessment* (eds. Ch. Körner and E.M. Spehn), published by Parthenon in 2002. The chapters of this volume have been selected in a peer-reviewing process from the presentations offered at two GMBA workshops, one in Africa (Moshi, Tanzania, August 19 to August 24, 2002) and the other one in the Andes (La Paz, Bolivia, August 20 to August 23, 2003). More than 50 researchers actively participated, sharing knowledge from all major mountain regions, with a particular focus on the Andes and the African mountains. The two workshops profited greatly from the hospitality of the African Mountain Association (AMA), which hosted the African workshop at its sixth international conference on sustainable mountain development in Africa. We would like to cordially thank Prof. Salome Misana of the Department of Geography, University of Dar es Salaam, Tanzania, for the organization of the conference and for her local support and input during the first workshop. The second workshop in the Andes was locally organized by Maximo Liberman, SERNAP, in Huarina at the shore of Lake Titikaka in Bolivia, under the auspices of the Andean Mountain Association (AMA).

Under the patronage of, and with support from, DIVERSITAS, these workshops have been underwritten by various agencies. The workshops and the synthesis process were generously funded by the Swiss Agency for Development and Cooperation. The Swiss Federal Office for Agriculture enabled the cooperation with the Swiss Federal Research Station of Agroecology and Agriculture (Agroscope Zürich–Reckenholz) on this project. The Food and Agriculture Organization (FAO) of the United Nations supported the preparation of this publication through the FAO/Netherlands Partnership Programme "Assessment of Agricultural Biodiversity." SERNAP (Servicia Nacional de Areas Protegidas de Bolivia)/ II Bolivia supported the Spanish edition of this volume, printed in Bolivia (SERNAP, La Paz, 2005).

We wish to thank the following persons who helped with the editing of this volume: Andreas Grünig of the Swiss Federal Research Station for

Agroecology and Agriculture (Agroscope Zürich–Reckenholz) for his valuable help in the process of editing submitted manuscripts; Annemarie Brennwald, Sylvia Martinez, and Susanna Pelaez-Riedl of the Institute of Botany, University of Basel, for text editing and graphic support; Emma Sayer, who translated chapters to English, and Cecile Belpaire (La Paz, Bolivia), who translated chapters to Spanish in the Spanish edition.

Under the auspices of the Swiss Academy of Natural Sciences, the GMBA office in Basel, Switzerland (Eva Spehn and Sylvia Martinez) were supported by the Swiss Federal Office of Science and Education 2001–2003 and the Swiss National Science Foundation (SNF) (2004–).

Eva Spehn, Maximo Liberman, and Christian Körner

Basel, Switzerland and La Paz, Bolivia
January 2005

Contributors

Bhupendra Singh Adhikari
Wildlife Institute of India
Dehradun, India

Khukmatullo Akhmadov
Tajik Forestry Research and Development Institute
Dushanbe, Tajikistan

Humberto Alzérreca Angelo
Programa Estralégico de Acción para la Cunca del
  Rio Bermejo (PEA-Bolivia)
Tarija, Bolivia

Roxana Aragón
Facultad de Agronomía
Universidad de Buenos Aires
Buenes Aries, Argentina

Yoseph Assefa
Department of Biology
Addis Ababa University
Addis Ababa, Ethiopia

Jan C. Axmacher
Lehrstuhl Biogeographie
Universität Bayreuth
Bayreuth, Germany

Khadga Basnet
Central Dept. of Zoology
Tribhuvan University
Kathmandu, Nepal

Erwin Beck
Lehrstuhl fur Pflanzenphysiologie
Universität Bayreuth
Bayreuth, Germany

Siegmar Breckle
Department of Ecology
University of Bielefeld
Bielefeld, Germany

Uta Breckle
Department of Ecology
University of Bielefeld
Bielefeld, Germany

Jorge Alberto Bustamante Becerra
Department of Ecology, Biosciences Institute
University of São Paulo
São Paulo, Brazil

Julietta Carilla
Laboratorio de Investigaciones Ecológicas de las
  Yungas
Universidad Nacional de Tucumán
Tucumán, Argentina

Luciana Cristóbal
Laboratorio de Investigaciones Ecológicas de las
  Yungas
Universidad Nacional de Tucumán
Tucumán, Argentina

Terry M. Everson
School of Biological and Conservation Sciences
University of KwaZulu–Natal
Pietermaritzburg, South Africa

Masresha Fetene
Department of Biology
Addis Ababa University
Addis Ababa, Ethiopia

Konrad Fiedler
Population Ecology
Institute for Ecology and Conservation Biology
University of Vienna
Vienna, Austria

Menassie Gashaw
Ethiopian Wildlife Organization
Addis Ababa, Ethiopia

Roger B. Good
National Parks and Wildlife Service
Queanbeyan, New South Wales, Australia

Steven M. Goodman
WWF
Anatananarivo, Madagascar
and
Field Museum
Chicago, Illinois

H. Ricardo Grau
Laboratorio de Investigaciones Ecológicas de
las Yungas
Universidad Nacional de Tucumán
Tucumán, Argentina

Ken Green
National Parks and Wildlife Service Snowy
Mountains Region
Jindabyne, New South Wales, Australia

Stephan R.P. Halloy
Instituto de Ecologia
Universidad Mayode San Andrés
La Paz, Bolivia

Andreas Hemp
Department of Plant Physiology
Bayreuth, Germany

Zulimar Hernández
Instituto de Ciencias Ambientales y Ecológicas
Universidad de Los Andes
Mérida, Venezuela

Christine Huovinen
WSL, Swiss Federal Institute for Snow and
Avalanche Research SLF
Davos Switzerland

Stuart W. Johnston
School of Resources, Environment and Society
Australian National University
Canberra, Australia

Christian Körner
Institute of Botany
University of Basel
Basel, Switzerland

Michael Kreuzer
Institute of Animal Science, Animal Nutrition
Swiss Federal Institute of Technology (ETH)
Zürich, Switzerland

Jorge C. Laura
Asociación de Ganaderos en Camélidos de los
Andes Altos (AIGACAA)
El Alto de La Paz, Bolivia

Maximo Liberman
Servicio Nacional de Areas Protegidas
La Paz, Bolivia

Freddy Loza
Asociación Boliviana de Teledetección y
Mediambiente
La Paz, Bolivia

Demetrio Luna
Asociación de Ganaderos en Camélidos de los
Andes Altos
El Alto de La Paz, Bolivia

Herbert V.M. Lyaruu
Botany Department
University of Dar es Salaam
Dar es Salaam, Tanzania

Agustina Malizia
Laboratorio de Investigaciones Ecológicas de
las Yungas
Universidad Nacional de Tucumán
Tucumán, Argentina

Andrea Corinna Mayer
Swiss Federal Institute for Snow and Avalanche
Research
Davos, Switzerland

Marcelo Fernando Molinillo
Instituto de Ciencias Ambientales y Ecológicas
Universidad de Los Andes
Mérida, Venezuela

Maximina Monasterio
Instituto de Ciencias Ambientales y Ecologicas
Universidad de los Andes
Mérida, Venezuela

Mariano Morales
Departamento de Dendrocronología e Historia
  Ambiental
IANIGLA-CRICYT
Mendoza, Argentina

Craig D. Morris
Range and Forage Institute
Agricultural Research Council
Pietermaritzburg, South Africa

Klaus Müller-Hohenstein
Lehrstuhl Biogeographie
Universität Bayreuth
Bayreuth, Germany

George Nakhutsrishvili
Institute of Botany
Georgian Academy of Science
Tbilisi, Georgia

Jonny Ortega
Asociación de Ganaderos en Camélidos de los
  Andes Altos
La Paz, Bolivia

Jesus Orlando Rangel Churio
Instituto de Ciencias Naturales
Universidad Nacional de Colombia
Bogotá, Colombia

Bernardin Pascal N. Rasolonandrasana
WWF
Ambalavao, Madagascar

Gopal S. Rawat
Wildlife Institute of India
Dehradun, India

Lina Sarmiento
Instituto de Ciencias Ambientales y Ecologicas
Universidad de los Andes
Núcleo la Hechicera, Facultad de Ciencias
Mérida, Venezuela

Ludger Scheuermann
Department of Zoology
State Museum of Natural History Karlsruhe
Karlsruhe, Germany

Marion Schrumpf
Max Planck Institute for Biogeochemistry
Jena, Germany

Anton Seimon
Earth Institute at Columbia University
New York, New York

Lisa A. Simpson
CRC Freshwater Ecology
University of Canberra
Canberra, Australia

Julia K. Smith
Instituto de Ciencias
Ambiental y Ecologicas
Universidad de los Andes
Menda, Venezuela

Eva M. Spehn
Global Mountain Biodiversity Assessment
Institute of Botany
University of Basel
Basel, Switzerland

Veronika Stöckli
Swiss Federal Institute for Snow and Avalanche
  Research
Davos, Switzerland

Alfredo Tupayachi
Facultad de Ciencias Biológicas
Universidad Nacional de San Antonio Abad de
  Cuzco
Cuzco, Perú

Ricardo Villalba
Departamento de Dendrocronología e Historia
  Ambiental
Mendoza, Argentina

Karsten Wesche
Institute of Geobotany and Botanical Garden
University of Halle-Wittenerg
Halle, Germany

Zerihun Woldu
Department of Biology
Addis Ababa University
Addis Ababa, Ethiopia

Walter Wucherer
Department of Ecology
University of Bielefeld
Bielefeld, Germany

Karina Yager
Department of Anthropology
Yale University
New Haven, Connecticut

Table of Contents

Part I

Introduction

1 High-Elevation Land Use, Biodiversity, and Ecosystem Functioning

Christian Körner, Gia Nakhutsrishvili, and Eva Spehn

ANTHROPOGENIC HIGHLAND ECOSYSTEMS

Humans have shaped much of the world's highlands over millennia. Landscapes of sustainable productivity, high biodiversity, and aesthetic attractiveness have developed through livestock grazing. These landscapes also exhibit high ecosystem stability, a key requisite for erosion control and catchment quality (Körner, 2000, 2004; Figure 1.1).

As a cultural heritage associated with traditional-knowledge-based land management, many of these high-elevation pasture landscapes, hayfields, marginal crop fields, and rangelands are of significant conservational and historical value. In some parts of the world, however, highland management had no tradition (e.g., New Zealand and Australia), and when abruptly introduced to an unadapted flora, often had disastrous consequences (e.g. Costin 1958).

Over the last 50 years, these anthropogenic highland biota have undergone dramatic changes associated with even more dramatic societal and economic changes, in addition to the atmospheric (climatic) changes underway. In the more wealthy parts of the world, much of the highlands have undergone extensivation of use or abandonment. In the less economically privileged parts, population growth and land use pressure have often caused an expansion of agricultural land use into less suitable regions and abandonment of traditional land use practices. Both of these facets of global change have had drastic influences on highland integrity and biodiversity. Unfortunately, both these departures from the traditional middle ground of sustainable land use have caused a loss of biological richness, and both tend to incur land degradation, though this is only a transitory risk in the case of abandonment (e.g. Tasser et al. 2003) but is often terminal in the case of overusing when soils are washed away.

In this overview of the ecological dimensions of highland grazing, we will follow the simple and common biogeographic nomenclature of elevational belts. We will use the altitudinal position of the natural upper-climatic *treeline*, defined as the line connecting the uppermost pockets of trees (i.e. below the tree species line but above the forest line; Körner, 2003), as a reference (irrespective of whether such forest patches are locally present or not). We will define the mountain slopes below as *montane* and the naturally treeless land above as *alpine*. In this sense "alpine" does not refer to the Alps but applies globally (following from its preIndo-Germanic meaning of "steep slopes"), with "Andean" and "Afroalpine" as synonyms. The climatic high-elevation treeline correlates worldwide with a seasonal mean temperature of $6.7 \pm 0.8°C$ (independent of season length; Körner and Paulsen, 2004). Somewhat lower threshold temperatures (5 to 6°C) can be found at the equator (treelines at 3800 to 4100 m), but the thresholds in the subtropics match with those at higher latitudes. In the humid and semihumid tropics and subtropics, much of the high-elevation pastureland is found between 500 and 800 m below and between 300 and 400 m above the treeline elevation (i.e. between 3000 and 4400 m), with lower elevations commonly used for crop production and higher elevations commonly carrying too little vegetation and not regularly grazed. In the Northern Hemisphere temperate zone, with treeline positions varying widely between 1500 and 3500 m

3

FIGURE 1.1 Fingerprints of millennia of land use in the highlands. Examples of anthropogenic grassland from (a) Bolivian altiplano, 4100 m; (b) Cayambe region, 3700 m, Ecuador; (c) Sajama region bofedales, Bolivia, 4100 m; and (d) Spiti Valley, the Himalayas, 3700 m, India.

depending on latitude and oceanic climate, the corresponding amplitudes are wider, namely, from at least 1000 m (1500 m in continental ranges) below to 400 m above treeline. These are the elevational ranges considered here and in the remainder of this volume when the term *highland* is used. As the focal elevations of GMBA are the upper-montane, treeline ecotone, and alpine belts, most of the contributions refer to these higher parts of what could be considered highlands in the widest sense.

According to an assessment by Kapos et al. (2000; cf. Körner, 2004), the global land area above 1000 m and below 4500 m represents 14.3% of the terrestrial area. Given that (1) in the subtropics and tropics, much of the lower part of this topography-based assessment falls outside the climatic range of interest here, and that (2) a great fraction of mountains falls in the largely bare polar and subpolar regions, a realistic estimate of the global land area fraction suitable for agricultural use in the highlands will be somewhere around 8%, with 3% falling in the alpine belt (Körner 1995), and the remaining (around 5%) in the montane belt. About 25% of the montane land area is still forested according to Kapos et al. (2000), and a similar area may

be arid or barren, so that the nonforested, potentially grazed montane and alpine highlands will cover roughly 5% of the global terrestrial area, an area as large as the polar tundra region (Körner, 1995). Approximately half of this area lies in the tropics and subtropics.

As small as this area may look on a global scale, it covers a very critical mountain zone. It has been estimated that nearly half of humanity depends directly or indirectly on the water yield from mountain catchments (Messerli and Ives, 1997; Messerli, 2004), with the vegetation-covered upper-catchment regions playing a key role for clean and steady discharge. In this sense, highlands control much of the so-called water towers of the globe, and the functional integrity of these highlands matters for land areas (and populations) by far exceeding their actual size (Figure 1.2). The slopes of these catchments are only as stable as their green cover. This cover needs a high functional diversity of plants to fulfill its protective role under all sorts of unpredictable environmental conditions. Thus the ecology and richness of highland biota are intimately linked to the welfare of a large fraction of human population, beyond their significance for local livelihoods (Körner and Spehn, 2002;

FIGURE 1.2 Nearly half of mankind depends in one way or the other on mountain water. Highland vegetation is the safeguard of catchment quality and yield. It cleans, stores, and channels water to the lowlands. Land use in these regions has far-ranging economic consequences. From top to bottom: upper catchment, Bolivia, 4000 m; montane transgression, Sichuan, west China; irrigation canal, lowland California.

Körner, 2004). Given that much of the tropical and subtropical mountain forelands is rather dry, this interdependency is even larger at low-latitude regions. The teleconnection between highland grazing grounds and metropolitan areas may be thousands of kilometers as, for instance, exists between the upper-Nile catchments and Cairo or between eastern Anatolia and what was Mesopotamia. Sustainable highland management, thus, has significant economic impact on people living far outside the mountains.

In this introductory chapter, we (1) aim to summarize a few general principles that govern the functioning of highland biota with special reference to low latitudes, (2) will provide a brief summary of previous observations on highland pasture systems, and (3) will, then, open the arena for the global change implications for biodiversity and ecosystem functioning in subtropical and tropical highlands, the main theme of this volume.

DRIVERS OF HIGHLAND ECOLOGY (WITH SPECIAL REFERENCE TO THE TROPICS AND SUBTROPICS)

The following is a brief reconsideration of the major forces that shape upland biota. These fall into topography-related and climatic drivers and biological determinants.

Compression of climatic zones. Mountains are inhabited by more species of plants, animals, and microbes as one would estimate from their land area and have often been called "hot spots" of biodiversity (Körner, 2004). This has several reasons intrinsically linked to topography and gravity. Due to the elevational range covered, mountains encapsulate several climatic life zones that would otherwise be separated by thousands of kilometers at low elevation (Barthlott et al. 1996). Hence, nowhere else on land can more biological richness be encountered on a 100-km^2 scale than on the slopes of a high tropical mountain. In relative terms, this effect also holds for mountains in extratropical regions.

Habitat diversity. The second important factor at smaller scales is topographic diversity. Exposure, steepness of slope, variation of substrate, and microclimate over short distances create a multitude of microhabitats, each nesting a different set of organisms. This habitat diversity again permits aggregation of rather diverse biota over otherwise short distances. Gravity is the primary force behind this geodiversity; where it lacks action, as in plains, irrespective of elevation, biological diversity declines. Because preferred grazing grounds are often flat and smooth, their biological inventory is commonly smaller than is found on the surrounding slopes. However, the species pool in such plains could be even lower without grazing because grazing often creates "structure" by patchy disturbance, dung deposition, food preference, etc. (Edwards et al. 2004). Such effects of inclination on biodiversity can even be seen at very large scales. The total flora of the arctic tundra of Eurasia and North America (much of it flat terrain) contains about 1000 species of flowering plants alone, a number found in the thermally similar alpine flora of the Caucasus Mountains, or the Alps plus the Pyrenees. Therefore, it is important to keep in mind that steep slopes and the gravitational forces that shape them are key to biodiversity but, at the same time, these are the most fragile parts of the high-elevation landscapes.

Microclimate. Tied to habitat diversity are climatic forces, which strongly differ from what meteorological stations report. Above the treeline, slope exposure and shelter are more important for the daytime climate that organisms experience than absolute elevation (Körner, 2003). What is even more important is that plants manipulate the microclimate. The stature and density of plants have a major influence on the climate that they experience. Grazing animals may change this structure and, hence, the effective climate, and humans may interfere by cutting, weeding, or burning, and by the grazing regime that they permit. Low-stature plants may experience outstanding high temperatures that would never be predicted from meteorological station data (Körner et al. 1983, for New Guinea; Diemer 1997, for the Ecuadorian Andes; Hofstede et al. 1995a). The reason why there are no trees above a certain elevation is not that trees have a poorer physiology than low-stature vegetation. It is only because of their architecture that trees cannot trap the needed warmth for growth, once saplings emerge from the protective grass or shrub layer above treeline. The management of highland vegetation always incurs a manipulation of microclimatic conditions on which plants, their microfauna, and microbial partners depend more and more, the higher the elevation (Figure 1.3).

Soils and slopes. The third of the topography-related drivers is the soil. Soils with their biota not only store and recycle nutrients, they also hold the moisture for dry periods and provide mechanical hold for roots; it is obvious that their depth and structure depends on topography and age. It has often been claimed that the mechanical strength of roots is weakened under grazing, but a broad literature survey does not support this (Milchunas and Lauenroth, 1993). On slopes, the overarching formula is

rather trivial: without soil, there is no vegetation, but without vegetation, there is also no soil. Only vegetation can secure the soil against gravity, and once the soil is eroded, the ecosystem is gone. Soil integrity, thus, is the number one driver of highland biota, but as with microclimate, vegetation is the key factor in soil preservation. From a chronological perspective, vegetation and soils developed jointly. As plants secured the initial substrate, fines and humus could accumulate, and vegetation succeeded into more mature stages, tying up more humus, and so on. Any land use regime is to be measured by how it interferes with this mutualistic system, and whether it permits a new and rapid succession once the system has been disturbed and reverted to initial stages, with fragile and loose substrate and open ground.

Water. Tropical and subtropical rangelands at high elevation are rather dry in many parts of Africa and South America (but not Southeast Asia, West Africa, and northwestern South America), because they are above the regional advection or condensation layer and thus receive much less moisture. Often there is a drastic decline in precipitation from a midmontane maximum (cloud forest climate) to a semiarid situation in the highest ranges and plateaus. However, whenever studied, the individual plants in these high-elevation drylands have not been found drought stressed (Geyger 1985, references in Körner 2003). This paradox finds a simple solution if one accounts for ground cover. It appears that ground cover (leaf area index, LAI) is controlled by unknown mechanisms in such a way that the transpiring leaf area per unit ground area matches the available ground moisture, and hence declines with declining precipitation. This has serious implications for land cover management. There may not be enough moisture to permit full ground cover year round. This is where the interplay between often dominant perennial grasses (mostly tussock grasses) and intertussock-space vegetation comes into play. It is also worth noting that the same LAI can be packed in few tall structures, leaving most of the ground unprotected or in low-stature structures spread over much of the surface, thus protecting it. A combination of moderate grazing and burning tends to favor tussock-type morphologies, whereas heavy grazing with or without fire was found to diminish the tussock contribution to

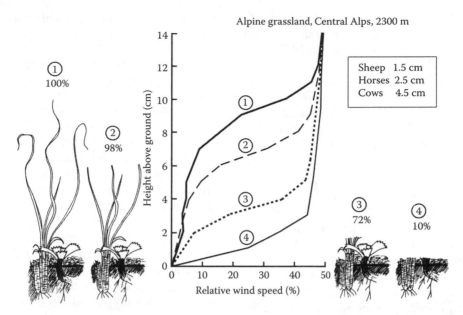

FIGURE 1.3 Examples of how grazing affects the microclimate in alpine grassland, in this case illustrated by the wind regime after partial removal of biomass in *Carex curvula* mini-tussocks in the Alps at 2300 m. The inserted box gives average grazing depth of different domestic animals on a uniform turf of grass. Note the dramatic effect of only minor removals of dead (last season) leaf ends, as happens as a consequence of light grazing.

biomass in a site comparison in Colombia (Hofstede, 1995b). In climates with very high humidity and high frequency of clouds or fog, long-leaved tussock grasses intercept significant amounts of moisture not found in rain collectors and thus have a profound influence on the water balance (Mark 1994 and references therein).

Nutrients. Nutrient availability is tied to topography, water availability, and soil age, as anywhere else in the world, but topography obviously plays a more important role in mountains, with a steady, physical translocation of nutrients from source areas (convex topography) to sink areas (concave topography). Old successional stages of vegetation on slightly inclined ground have commonly arrived at a steady-state nutrient capital that is recycled. These systems may be self-sufficient in N, but depleted in P. Young systems and those on active slopes depend on a new input of N, but may tap sufficient P from fresh mineralization. Hence, successional stage and slope play key roles. Animals recycle and retranslocate nutrients and can "engineer" a new nutrient landscape, in which a small fraction of the land may be sinks (dung deposits), and large fractions are sources for nutrients (intake of forage). This depends much on animal type and animal behavior. At sustain-

able stocking rates, cattle (similarly in yak or camels) commonly dump dung on around 2% of the landscape per year (Körner, 2000; Edwards et al. 2004), whereas sheep, goats, llamas, vicuñas, and guanacos spread dung over wider areas. A resume of grazing consequences for alpine biodiversity arrived at the conclusion that dung deposition is more influential than biomass removal (grazing) per se (Erschbamer et al. 2003). High-elevation plants are commonly well supplied with nutrients, possibly because they grow and use nutrients in a way that permits high-tissue concentrations ("luxurious" consumption, cf. Chapin et al. 1986; Körner, 1989, 2003; Bowman, 1994). However, the abundance of species that produce low N leaves wherever they occur is often higher in high-elevation grasslands (all long-lived, rigid leaves), in part, perhaps, as a response to grazing pressure. Animal grazing can influence forage quality in tussock grasslands (e.g. Mark, 1994; Chaneton et al. 1996), although this is not necessarily the case with natural plant herbivory (examples for cold climates in Bliss et al. 1981; Jonasson et al. 1986). There is also a clear trend of nutrient depletion in alpine plant leaves as season length increases from polar to tropical latitudes (Körner 1989). Even small amounts of nutrient (N) addition to alpine grasslands have been found to stimulate growth significantly, but responses depend on moisture availability and plant type (Bowman et al. 1995). A 4-year addition of 40 kg N ha^{-1} a^{-1} in the Alps doubled biomass (Körner et al. 1997). However, even a 15 kg N ha^{-1} a^{-1} addition produced a significant stimulation in the year of application (E. Hiltbrunner, unpublished), and 100 kg ha^{-1} a^{-1} can convert a glacier forefield into a hay meadow (Heer and Körner 2002).

Biomass and productivity. Provided moisture permits, primary production at high elevation is, in large part, a matter of time. When green leaves are present, neither photosynthesis nor growth is commonly restricted during the day when the canopy warms up. However, the time to grow may be limited by the length of the active season, either by periods with subfreezing temperatures or periodic drought, or their combination. On a 24-h scale, the formation of new tissue may be limited by subzero temperatures during the night and otherwise warm daytime temperatures. When rated by the length of the growing season, biomass accumulation in the temperate alpine zone (over 2 to 3 months) has been found indifferent from humid lowland tropical productivity (during 12 months; Körner 2003). Tropical alpine productivity has not been studied to date, but if one takes the peak biomass (Table 1.1) as a surrogate for annual production, the ca. 750 g m represents little more than half of mean humid tropical lowland productivity, despite a common 12-month season. Regular low nighttime and early-morning temperatures may effectively reduce the growth period to half a year or less, and ground cover may be reduced by needle ice formation, water shortage, or overgrazing. Because there is no structural growth below 0°C (and hardly any below 5°C) but positive net photosynthesis of leaves down to 5°C and at least a third of maximum carbon uptake at 5°C, the investment of carbon in new structures will always be more restricted by low temperature than its acquisition. It is important to bear in mind that night-time temperatures at the level of grass leaf meristems (which is several centimeters below the ground) is co-determined by the density and insulation of the ground cover. At the same time, a dense ground cover reduces daytime soil heat flux. Hence, there is a delicate balance between the two effects of ground cover, which can be dipped by grazing pressure.

GRAZING THE HIGHLANDS: THE TWO SIDES OF THE COIN

We may look at grazing as something that may harm or help, depending on the dose, i.e. its rate and duration. All natural vegetation is grazed or browsed, and plant–animal interactions commonly shape vegetation as we see it. Biomass removal can induce compensatory growth (e.g. McNaughton, 1983; Trlica and Rittenhouse, 1993; McIntire and Hik, 2002), i.e. an overall increase in productivity, although this has not been studied in highlands and is questioned by some authors (Belsky 1986). However, it also opens niches to plant species that were otherwise suppressed. Moderate grazing tends to reduce dominance of a few species and to open space for many minor species. Many years of trampling may also terrace slopes, which commonly reduces surface runoff and erosion. It also increases habitat niches for certain species and accelerates the nutrient cycle.

In the Alps, moderate grazing by cattle and sheep at alpine elevations, several hundred meters above treeline, commonly exerts no destructive impact on vegetation. In fact, fencing cattle out from what is believed to be natural pristine alpine grasslands for 6 years during the 6-week grazing season (out of the 10- to 12-week total growing season) leads to a 16% reduction in standing-crop biomass at peak season and a reduction of the contribution of minor species to biodiversity. The negative effect of the prevention of grazing was even visible, with the area inside the fence appearing less lush and with significantly fewer flowers (Körner, 2000). A very similar observation was made by Pucheta et al. (1998) in montane grasslands of central Argentina, where the decline in species richness after fencing was evident after 4 years and continued over the 15-year observation period. At a consumption of 35%, the biomass productivity was not affected. Sundriyal (1992) reports 32% consumption of biomass along the southern slopes of the Himalayas with no harm to vegetation. Even in Arctic sedge communities, grazing was found to be stimulating (Henry and Svoboda 1994). Dense- and short-grazed grass mats also

TABLE 1.1
Standing crop (aboveground) life biomass in subtropical and tropical tussock grasslands in Upper Montane (close to potential treeline) or above elevations, and comparative numbers for New Zealand "snow tussock" and Northern Hemisphere temperate alpine grasslands (from various sources)[*]

Sampling Region (n Locations)	Latitude	Altitude (m)	Biomass (g m^{-2}) Min (mean) Max	Source[*]
Tussock grassland				
Colombia	5° N	3300–3400	314 () 1854	a
(3)		3620–3670	603 (978) 1374	a
(4)		3950–4100	440 (720) 860[**]	c
New Guinea (3)	6° S	3400–4350	490 (606) 722[***]	a
New Zealand and subantarctic islands	44° S,	30–1260	363 (732) 918	a
(4)	52–54° S			
Nontussock grassland				
Venezuela (5)	9° N	3530–4700	149 (273) 427	d
Indian Himalayas (8)	30° N	3100–4200	90 (223) 402	e, f, g
Various temperate alpine grasslands (9)	40–47° N	2000–3650	150 (250) 470	b

[*](a), (b) compilations in Hofstede et al. 1995 (Colombia) and Körner 2003 (New Guinea, Alps, Caucasus, Rocky Mountains, New Zealand); original data by (c) Hofstede et al. 1995b; (d) Smith and Klinger 1985; (e) Sundriyal 1992 (two ungrazed plots); (f) Ram 1992 (one unclipped site); (g) Rikhardi et al. 1992 (five sites).

[**] These numbers exclude the trunks of giant rosettes (Espeletia sp.) but include all other life mass of Espeletia such as green leaves, flowers, meristems, and a mean of 44 g m^{-2} for small shrubs and cryptograms.

[***] Means under parentheses for the given min + max only, all other means for replicate sites/sampling areas.

create a surprisingly warm microclimate (Cernusca and Seeber, 1981), on which many species depend (Kikvidze and Nakhutsrishvili, 1998; Callaway et al. 2002).

On the other hand, heavy grazing on highly weathered soils is well known to compact the soil, reduce infiltration, increase runoff, and increase erosion and sediment yield (e.g. Trimble and Mendel, 1995; Heitschmidt and Stuth, 1991), and often it depauperates vegetation, with many examples from around the globe (e.g. Mahaney and Linyuan, in 1991 for northwest China at 2800 to 3300 m, with overgrazing by yaks and horses; Körner 2000). On poorly weathered, coarse, and young (often volcanic) substrate, overgrazing prevents soil stabilization and the establishment of a protective plant cover, a frequent situation in tropical highlands. Heavy grazing may massively reduce highland productivity, as was shown by Taddesse et al. (2003) for lower montane pastures in Ethiopia, which lost two thirds of their productivity unless receiving extra manure. In this respect, highlands are no exception from lowlands, where these overgrazing effects have been studied much more widely (e.g. Tongway and Ludwig, 2000).

When uncontrolled herds are allowed into the pristine upper-alpine area, there is danger of a negative outcome, simply because the ground cover is not complete and the substrate is unstable. Furthermore, grazing tends to remove the reproductive parts first (an estimate of 80% loss of seed for the Rocky Mountains by Galen 1990). However, a comparative long-term test of mown vs. grazed upper-montane grassland revealed a big surprise. Although, indeed, more than 80% of the seed mass was removed by grazing, the density of seedlings or juvenile plants was >80% higher in grazed compared to mown grassland (Figure 1.4). It appears that grazing by far overcompensated the loss in diaspores by facilitating recruitment, most likely by opening regeneration niches and by mechanical disturbance of the ground.

In the long run, regularly mown meadows (where hay-making is still practiced in seasonal upper-montane grasslands) lead to a greater diversity of dicotyledonous herbaceous species, mostly rosette-forming species, whereas graminoids are not affected by either treatment (Figure 1.5 and Figure 1.6). Of course, this is of less relevance for nonseasonal climates, when no fodder reserves are needed.

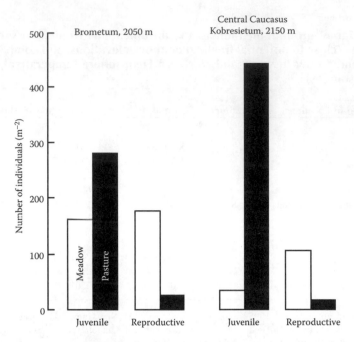

FIGURE 1.4 Although grazing removes most of the reproductive investments of plants as compared to a high-elevation hayfield, the chances for recruitment are far higher in a pasture, as exemplified here for a long-term fencing trial in the Central Caucasus at 2050 m near Kasbegi.

FIGURE 1.5 One of the oldest test sites of mowing-vs.-grazing effects on high-elevation grasslands (here at 2050 m elevation) in the Central Caucasus near Kasbegi, with the Kasbek summit (5047 m) in the back. Details illustrate the grazing effect on the height and density of plant cover.

Plants undergo characteristic adjustments of their stature when being grazed instead of being mown or remaining untouched (Diaz et al. 1992; Nakhutsrishvili, 1999). Some of these adjustments may even be ecotypic. Figure 1.7 illustrates the changes seen in the same species, but sampled from adjacent meadow or pasture habitats. Grazing leads to stunted stature, flat leaf position, fewer and smaller leaves (Table 1.2) and, as a consequence, to a lower leaf area index (LAI). These adjustments resemble sun-vs.-shade modifications and reflect the reduced mutual shading within the canopy when much of the biomass is removed by regular grazing. Metabolism and gas exchange are far more intense in the foliage of such pastures (Körner and Nakhutsrishvili, 1987) and so is light consumption per unit of foliage area. There is little

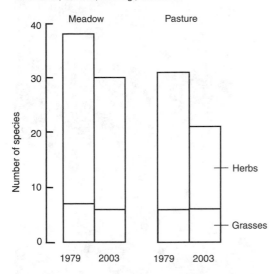

Brometum, 2050 m, Kasbegi, Caucasus

FIGURE 1.6 The effect of grazing vs. mowing on plant species diversity, 9 and 33 years after fencing out sheep at the site shown in Figure 1.5.

information from highlands for below-ground responses to grazing. Hofstede and Rossenaar (1995) found no difference in Colombian páramo grasslands that was either ungrazed or grazed in combination with fire (about 1.2 kg of roots per m²). However, a site with very heavy

grazing without burning showed a significantly higher root mass of 2.1 kg m², a remarkable effect in light of the importance of belowground structures for soil stabilization.

The removal of leaf area by grazing or mowing has one important secondary consequence. When it reduces LAI below about 2, evapotranspiration becomes reduced. In two independent tests with weighing lysimeters in alpine (Austrian Alps) and upper-montane (Central Caucasus) grassland as shown in Figure 1.8, the water loss was reduced by about 10% despite the better aerodynamic coupling of vegetation (Figure 1.3) and intensified transpiration per unit leaf area. Reduced evapotranspiration at equal precipitation increases runoff and catchment yield. It had been estimated that a short-grazed, intact, alpine pasture could add to the ungrazed reference an equivalent of water and electric energy that corresponds to a value of about \$150 ha⁻¹ yr⁻¹ at a 2-km difference in altitude (Körner et al. 1989). It was also shown that land abandonment could transitorily reduce evapotranspiration from high-elevation grasslands (Tappeiner and Cernusca, 1998). Costin (1958) demonstrated that inappropriate alpine heathland grazing that induces a net loss of only 2 to 5% of catchment value repre-

TABLE 1.2
Morphological and biomass differences in a selection of important grassland species from untouched and grazed grasslands in Upper Montane, Central Caucasus

Species	Leaf Area (cm²)			Leaf Number			Plant Height (cm)			Aboveground Biomass (g)		
	M	P		M	P		M	P		M	P	
Ranunculus oreophilus	4.5	2.2	(51)	5.6	3.7	(34)	20.2	5.2	(74)	0.165	0.022	(87)
Leontodon hispidus	6.6	4.7	(29)	10.0	9.1	(9)	20.6	9.6	(53)	0.400	0.083	(79)
Veronica gentianoides	2.2	1.4	(36)	12.3	9.6	(22)	22.5	10.6	(53)	0.162	0.057	(65)
Plantago caucasica	6.9	3.2	(54)	34.2	21.1	(38)	12.8	6.4	(50)	1.175	0.270	(77)
Potentilla crantzii	5.0	1.3	(74)	23.8	11.3	(53)	7.6	3.9	(49)	0.195	0.022	(89)
Alchemilla sericata	9.0	3.6	(60)	12.5	12.1	(3)	9.4	4.4	(53)	0.412	0.127	(69)
All species mean	5.7	2.7	(52)ᵃ	16.4	11.2	(32)	15.5	6.7	(57)	0.418	0.097	(77)
±S.D.	2.3	1.3		10.6	5.7		6.4	2.8		0.388	0.094	

Note: Either growing in a 30-cm-tall, fenced meadow (M) or an adjacent 3-cm-high pasture (P) turf under regular sheep grazing, as shown in Figure 1.5 (in brackets, the percentage difference of P vs. M).

ᵃMean percentage difference calculated from all-species mean; M vs. P differences are significant at $p < 0.01$, except for leaf number ($p = 0.08$).

FIGURE 1.7 Differences in morphology and size of plants in grassland species from either mown or grazed sites (location as in Figure 1.5).

FIGURE 1.8 Weighing lysimeters as successfully used in high-elevation grasslands in the humid temperate zone. (cf. Körner et al. 1989; Körner and Nakhutsrishvili 1987) Regular summer rains ensure that plants do not depend on moisture from deeper than the lysimeter soils. About 90% of the roots are commonly found in the top 15 cm. Grazed vegetation consumes about 10% less water. (a) Central Alps, 2300 m (Hohe Tauern National Park, Austria), (b) at the site shown in Figure 1.5.

sents an economic loss greater than the economic gain from the associated animal husbandry. These four examples illustrate that grazing regimes can have a profound influence (both negative and positive) on regional hydrology. The economic disadvantages easily exceed the immediate land use benefit. The advantages could add to the land manager's profit if acknowledged by those who benefit from it.

THE CHALLENGE OF STUDYING HIGH-ALTITUDE RANGELANDS IN THE TROPICS AND SUBTROPICS

The Nature and Cause of Tussock Grass Dominance

Most temperate high-elevation grasslands differ from the dominant forms of grassland in tropical and subtropical highlands. A major difference is the growth form of the dominant grasses

and their leaf longevity (Figure 1.9 and Figure 1.10). Given the long or even year-round season length at low latitudes, there are no constraints to leaf longevity, and there is a selective advantage with respect to grazing resistance by rigid, sclerenchymatous, poor-quality, long-lived leaves of sizes sometimes exceeding 1 m (Cabrera, 1968). The close relationship between leaf longevity, high sclerophylly, and low-nutrient concentration is well established (Chapin, 1980; Reich et al., 1992). With their shoots forming solid tussocks, these grasses are mechanically extremely robust, escape trampling, and do well under fire, because meristems are protected by a tunica of stumps, litter, and substrate. There is a rich literature describing the floristics and biology of such high-elevation tropical tussock grasslands and heathlands (e.g., Hedberg, 1964; Vareschi, 1970;

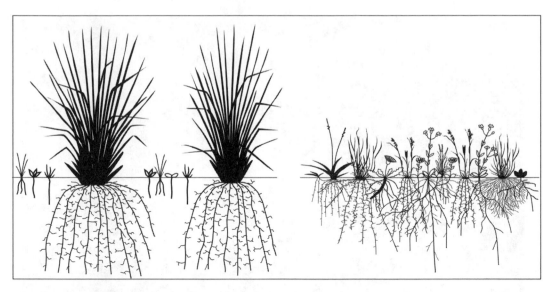

FIGURE 1.9 A schematic representation of typical temperate and cool subtropical zone mat-forming alpine grassland as it may be found anywhere in Eurasia or North America from 30° to 70°N (but also at corresponding latitudes in the Southern Hemisphere) as compared to the warm subtropical and tropical highlands, as well as in the oceanic temperate south, where tall tussock grasses dominate.

Smith, 1977; Cleef et al. 1983; Vuilleumier and Monasterio, 1986; Balslev and Luteyn, 1992; Rundel et al. 1994; Miehe and Miehe, 1994; Safford, 1999).

Water shortage does not seem to be a major selective driver of the dominance of the tall tussock growth form, because it is also abundant in the wet tropics, as for instance in Papua, New Guinea (Hnatiuk, 1978). Given the dominance of very similarly structured tussock grasslands in oceanic New Zealand and on the subantarctic islands, one cannot escape the conclusion that short season, sharp frost, and perhaps long snowpack exclude this leaf life history, giving way to low-stature vegetation, with short-lived, aboveground leaf parts as they dominate the temperate mountain grasslands and arctic grasslands (Hnatiuk, 1978; Mark et al. 2000). The absence of tall tussocks also means that there is even more mechanical impact on vegetation and no channeling of trampling trails. There are certain conditions in the tropics in which short-stature grasslands do develop, as for instance in the humid bofedales of the Andean altiplano. In other cases, shrubs take over without land use, and tall tussocks dominate at intermediate grazing and burning intensities but nearly disappear at very frequent burning (Suarez and Medina, 2001). For

the Colombian páramos, Hofstede et al. (1995b) consider fire as the single-most important selective force that induces the transgression from grassland mats to tussock dominance, but they also noted that the heaviest forms of land use diminished abundance of tall tussock grasses.

Although it is not questioned for the temperate zone that montane pastures are man-made substitutes for forest, the situation is less clear in the tropics. In certain areas of the Andes, continuous *Polylepis* forests, which used to be widespread up to an altitude of about 4500 m, were destroyed by felling, fire, and grazing, and were replaced by grasslands or heathlands. However, large areas of the puna, jalca, and páramo below the upper boundary of tree-growth in the South American Andes seem to be naturally treeless due either to aridity or regular natural fires (Lauer et al. 2001). In the Colombian páramos, the biological richness and endemism (Vuilleumier and Monasterio 1986), as well as genetic age of specialist taxa in the treeless páramos suggest a pristine nature (Cleef et al. 1983; van der Hammen and Cleef, 1986). The influence of fire is very obvious in African high mountains, where the treeline is depressed substantially below its climatic high-elevation limit (Wesche et al. 2000; Hemp and Beck, 2001).

FIGURE 1.10 High-elevation tussock grassland as found in tropical mountains but also in some temperate regions, as for instance in New Zealand and New Guinea. (a) Bolivia (Sajama region), 4150 m; (b) Mexico (Pico di Orizaba), 4050 m; (c) Ecuador (Páramos de La Virgen), 4000 m; (d) tussock-shrubland, Tanzania (Mt. Kilimanjaro), 3900 m; (e) Papua New Guinea (Mt. Willhelm), 4420 m; (f) New Zealand, southern Alps (Mt. Brewster), 1100 m.

BIOMASS AND PRODUCTIVITY

Several research teams have examined the biomass storage in these high-elevation grasslands (Table 1.1). Although there is much variation, the green (life) aboveground part of the dry matter per unit land area is often around 750 g m^{-2} (400 to 1300 g m^{-2} may be a range commonly found, disregarding extremes), which is three times the amount found in alpine grasslands of the temperate zone. From Table 1.1, it can further be concluded that high-elevation

tussock grasslands store on average about three times more life biomass than mat-forming low-stature grasslands, irrespective of latitude. This may simply reflect a three-times-greater leaf longevity in vegetation that includes tall tussock grasses, a field to be explored. Given the common lower palatability of tussock grass leaves (low specific leaf area, low N concentration), leaf functional traits theory would predict this (Reich et al. 1992).

Whereas there is a wealth of biomass data, the productivity of tropical and subtropical grassland has never been explored. This has to do with the great difficulty of assessing growth rates in a close-to-nonseasonal climate with long-lived tillers and leaves. Biomass (by definition, the life part) often composes only 20% of total aboveground phytomass (which includes dead parts), so great fuel loads may accumulate, which facilitate burning during dry periods. Hofstede et al. (1995b) made an interesting observation in the fairly humid Colombian páramos, namely that fire and grazing reduce the amount of litter and attached dead structures, but the life biomass examined at one point in time in areas of contrasting management history remained fairly unaffected. This leads to a key question that awaits careful analysis: How often is biomass recycled during a year?

Take a standing green crop of 600 g m^{-2}. This could, for example, represent the accumulated biomass of 2-years (very long-lived tillers and/or leaves), or it could be the steady-state, mean crop through which three tiller/leaf generations had "cycled" per year. The aboveground net primary production per year could hence be 300 or 1800 g m^2, i.e. it could vary sixfold. Because cutting affects regrowth, the problem cannot be solved by regular harvests. The only feasible procedure is a study of tiller dynamics, of birth and death of tillers, as was done by Diemer (1998) for small herbaceous species in the Ecuadorian páramo at 4000-m elevation. Depending on species, he found a mean leaf duration of 3 to 22 months in the herbaceous ground cover between tussocks. Therefore, there is a wide spectrum of possibilities, not permitting any prediction. Cutting treatments would add an interesting applied facet to such a study of tiller dynamics (the issue of compensatory growth potential) but are no substitute to the demographic approach. The only data for tall tussock grass leaf longevity come from New Zealand and range from 2.6 to 3.2 year (Meurk, 1978). From circumstantial evidence, Hnatiuk (1978) arrives at life spans somewhere between 7 and 16 months for tussocks near the treeline in New Guinea, so a year may be a reasonable first approximation for such wet tropical conditions. This would be 4 to 6 times the life span of leaves in the Alps (Körner 2003).

FUNCTIONAL DIVERSITY

Beyond the presence of taxa, the presence of certain functional types of plants is key to ecosystem integrity and land use value. Animals may have a profound influence on the balance among such functional groups, of which six are of particular importance:

- Tussock-forming grasses
- Low-stature shrubs
- Mat-forming graminoids
- Legumes with N-fixing symbionts
- Rosette-forming, non-legume herbaceous species
- Cryptogams

According to the analysis by Hofstede et al. (1995b) in Colombia, the nontussock, nonshrub fraction (herbs and short grasses) accumulate between 60 and 94% of total green phanerogam biomass at the end of the humid season. The cryptogam fraction varies from zero in heavily used to 15% in undisturbed grassland. The high fraction of short herbaceous, semi-herbaceous, and graminoid species comes at a surprise, and its fraction was highest in the most heavily burned and grazed plots, whereas the tussock fraction at harvest arrived at only 6% of life mass. Whether this reflects a special situation in these Colombian sites, or applies more generally, awaits study. The data at least contrast the view that land use is in favor of sturdy tussocks under all conditions. The intertussock vegetation thus plays a dominant role in ground coverage, is also far more diverse (Figure 1.11) than the tussock and shrub component, and is richer in nutrients.

Intertussock ground cover may, however, not be present year round and may include many ephemeral species emerging during wet periods, whereas tussocks are perennial. Over a large fraction of the year, the occupancy of the intertussock space, and thus erosion control, depends on the abundance of short, long-lived, mat-forming grasses and sedges. The legume fraction often depends on phosphate availability and may be enhanced by P-fertilizer without negative effects on the remainder species assemblage, as has often been shown for marginal, conservationally precious land in the temperate zone. Again, this is a pre field that needs careful exploration in the tropics.

FIGURE 1.11 The intertussock space can be occupied by a highly diverse flora of small stature species, which greatly contribute to soil stabilization, biodiversity, and biomass: *Festuca orthophylla* tussocks with *Cajophora* sp., *Perezia* sp., *Lesquerella* sp., and short *Aciachne* sp. grass mats in the intertussock space (photographs composed from north Argentina and west Bolivia).

Pasture encroachment by unpalatable species is a worldwide problem. In the tropics, two forms of intrusions are most widespread: poor-forage-quality graminoids and shrubs. When such encroachments happen, not only does pasture quality drop dramatically but also biological richness. The net land area for livestock grazing can become much reduced (Figure 1.12). Fire is one way to cope with this. The use of shrubs as fuel in these inherently fuel-poor regions is an alternative that could profit from the adoption of some simple heating technology and would release pressure from the remaining woodlands.

NEEDED NEXT STEPS

Decades of research in tropical and subtropical highlands established a firm picture of the diversity of plant species and, less so, for other groups of organisms. However, there is still an open arena for a suite of key studies needed to understand the diversity, functioning, and responsiveness to land uses of these biota. The Moshi-La Paz research agenda at the end of this volume is a condensate of the important needs arrived at by a group of 40 researchers from 20 different countries.

Therefore, it seems the most urgent biological questions are:

1. What is the productivity of these highlands, and how does it respond to grazing and burning?
2. What is the role played by intertussock vegetation in terms of biodiversity, productivity, and erosion control?
3. What is the role of legumes, and how can their contribution to biodiversity and productivity be enhanced?
4. How does land use affect forage quality, and how may it be improved without or with low-frequency fire management?
5. What land use regimes need to be applied to achieve year-round ground cover (e.g. more mat-forming, grazing-only [rather than fire] driven vegetation)?

and make available existing but scattered knowledge from biological sciences that could assist in decision making on land care issues, irrespective of other implications. The community of researchers that met for the first time during these two occasions, and now presents some of the results discussed, felt that there is not enough time for trial-and-error or *laissez faire* approaches, and decisions should be based on knowledge. It also became clear that the priority research themes had neither been defined nor were there any concerted research programs or activities along common questions. Not only for the human needs in and around these fragile highlands, but also for the degrading biological richness itself and the irreversible loss of ecosystem functioning, research on biological implications of highland management in the subtropics and tropics is highly needed. This synthesis project aims to be a stepping stone in this process. The Global Mountain Biodiversity Assessment (GMBA) program, with its office in Basel, Switzerland, a cross-cutting network of DIVERSITAS (Paris), operates as a coordination platform, think tank, and patronage institution for this endeavor.

FIGURE 1.12 The net land area for grazing can shrink dramatically when thorny scrub intrudes or when unpalatable tussock grasses occupy much of the space. (a) Burning tall tussock grass (*Festuca orthophylla*), Bolivia, 4150 m; (b) recently burned tussocks, Pico de Orizaba, Mexico, 4050 m; and (c) yak pasture encroached by *Berberis* sp., Langtang, Nepal, 3200 m.

Ultimately, such investigations should form a sound basis for sustainable use of these highlands.

Beyond such biological approaches, there are a suite of pedological, hydrological, and animal husbandry questions and, ultimately, a suite of sociological and economical questions to be solved. However, the workshops that form the basis for this volume were organized to gather

SUMMARY

In this introductory overview, we highlight the significance of biodiversity and ecosystem functioning in the non-forested highlands of the world, with special reference to tropical and subtropical regions. After defining the terms and the worldwide extent of the considered biota, we recall some key drivers of highland diversity and ecosystem functioning, such as quaternary history, topography, microclimate, water, and nutrients, and summarize basic knowledge on biomass production and responses to land use, grazing, and fire, in particular. Animal husbandry in the highlands, temperate as well as tropical and subtropical, can exert positive and negative impacts, which range from increasing biodiversity and soil stability to dramatic losses of biological richness and soil destruction. Land use in the highlands has profound consequences for catchment value and, thus, the forelands, which depend on steady and clean water supplies from mountains in various ways. We show how land use can shape highland biota, with examples from our own works in the Alps and the Caucasus and with reference to the published evidence from tropical highlands. It is concluded that the productivity of low-latitude highlands is largely unknown and that future research should focus on the yield, quality, and functional as well as taxonomic diversity of these biota in the light of land management and sustainability of land use and catchment value.

ACKNOWLEDGMENTS

A.M. Clccf provided helpful comments on the manuscript. We thank the Swiss Academy of Sciences (SANW); the Swiss National Science Foundation (SNF); the Swiss Office of Science and Education (BBW); the Swiss Agency of the Environment, Forest, and Landscape (BUWAL); the Swiss Agency of Agriculture (BLW) (all in Berne); the United Nations University of UNEP (Tokyo), and DIVERSITAS (Paris) for its past and continuous support of GMBA; and the Swiss Agency of Development and Cooperation SDC and FAO (Rome) for specific funds for this project.

References

Balslev, H., Luteyn, J.L. (1992). *Páramo: An Andean Ecosystem under Human Influence*. Academic Press, San Diego.

Barthlott, W., Lauer, W., Placke, A. (1996). Global distribution of species diversity in vascular plants: towards a world map of phytodiversity. *Erdkunde* 50: 317–327.

Belsky, A.J. (1986). Does herbivory benefit plants? — a review of the evidence. *Am Naturalist* 127: 870–892.

Bliss, L.C., Heal, O.W., Moore, J.J. (1981). *Tundra Ecosystems: A Comparative Analysis*. Cambridge University Press, Cambridge.

Bowman, W.D. (1994). Accumulation and use of nitrogen and phosphorus following fertilization in two alpine tundra communities. *Oikos* 70: 261–270.

Bowman, W.D., Theodose, T.A., Fisk, M.C. (1995). Physiological and production responses of plant growth forms to increases in limiting resources in alpine tundra: implications for differential community response to environmental change. *Oecologia* 101: 217–227.

Cabrera, A.L. (1968). Ecologia vegetal de la Puna. In Geo-ecology of the mountainous regions of the tropical Americas. Proceedings of the UNESCO, Mexico Symposium, 1966. *Colloquium Geographicum (Bonn)*, 9: 91–116.

Callaway, R.M., Brooker, R.W., Choler, P., Kikvidze, Z., Lortie, C.J., Michalet, R., Paolini, L., Pugnaire, F.L., Newingham, B., Aschehoug, E.T., Armas, C., Kikodze, D., Cook, B.J. (2002). Positive interactions among alpine plants increase with stress. *Nature* 417: 844–848.

Cernusca, A., Seeber, M.C. (1981). Canopy structure, microclimate and the energy budget in different alpine plant communities. In Grace, J., Ford, E.D., Jarvis, P.G. (Eds.). *Plants and Their Atmospheric Environment*. Symposium of the British Ecological Society. Blackwell, Oxford, pp. 75–81.

Chaneton, E.J., Lemcoff, J.H., Lavado, R.S. (1996). Nitrogen and phosphorus cycling in grazed and ungrazed plot in a temperate subhumid grassland in Argentina. *J Appl Ecol* 33: 291–302.

Chapin, F.S. III. (1980). The mineral nutrition of wild plants. *Annu Rev Ecol Syst*, 11: 233–260.

Chapin, F.S. III, Vitousek, P.M., Van Cleve, K. (1986). The nature of nutrient limitation in plant communities. *Am Naturalist* 127: 48–58.

Cleef, A.M., Rangel, J.O., Salamanca, S. (1983). Reconocimiento de la vegetación de la parte alta del transecto Parque Los Nevados. In Van der Hammen, T., Pérez, P., Pinto, E. (Eds.). *La Cordillera Central Colombiana. Estudios de ecosistemas tropandinos I*. Cramer, Vaduz, pp. 150–173.

Costin, A.B. (1958). The grazing factor and the maintenance of catchment values in the Australian Alps. *CSIRO Div Plant Ind Tech Paper* 10: 3–13.

Diaz, S., Acosta, A., Cabido, M. (1992). Morphological analysis of herbaceous communities under different grazing regimes. *J Veg Sci* 3: 689–696.

Diemer, M. (1997). Plant microclimate and growth conditions in the páramo zone of Ecuador. In Valencia, R., Balslev, H. (Eds.). *Estudios sobre diversidad y ecologia de plantas*. Pontificia Universidad catolica del Ecuador, pp. 255–262.

Diemer, M. (1998). Leaf lifespans of high-elevation, aseasonal Andean shrub species in relation to leaf traits and leaf habit. *Global Ecol Biogeogr Lett* 7: 457–465.

Edwards, P.J., Berry, N.R., Güsewell, S., Jewell, P.L., Kreuzer, M. (2004). Long-term effects of cattle grazing upon the nutrient status of alpine pastures. In Lüscher, A. (Ed.). *Land Use Systems in Grassland Dominated Regions.* European Grassland Congress 2004, Luzern, in press.

Erschbamer, B., Virtanen, R., Nagy, L. (2003). The impacts of vertebrate grazers on vegetation in European high mountains. In Nagy, L., Grabherr, G., Körner, C., Thompson, D.B.A. (Eds.). *Alpine Biodiversity in Europe. Ecological Studies,* 167: 377–396, Springer-Verlag, Berlin.

Galen, C. (1990). Limits to the distributions of alpine tundra plants: herbivores and the alpine skypilot, *Polemonium viscosum. Oikos,* 59: 355–358.

Geyger, E. (1985). Untersuchungen zum Wasserhaushalt der Vegetation im nordwestargentinischen Andenhochland. *Dissertationes Botanicae,* 88: 176, Cramer, Berlin.

Hedberg, O. (1964). Features of afroalpine plant ecology. *Acta Phytogeographica Suecica,* 1–149, Uppsala.

Heer, C., Körner, C. (2002). High elevation pioneer plants are sensitive to mineral nutrient additon. *Basic Appl Ecol* 3: 39–47.

Heitschmidt, R.K., Stuth, J.W. (1991). *Grazing Management: An Ecological Perspective.* Timber Press, Portland.

Hemp, A., Beck, E. (2001). *Erica excelsa* as a fire-tolerating component of Mt. Kilimanjaro's forests. *Phytocoenologia* 31: 449–475.

Henry, G.H.R., Svoboda, J. (1994). Comparisons of grazed and non-grazed high-arctic sedge meadows. In Svoboda, J., Freedman, B. (Eds.). *Ecology of a Polar Oasis.* Captus Press, New York, pp. 193–194.

Hnatiuk, R.J. (1978). The growth of tussock grasses on an equatorial high mountain and on two sub-antarctic islands. In Troll, C., Lauer, W. (Eds.). *Geoecological Relations between the Southern Temperate Zone and the Tropical Mountains.* Erdwiss Forschung 11, Franz Steiner Verlag, Wiesbaden, pp. 159–188.

Hofstede, R.G.M., Chilito, E.J., Sandovals, E.M. (1995a). Vegetative structure, microclimate, and leaf growth of al páramo tussock grass species, in undisturbed, burned and grazed conditions. *Vegetatio* 119: 53–65.

Hofstede, R.G.M., Mondragon Castillo, M.X., Rocha Osorio, C.M. (1995b). Biomass of grazed, burned, and undisturbed Páramo grasslands, Colombia. I. Aboveground vegetation. *Arct Alp Res* 27: 1–12.

Hofstede, R.G.M., Rossenaar, A.J.G.A. (1995). Biomass of grazed, burned, and undisturbed Páramo grasslands, Colombia. II. Root mass and aboveground: belowground ratio. *Arct Alp Res,* 27: 13–18.

Jonasson, S., Bryant, J.P., Chapin, F.S. III, Andersson, M. (1986). Plant phenols and nutrients in relation to variations in climate and rodent grazing. *Am Naturalist* 128: 394–408.

Kapos, V., Rhind, J., Edwards, M., Price, M.F., Ravilious, C. (2000). Developing a map of the world's mountain forests. In Price, M.F., Butt, N. (Eds.). *Forests in Sustainable Mountain Development (IUFRO Research Series 5).* CABI Publishing, Wallingford Oxon, pp. 4–9.

Kikvidze, Z., Nakhutsrishvili, G. (1998). Facilitation in subnival vegetation patches. *J Veg Sci* 9: 261–264.

Körner, Ch., Allison, A., Hilscher, H. (1983). Altitudinal variation in leaf diffusive conductance and leaf anatomy in heliophytes of montane New Guinea and their interrelation with microclimate. *Flora* 174: 91–135.

Körner, Ch. (1989). The nutritional status of plants from high altitudes. A worldwide comparison. *Oecologia* 81: 379–391.

Körner, Ch. (1995). Alpine plant diversity: a global survey and functional interpretations. In Chapin, F.S. III, Körner, Ch. (Eds.). *Arctic and Alpine Biodiversity: Patterns, Causes and Ecosystem Consequences. Ecol Studies,* 113: 45–62, Springer-Verlag, Berlin.

Körner, Ch., Nakhutsrishvili, G. (1987). Der Einfluss der Schafbeweidung auf den Wasserhaushalt der Vegetation (russ). In Rabotnov, T.A. (Ed.). *Ekologitseskie issledowanija wisokogornich lugov Kasbegi (Ökologische Untersuchungen der Hochgebirgsweiden von Kasbegi).* Academia nauk GSSR, Mezniereba, Tbilissi, pp. 99–120.

Körner, Ch., Diemer, M., Schäppi, B., Niklaus, P., Arnone, J. (1997). The responses of alpine grassland to four seasons of CO_2 enrichment: a synthesis. *Acta Oecologica,* 18: 165–175.

Körner, Ch., Wieser, G., Cernusca, A. (1989). Der Wasserhaushalt waldfreier Gebiete in den österreichischen Alpen zwischen 600 und 2600 m Höhe. In Cernusca, A. (Ed.). *Struktur und Funktion von Graslandökosystemen im Nationalpark Hohe Tauern*. Veröff Oesterr MaB-Hochgebirgsprogramm Hohe Tauern Band 13. Universitätsverlag Wagner, Innsbruck and Austrian Acad Sci, Vienna, pp. 119–153.

Körner, Ch., Spehn, E.M. (2002). *Mountain Biodiversity, A Global Assessment*. The Parthenon Publishing Group, Boca Raton.

Körner, Ch. (2000). The alpine life zone under global change. *Gayana Bot (Chile)*, 57: 1–17.

Körner, Ch. (2003). *Alpine Plant Life* (2nd ed.). Springer-Verlag, Berlin.

Körner, Ch. (2004). Mountain biodiversity, its causes and function. *Ambio*, special report 13: 11–17.

Körner, Ch., Paulsen, J. (2004). A worldwide study of high altitude treeline temperatures. *J Biogeogr* 31: 713–732.

Lauer, W., Rafiqpoor, M.D., Theisen, I. (2001). *Physiogeographie, Vegetation und Syntaxonomie der Flora des Páramo de Papallacta (Ostkordillere Ecuador)*. Franz Steiner Verlag, Stuttgart.

Mahaney, W.C., Linyuan, Z. (1991). Removal of local alpine vegetation and overgrazing in the Dalijia mountains, northwestern China. *Mount Res Dev* 11: 165–167.

Mark, A.F. (1994). Effects of burning and grazing on sustainable utilisation of upland snow tussock (*Chionochloa* spp.) rangelands for pastoralism in South Island, New Zealand. *Aus J Bot* 42: 149–161.

Mark, A.F., Dickinson, K.J.M., Hofstede, R.G.M. (2000). Alpine vegetation, plant distribution, life forms, and environments in a perhumid New Zealand region: oceanic and tropical high mountain affinities. *Arct Antarct Alp Res* 32: 240–254.

McIntire, E.J.B., Hik, D.S. (2002). Grazing history versus current grazing: leaf demography and compensatory growth of three alpine plants in response to a native herbivore (Ochotona collaris). *J Ecol* 90: 348–359.

McNaughton, S.J. (1983). Compensatory plant growth as a response to herbivory. *Oikos* 40: 329–336.

Messerli, B., Ives, J.D. (Eds.) (1997). *Mountains of the World: A Global Priority*. Parthenon Publishing Group, New York.

Meurk, C.D. (1978). Alpine phytomass and primary productivity in Central Otago, New Zealand. *NZ J Ecol* 1: 27–50.

Miehe, S., Miehe, G. (1994). *Ericaceous Forests and Heathlands in the Bale Mountains of South Ethiopia. Ecology and Man's Impact*. Stiftung Walderhaltung in Afrika and Bundesforschungsanstalt für Forst und Holzwirtschaft, Hamburg.

Milchunas, D.G., Lauenroth, W.K. (1993). Quantitative effects of grazing on vegetation and soils over a global range of environments. *Ecol Monogr* 63: 327–366.

Nakhutsrishvili, G. (1999). The vegetation of Georgia (Caucasus). *Braun-Blanquetia* 15.

Pucheta, E., Cabido, M., Diaz, S., Funes, G. (1998). Floristic composition, biomass, and aboveground net plant production in grazed and protected sites in a mountain grassland of central Argentina. *Acta Oecol* 19: 97–105.

Ram, J. (1992). Effects of clipping on aboveground plant biomass and total herbage yields in a grassland above treeline in central Himalaya, India. *Arct Alp Res* 24: 78–81.

Reich, P.B., Walters, M.B., Ellsworth, D.S. (1992). Leaf life-span in relation to leaf, plant, and stand characteristics among diverse ecosystems. *Ecol Monogr* 62: 365–392.

Rikhari, H.C., Negi, G.C.S., Pant, G.B., Rana, B.S., Singh, S.P. (1992). Phytomass and primary productivity in several communities of a central Himalayan alpine meadow, India. *Arct Alp Res* 24: 334–351.

Safford, H.D. (1999). Brazilian páramos I. An introduction to the physical environment and vegetation of the campos de altitude. *J Biogeogr* 26: 693–712.

Smith, J.M.B. (1977). Origins and ecology of the tropical pine flora of Mt Wilhelm, New Guinea. *Biol J Linnean Soc* 9: 87–131.

Smith, J.M.B., Klinger, L.F. (1985). Aboveground:belowground phytomass ratios in Venezuelan páramo vegetation and their significance. *Arct Alp Res* 17: 189–198.

Suarez, E., Medina, G. (2001). Vegetation structure and soil properties in Ecuadorian páramo grassland with different histories of burning and grazing. *Arct Antarct Alp Res* 33: 158–164.

Sundriyal, R.C. (1992). Structure, productivity and energy flow in an alpine grassland in the Garhwal Himalaya. *J Veg Sci* 3: 15–20.

Taddesse, G., Peden, D., Abiye, A., Wagnew, A. (2003). Effect of manure on grazing lands in Ethiopia, East African highlands. *Mount Res Dev* 23: 156–160.

Tappeiner, U., Cernusca, A. (1998). Effects of land-use changes in the Alps on exchange processes (CO_2, H_2O) in grassland ecosystems. In Kovar, K., Tappeiner, U., Peters, N.E., Craig, R.G. (Eds.). *Hydrology, Water Resources and Ecology in Headwaters.* IAHS Publ., 248: 131–138.

Tasser, E., Mader, M., Tappeiner, U. (2003). Effects of land use in alpine grasslands on the probability of landslides. *Basic Appl Ecol* 4: 271–280.

Tongway, D.J., Ludwig, J.A. (2000). The nature of landscape dysfunction in rangelands. In Ludwig, J.A., Tongway, D.J., Freudenberger, D.O., Noble, J.C., Hodgkinson, K.C., Griffin, G.F., MacLeod, N.D., Brown, J.R. (2000) *Landscape Ecology, Function and Management: Principles from Australia's Rangelands.* CSIRO Publishing, Collingwood, pp. 49–61.

Trimble, S.W., Mendel, A.C. (1995). The cow as a geomorphic agent — a critical review. *Geomorphology,* 13: 233–253.

Trlica, M.J., Rittenhouse, L.R. (1993). Grazing and plant performance. *Ecol Appl* 3: 21–23.

van der Hammen, T., Cleef, A.M. (1986). Development of the high Andean paramo flora and vegetation. In Vuilleumier, F., Monasterio, M. (Eds.). *High Altitude Tropical Biogeography.* Oxford University Press, New York, pp. 153–201.

Vareschi, V. (1970). *Flora de los Páramos de Venezuela.* Universidad de los Andes, Merida, Venezuela.

Vuilleumier, F., Monasterio, M. (1986). *High Altitude Tropical Biogeography.* Oxford University Press, New York.

Wesche, K., Miehe, G., Kaeppeli, M. (2000). The significance of fire for Afroalpine ericaceous vegetation. *Mount Res Dev* 20: 340–347.

Part II

Effects of Fire on Mountain Biodiversity

2 Diversity of Afroalpine Vegetation and Ecology of Treeline Species in the Bale Mountains, Ethiopia, and the Influence of Fire

Masresha Fetene, Yoseph Assefa, Menassie Gashaw,
Zerihun Woldu, and Erwin Beck

INTRODUCTION

Uplift and volcanism in the Miocene and Oligocene geological periods (between 38 and 7 million BP) resulted in the covering of all the underlying rocks and the formation of the East African mountains that rest like islands on the surrounding hills and plains. These Afromontane archipelagos are distributed on both sides of the East African Rift Valley.

The Bale Mountains lie in the southeastern part of the Ethiopian highlands, about 850 km north of the equator. The highest peak in Bale, Tulu Dimtu, is the second highest peak in Ethiopia and the seventh in Africa (see Figure 2.1). The East African mountain nearest to the Bale mountains is Mt. Kulal, 550 km south in the Turkana Depression.

The vegetation of the Bale Mountains has been the subject of studies by a number of botanists and ecologists. A full account of the history of botanical exploration of the Bale Mountains has been provided by Miehe and Miehe (1994). In a series of publications, Hedberg (1975, 1986) made important analyses of the vegetation and ecology of Afroalpine regions in Ethiopia. Weinert (1981), Weinert and Mazurek (1984), and Uhlig (1988) also conducted ecological research on the vegetation of the Bale Mountains. Miehe and Miehe (1994) presented a detailed study on ericaceous vegetation and on the plant communities within the ericaceous

zones of the Bale Mountains. The present study attempts to provide a description of plant communities in the entire altitudinal range of the Afroalpine and ericaceous zones.

The ericaceous belt of the Bale Mountains is a region most seriously affected by the progressive increase of human activities. Cattle and horses put heavy pressure on the vegetation, especially at the lower altitudes. The ericaceous bushes are cut for fuel wood and are frequently burned by the local people for various reasons. This results in the destruction of the vegetation and in the disappearance of the fauna, and hence leads to a reduction of the region's biodiversity.

The present study aims at (1) describing the plant communities of the Afroalpine and ericaceous zones, (2) documenting the distribution patterns of treeline species and the changes in the structure of ericaceous vegetation with altitude, and (3) assessing the incidence and influence of fire on the diversity and composition of vegetation in the ericaceous belt.

MATERIAL AND METHODS
DESCRIPTION OF THE STUDY AREA

Geology and Climate

The study area is the Harenna Escarpment, located at the southern slopes of the Bale Mountains between 6°45 and 7° N and 39°45 and

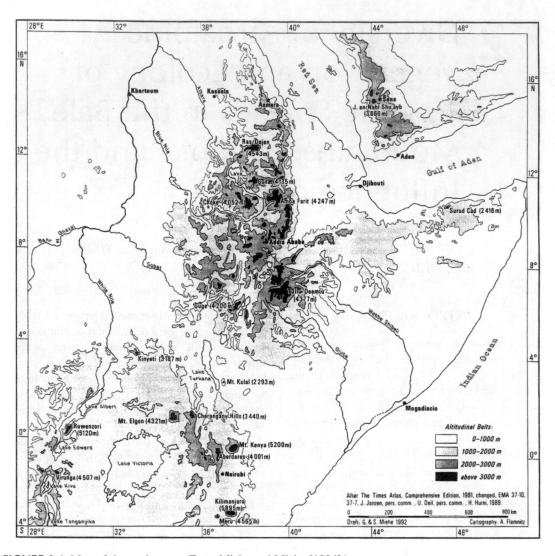

FIGURE 2.1 Map of the study area. (From Miehe and Miehe [1994].)

39°40 E. The rocks of the volcanic outpourings are predominantly trachytes but also include rhyolites, basalts, and associated agglomerates and tuffs. Although adequate information about glaciations is lacking, the current landforms in the mountains appear to have resulted from actions of tectonics and glaciations. At least two glacial periods are documented in the mountains (18,000 BP and 2,000 BP, Bonnefille, 1993).

In contrast to the northern highlands, southern Ethiopia is within the East African climatic domain, which is highly influenced by south-easterlies from the Indian Ocean during most of the year. As in most Ethiopian highlands, the intertropical convergence zone (ITCZ) and local altitudinal and topographic influences affect the distribution of the precipitation in the Bale Mountains. Annual rainfall in the Bale Mountains ranges between 600 and 1500 mm depending on the relief (see Table 2.1).

The diurnal variability in temperature in the Bale Mountains is higher than the seasonal variation. A minimum temperature of −15°C was recorded by Hillman (1986) on the Sanetti Plateau (3850 m), whereas Miehe and Miehe (1994) recorded a nocturnal minimum temperature of −3°C in sparsely vegetated areas of the ericaceous belt. Solifluction is common in the

Afroalpine area and in the upper parts of the ericaceous vegetation.

Recently, the ericaceous and the Afroalpine areas have been subjected to increasing grazing pressure. The number of livestock varies in the wet and dry seasons (the maximum is 46/km^2 in the plateau and minimum is less than 2/km^2) (Hilman, 1986; Gottelli and Sillerio-Zubiri, 1992). Poaching of mountain nyalas and small antelopes is also common in the area. These activities are accompanied by deliberate setting of bush fires for hunting, and clearing and improvement of pastures (Miehe and Miehe, 1994).

There is evidence of early settlements in some valleys and plains in the area. Recently, with the construction of an all-weather road traversing the plateau, there is an increase in barley cultivation in the ericaceous and Afroalpine vegetation. However, the highlands of the Bale Mountains are still less densely populated than the Semien Mountains of northwestern Ethiopia (see Table 2.2). For instance, barley is cultivated in Bale at 600 to 800 m lower than in Semien. This is due to the transhumant mode of living in the Bale Mountains.

Vegetation Sampling

The current study considers vegetation in the ericaceous belt of the Bale Mountains along an altitudinal gradient ranging from 3000 to 4200 m. Transects were laid out based on homogeneity of the vegetation (Mueller-Dombois and Ellenberg, 1974). Relevés of 15 m × 15 m size were established at 50-m vertical distance (altitude). Within each altitudinal level, replicate relevés were put with minimum lateral distance

of 20 m. Within each relevé, a subplot of 2 m × 2 m was made for the herbaceous vegetation. All vascular plants in each relevé were recorded. We estimated abundance for single species using the 9-level ordinal cover abundance scale following Braun Blanquet as modified by Van der Maarel (Van der Maarel, 1979). The height of trees and shrub species, diameter at breast height (DBH) for trees, and the diameter at stump height (DSH) for shrubs were also recorded in all relevés.

Soil and Environmental Data

The rainfall measurements were compiled for the time of fieldwork and for the previous 11 months. Climate data of the area from previous studies were also considered (Miehe and Miehe, 1994). For each plot, information on altitude, slope, inclination, soil surface, and vegetation cover, etc., were collected. Soil samples were collected from the topsoil and at a depth of 30 cm from the surface of each relevé. Soil moisture, texture, pH, and total nitrogen were determined for each sample.

Incidence of Fire

Records on incidence of recent fires were gathered. In addition to the information obtained from the local people, the incidence of fire was assessed from the presence or absence of *Bryum argenteum* (a moss that grows after fire), charcoal, and remnants of charred twigs and ligno-tubers. The presences of each of these indicators were summed for each relevés, yielding a combined index of fire incidence.

TABLE 2.1
Annual rainfall for northern (n) and southern (s) slopes of Bale Mountains and on Sanetti Plateau (P)

Locality	Altitude (masl)	Rainfall (mm)	Years
Chorchora (n)	3500	1086	1985–1991
Goba (n)	2720	925	1968–1980
Koromi (P)	3850	1051	1985–1991
Mena (s)	1250	387	1983–1988
Rira (s)	3000	848	1987–1990
Tulu Konteh (P)	4050	852	1985–1991

Source: From Hilman (1986); Miehe and Miehe (1994).

TABLE 2.2
Population density in Semien and Bale Mountains (persons/km²) based on the census data taken for each zone (district) and the woredas (subdistricts) circumscribed by the mountains

	Zone/Woreda	Year		
		1998	1999	2000
Semien Mountains	North Gondar zone	49.8	51.2	52.6
	Debark	91.7	94.4	97
Bale Mountains	Bale zone	22.1	22.7	23.4
	Kokosa	160	164.5	169
	Dodola	91.2	94	96.9
	Adaba	52.2	53.7	55.3
	Sinana Dinsho	90.3	93.2	96.2
	Goba	44	45.8	47.6
	Menana Harenna Bulqi	14.1	14.5	14.9

Source: Central Statistical Authority (2001).

DATA ANALYSIS

Vegetation data were analyzed with hierarchical syntax clustering using agglomerative method with optimization (Podani, 2000). A resemblance matrix was calculated with the similarity ratio:

$$S_{ij} = 1 - \Sigma_{i\ xij\ xik} / (\Sigma_{i\ xij2} + \Sigma_{i\ xik2} - \Sigma_{i\ xij\ xik})$$

where $S(i, j)$ in row i and column j is the distance between observations i and j. Species-wise cover abundance values were used to classify vegetation communities. In classifying the communities, the subject group averages were used to evaluate the degree of dissimilarities among the relevés. Both the vegetation data and the environmental variables were analyzed with canonical correspondence analysis (CCA) using CANOCO (ter Braak and Smilaur, 1998) to explore the correlation between vegetation and environmental variables. Species richness and relative abundance were analyzed using the Shannon–Weaver index of diversity (Krebs, 1989).

RESULTS AND DISCUSSION

PLANT COMMUNITIES

The southern slope of the Harenna Escarpment with its montane forest between 1500 and 2800 m is more gentle than the ericaceous vegetation

above this altitude. The Bale Mountains have high floral and faunal diversity as well as endemicity. The floristic composition of the area has been reported by Friis (1986); Hedberg (1986); Negatu and Tadesse (1986); Woldu et al. (1989); Gashaw and Fetene (1996); and Bussman (1997).

A total of 60 relevés were sampled at the northwestern side of the Bale Mountains. The hierarchical classification gave six major plant communities. The first of these is the *Kniphofia– Euphorbia–Alchemilla* community (3400 to 3500 m). In this community, *Kniphofia foliosa*, *Euphorbia dumalis*, and *Alchemilla abyssinica* were the characteristic species. At the next altitudinal level, we find the *Alchemilla haumannii* community (3700 to 4000 m). This community is dominated by *A. haumannii*, which sometimes forms pure stands.

On the southeastern side of the Bale Mountains, a total of 110 relevés were sampled, in which 84 species of vascular plants were encountered. Eight of these were trees and shrubs, and the rest were herbaceous plants. The ericaceous vegetation was grouped into three altitudinal subzones following previous works: lower subzone (3000 to 3400 masl), central subzone (3400 to 3600 masl), and the upper subzone (3600 to 4000 masl) (see also Hedberg, 1951; Miehe and Miehe, 1994). Thirteen community types were identified from the cluster analysis. The communities were named based on the species with the highest cover abundance. The distribution of the communities varied in the lower (3000 to 3400 m), central (3400 to 3600 m), and upper (3600 to 4200 m) subzones of the ericaceous belt. Some of the community types occurred in the entire altitudinal range (3000 to 4200 m), whereas others were restricted to certain ranges. Plant diversity showed an inverse bell-shaped pattern. The upper and the lower subzones had higher diversities than the central one. The complete list of the communities and their respective distribution, diversity, and evenness in the three subzones are given in Table 2.3.

The *Schefflera volkensii–Erica trimera–Discopodium penninervium* community (altitude range, 3100 to 3300 m) is found at the lowermost part of the ericaceous subzone. The emergent tree in this community is *Schefflera volkensii*. Higher up in the lower ericaceous subzones, the *Erica trimera–Hagenia abyssinica–Hypericum revolutum* community occurs. The characteristic species for this community are *Erica trimera, Trifolium acaule, Hypericum revolutum, Hagenia abyssinica*, and *Discopodium penninervium*. At lower altitudes (between 3000 and 3200 m), this community forms a subcommunity that is characterized by the dominance of *Hagenia abyssinica* and *Hypericum revolutum*. Another community also common at the lower subzone of the ericaceous belt is the *Erica trimera–Polystichum–Hypericum revolutum* community (Plate 2.1a). The characteristic species of this community include *Erica trimera*, the codominant tree *Hypericum revolutum*, the most common fern *Polystichum* sp., *Discopodium penninervium*, and *Cynoglossum amplifolium*.

At the central subzone of the ericaceous belt, we find the *Erica trimera–Hypericum revolutum–Alchemilla abyssinca* community. In this community, the dominance of *Erica trimera* is conspicuous in the upper layer of the canopy. Another community of the central subzone is the *Erica trimera–Cynoglossum amplifolium–Discopodium penninervium* community.

Among communities of the upper subzone, we find the *Haplocarpha rueppellii–Alchemilla microbetula–Alchemilla pedata* community (3300 to 3900 m) and the *Satureja paradoxa–Asplenium aethiopicium-Geranium arabicum* community. In the former, *Haplocarpha rueppellii, Alchemilla microbetula, Alchemilla*

pedata, Myosotis abyssinica, and *Discopodium penninervium* are the characteristic species, whereas the characteristic species in the latter community are *Satureja paradoxa, Asplenium aethiopicium, Geranium arabicum, Crepis rueppellii*, and *Stachys aculeolata*. The upper part of the ericaceous belt had a patchy appearance with more openings. Depending on the microsite factors, the diversity was comparable with the lower part (2.35 ± 0.048 for the lower and 2.10 ± 0.05 for the upper) and was greater than in the central subzones.

DENSITY AND FREQUENCY OF TREELINE SPECIES

A total of eight tree and shrub species were recorded, out of which *Erica trimera* was found in almost all relevés, whereas one species (*Pittosporum viridiflorum*) was recorded in one relevé only and is not shown in Figure 2.2. *Erica trimera* and *Hypericum revolutum* showed similar trends in frequency in the lower and central subzone (Figure 2.2). *H. revolutum* was absent in the upper part of the ericaceous subzone. At the lower ericaceous subzone, the frequency of *E. trimera* was lower because of the competitive strength of the other montane woodland species (Miehe and Miehe, 1994). However, it is an important component of all three subzones of the ericaceous belt and no other species, including *Erica arborea*, showed such a wide distribution. *Erica arborea* was not found below 3200 m. *Rapanea melanophloeos, H. revolutum*, and *D. penninervium* were constituents of both the lower and central subzone but not of the upper subzone. *Schefflera volkensii* is restricted to the lower part of the ericaceous belt, and *Hagenia abyssinica* attained its highest frequency in the lower subzone. The density of the treeline species showed a similar trend as the frequency.

The height of treeline species decreased with increasing altitude (Table 2.4). The most notable change was observed for *E. trimera*. The regression analysis (Figure 2.3) showed a strong inverse relation between altitude and height ($R^2 = 0.60$). This could be attributed to the decrease in temperature with increasing altitude.

TABLE 2.3
The distribution of the diversity (H), evenness of the 13 community types, and average value for incidence of fire in lower, central, and upper subzones of ericaceous vegetation

| | Community Types | Distribution | | | Species Number | Diversity | | Fire Incidence Index: |
		Lower (3000–3400)	Central (3400–3600)	Upper (3600–4200)		Shannon Index	Evenness	
1	Erica trimera–Hagenia abyssinica–Hypericum revolutum	+	+		17	2.23	0.96	1.3
2	Erica trimera–Polystichum sp.–Hypericum revolutum	+	+	–	13	2.11	0.95	0.5
3	Erica trimera–Hypericum revolutum–Alchemilla abyssinica	+	+	–	13	2.36	0.97	0.6
4	Erica trimera–Cynoglossum amplifolium–Discopodium penninervium	–	+	–	8	2.43	0.99	1.0
5	Schefflera volkensii–Erica trimera–Discopodium penninervium	+		–	5	2.54	0.97	1.2
6	Senecio fresenii–Alchemilla abyssinica–Cynoglossum amplifolium	+	+	–	2	2.47	0.99	0.0
7	Erica trimera–Luzula johnstonii–Geranium arabicum	+	+	–	9	1.89	0.94	0.8
8	Lotus discolor–Polystichum sp.–Schefflera volkensii	+	–	–	2	2.31	0.96	0.0
9	Haplocarpha rueppellii–Alchemilla microbetula–Alchemilla pedata	+	+	+	12	2.11	0.95	0.8
10	Alchemilla fischeri–Luzula abyssinica–Cineraria abyssinica	+	+	+	15	2.21	0.97	0.1
11	Festuca richardii–Dryopteris inaequalis–Alchemilla haumanni	–	+	+	3	2.42	0.97	1.3
12	Alchemilla pedata–Asplenium aethiopicum–Alchemilla abyssinica	–	+		9	2.29	0.97	0.1
13	Satureja paradoxa–Asplenium aethiopicum–Geranium arabicum	–	–	+	2	0.81	0.42	2.0

Note: + indicates presence and – is absence.

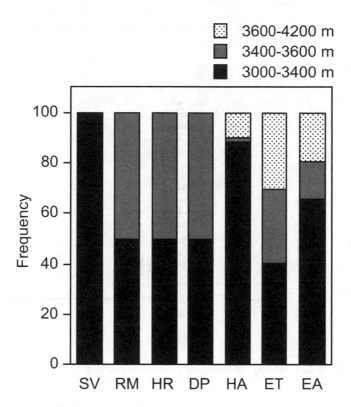

FIGURE 2.2 Frequency of treeline species in three ericaceous subzones: *Schefflera volkensii* (SV); *Rapanea melanophloeos* (RM); *Hypericum revolutum* (HR); *Discopodium penninervium* (DP); *Hagenia abyssinica* (HA); *Erica trimera* (ET); *Erica arborea* (EA). The three ericaceous subzones are the lower (3000 to 3400 masl); middle (3400 to 3600 masl); and upper (3600 to 4000 masl) zones.

TABLE 2.4
Height and DBH of five treeline species at (1) lower, 3000–3400 m, (2) central, 3400–3600 m, and (3) upper, 3600–4200 m, subzones of the ericaceous belt in Harenna Escarpment, Bale Mountains

Species	Subzones	DBH	Height	Number of Stems
D. penninervium	1		2.23 ± 1.90	119
	2		1.75 ± 1.28	
	3	—	—	—
E. arborea	1	—	1.62 ± 1.20	106
	2	4.04 ± 0	1.12 ± 0.80	
	3	—	0.95 ± 0	
E. trimera	1	23.34 ± 7.44	10.19 ± 2.20	255
	2	12.32 ± 12.38	5.30 ± 4.53	
	3	10.00 ± 8.20	2.08 ± 1.98	
H. revolutum	1	26.75 ± 8.83	13.60 ± 3.37	96
	2	16.74 ± 10.11	7.37 ± 5.86	
	3	—	—	
R. melanophloeos	1	25.02 ± 9.49	16.50 ± 6.87	91
	2	14.34 ± 7.81	8.96 ± 6.60	
	3	—	—	

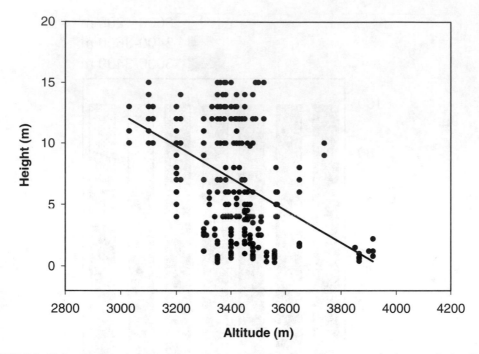

FIGURE 2.3 Biplot diagram showing the correlations of environmental parameters in the canonical ordination space.

TABLE 2.5
Pearson's correlation coefficient matrix for the nine environmental variables

	Altitude	Slope	Aspect	Moisture	pH	N	Fire	Sand	Clay
Slope	**0.725**								
Aspect	−0.001	−0.216							
Moisture	−0.33	−0.299	0.668						
pH	**−0.785**	0.629	−0.206	−0.375					
N	0.028	0.129	−0.066	0.013	−0.196				
Fire	**0.512**	0.411	0.476	0.224	0.100	−0.564			
Sand	0.638	**0.871**	−0.349	−0.457	**0.637**	0.226	0.150		
Clay	−0.201	−0.469	**0.554**	0.202	−0.480	−0.338	0.186	−0.418	
Silt	**−0.605**	−0.733	0.126	0.41	−0.475	−0.089	−0.248	**−0.902**	−0.014

Note: The magnitude indicates the degree of correlation. Positive signs indicate positive correlation and negative signs indicate inverse relation. Numbers in bold indicate significant correlation at $p < 0.05$.

RELATIONS BETWEEN DISTRIBUTION AND ECOLOGICAL CHARACTERS OF TREELINE SPECIES AND THEIR ENVIRONMENTAL FACTORS

The Pearson correlation analysis revealed a strong positive correlation between altitude and slope (0.8) and an even stronger negative correlation between altitude and pH. Percent silt and clay showed negative correlations at $r = −0.6$ and −0.4, respectively. The correlation coefficients of the environmental parameters are given in Table 2.5.

An ordination biplot was made for all environmental variables. The biplot diagram of the

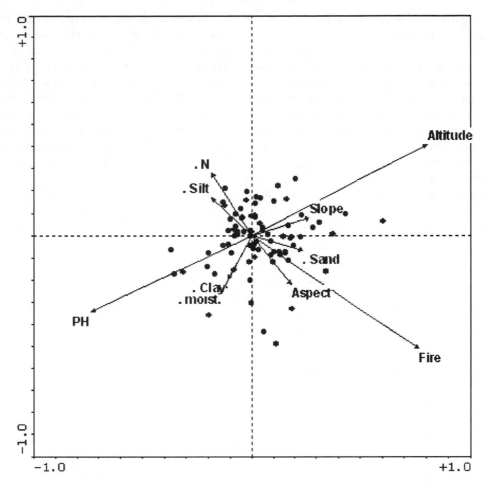

FIGURE 2.4 Regression analysis of the correlation of average height of *E. trimera* with altitude.

environmental variables reflects approximately the Pearson's correlation coefficients (Figure 2.4).

RECENT INCIDENCE OF FIRE

Recent incidence of fire showed an increasing tendency with increasing altitude (Figure 2.5). Fire incidence was not common in rocky areas with big boulders. The incidence was lower in areas with high cover of epiphytes, due, perhaps, to the convective cloud from Harenna that leads to the formation of thick epiphytic cover, playing a crucial role in insulation. Highly disturbed sites were avoided intentionally in this study. However, even in the relatively less disturbed vegetation, there was some evidence for recent occurrence of fire, especially at the upper subzone of the ericaceous vegetation. Incidence

of fire was more common at the upper part of the ericaceous vegetation. This is an indication that fire had little influence on the physiognomy of the lower part of ericaceous vegetation. This is in agreement with other investigations (Wesche, 2002). The highest incidence of fire was recorded in the *Satureja paradoxa–Asplenium aethiopicum–Geranium arabicum* community at the upper subzone of the ericaceous vegetation. The absence of indicators for fire incidence in the *Senecio fresenii–Alchemilla abyssinica–Cynoglos-sum amplifolium* community does not necessarily show the complete absence of fire in those localities. Alternatively, it may indicate the disappearance of the indicators of fire, which might be due to more severe disturbance.

IMPLICATIONS OF INCIDENCE OF FIRE TO VEGETATION DYNAMICS

Data collected from various sources (see Table 2.6) indicate that occurrence of fire in the Bale Mountains is very frequent. In the present study, samples were taken preferentially where there was a more or less undisturbed, continuous, and homogeneous vegetation. Thus, highly disturbed sites were avoided, as indicated earlier. However, even in the relatively less-disturbed vegetation, there was clear evidence of recent occurrences of fire, especially at the upper subzone of the ericaceous vegetation.

Incidence of fire is more common at the upper subzone of the largely shrubby ericaceous vegetation. The change in the stand structure of the vegetation due to high frequency of fire is observed more clearly in the northwestern slope of Bale Mountains (see Plate 2.1c).i Due to its proximity to towns, this zone has become more susceptible to human pressure. Thus, frequent fires and high grazing pressure seem to have nduced a transition in the plant community composition (see Plate 2.1b). This transition is usually from the *Erica trimera–Helichysum citrispinum* community to the *Euphorbia dumalis–Kniphofia foliosa* community as a result of

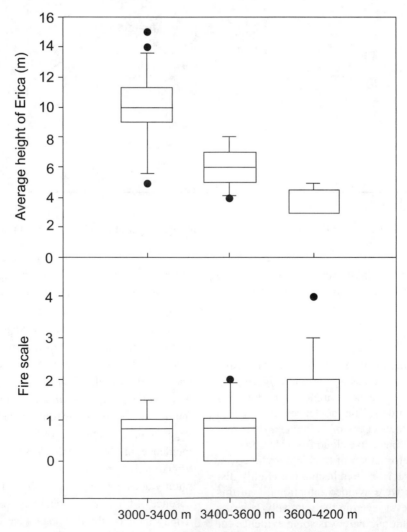

FIGURE 2.5 Box and Whisker plots of average height of *E. trimera* at three subzones, and the incidence of fire for southeastern transect (median and interquartile range).

TABLE 2.6
History of Fire Incidence in the Bale Mountains Area

Year	Locality	Burnt Area	Months
1963	Northwestern escarpment	Unmeasured	Unstated
1969	Around Finchaya Habra and other areas	Unmeasured	December
1971	Unstated	Unmeasured	Unstated
1973	Darkeena, Morabawa, and Worgona	500 ha	January/February
1973	Ukamasa, Layncha, and Danka Valley	35 kmΣ	March
1984	Different areas	195 kmΣ	December/March
1991	Northwestern escarpment (50 to 60 km by road from Goba town)	60 ha	December
1991	Northwestern escarpment	210 ha	December/January
1992	Along Kotera track (Simbirro) and other areas	Not measured	January/March
1992	Different areas, Toroshoma: Adelay and Gasuray	Not measured	Early April
1992	Gajera	Not measured	Early April
1993–1994	Batu Tiko and Gurari around two spots on Sanetti plateau, Adely ridge around Simbirro, and other areas	But less than 1992 and 1993	December/February
1998	Northeastern and southwestern parts of the Bale Mountains and Borena	150,000 ha	December/April

[a]Data gathered from archival materials in Bale Mountains National Park Dinshu and Ministry of Agriculture, Addis Ababa, Ethiopia.

high grazing intensity and frequent fire incidence (see Plate 2.1c). This transition may be unique to the Bale Mountains as grazing intensity in the ericaceous belt is not observed at other Afroalpine regions of East Africa.

Fire may have influenced the establishment of some treeline species, such as *Hagenia abyssinica* and *Juniperus procera*, which have been shown to require an open environment to establish (Fetene and Feleke, 2002). There are no seedlings under their canopies, suggesting that fire could play a significant role in the dynamics of the high-mountain vegetation.

CONCLUSION

The ericaceous vegetation of the Bale Mountains is clearly affected by fire (and grazing). At lower altitudes (below 4000 m) in the ericaceous belt, we found a higher vegetation cover, a larger number of species per community, and better edaphic conditions (as shown by higher organic carbon, nitrogen, and soil moisture content, and better climatic conditions) than at higher altitudes. The most notable change in vegetation due to human influence is the replacement of the *Erica*-dominated vegetation

by the *Helichrysum*-dominated community due to fire and grazing.

With increasing population density, the relatively less disturbed southern slope is being converted to less woody formation, as in the northwestern slope. This would have a serious effect on the water catchment of the area. The Bale Mountains are the water catchment area for eight major rivers and the source of a large number of streams. Conservation of this fragile ecosystem requires study of the ecology of the area using permanent plots and awareness creation in the local population and participation of people in the management of the area. It also calls for a full commitment on the part of the government.

SUMMARY

The Bale Mountains in Ethiopia constitute the largest expanse of Afroalpine vegetation in Africa. The diversity of the vegetation and life-form types, climatic features, and influence of fire on the vegetation were studied with particular emphasis on the distribution and ecology of treeline species. The major plant community types on the northwestern side of the mountain are the *Kniphofia–Euphorbia–Alchemilla* scrub

PLATE 2.1A (a) The relatively little-disturbed ericaceous forest at the southern slope of the Bale Mountains; (b) expansion of the cushion form *Helichrysum* sp. community replacing the ericaceous forest; (c) a community with less palatable species (*Kniphofia–Euphorbia* community) dominates with severe grazing and fire. (a) is taken on the southern slope, whereas (b) and (c) are from the northwestern slope of the mountain.

PLATE 2.1B (a) The relatively little-disturbed ericaceous forest at the southern slope of the Bale Mountains; (b) expansion of the cushion form *Helichrysum* sp. community replacing the ericaceous forest; (c) a community with less palatable species (*Kniphofia–Euphorbia* community) dominates with severe grazing and fire. (a) is taken on the southern slope, whereas (b) and (c) are from the northwestern slope of the mountain.

PLATE 2.1C (a) The relatively little-disturbed ericaceous forest at the southern slope of the Bale Mountains; (b) expansion of the cushion form *Helichrysum* sp. community replacing the ericaceous forest; (c) a community with less palatable species (*Kniphofia–Euphorbia* community) dominates with severe grazing and fire. (a) is taken on the southern slope, whereas (b) and (c) are from the northwestern slope of the mountain.

community, *Alchemilla haumannii* meadow, *Helichrysum citrispinum–Alchemilla abyssinica* community, *Festuca–Haplocarpha–Helichrysum gofense* community, *Haplocarpha–Carex monostachya* community, and the *Helichrysum splendidum-Festuca abyssinica* community. On the southeastern side, the major plant communities include the *Schefflera volkensii-Erica trimera-Discopodium penninervium* community, *Erica trimera–Polystichum–Hypericum revolutum* community, *Erica trimera–Hypericum revolutum–Alchemilla abyssinica* community, and the *Erica trimera– Cynoglossum amplifolium–Discopodium penninervium* community at the central and upper altitudes. Microclimate and site conditions had high influence on diversity and life-form features. Fire incidence was very frequent and had a serious influence on the diversity. Incidence of recent fire increased with increasing altitude. Fire caused the expansion of secondary vegetation and encroachment of weeds such as *Euphorbia dumalis*, *Kniphofia foliosa*, and *Solanum aculeatum* into previous ericaceous vegetation zones.

ACKNOWLEDGMENT

Financial support for the study from Volkswagen–Stiftung, Germany, is gratefully acknowledged. We thank the staff of the National Herbarium, Addis Ababa, for their assistance in plant identification.

References

Bonnefille, R. (1993). Evidence for a cooler and drier climate in the Ethiopian uplands, 1.5 million years ago. *Nature* 303: 487–491.

Bussman, R.W. (1997). The forest vegetation of the Harenna escarpment (Bale Province, Ethiopia). Syntaxonomy and phytogeographical affinities. *Phytocoenologia* 27: 1–23.

Central Statistical Authority (2001). Federal Democratic Republic of Ethiopia Statistical Abstract Annual Report. Addis Ababa, Ethiopia.

Friis, I. (1986) Zonation of forest vegetation on south slopes of Bale Mountains, South Ethiopia. *Sinet: Ethiop J Sci* 9: 29–44.

Gashaw, M. and Fetene, M. (1996). Plant communities of the Afroalpine vegetation of Sanetti plateau. *Sinet: Ethiop J Sci* 19: 65–86.

Hedberg, O. (1951). Vegetation belts of the East African mountains. *Svensk Botanisk Tidskrift* 451: 140–204.

Hedberg, O. (1986). The Afroalpine flora of Ethiopia. *Sinet: Ethiop J Sci* 9(Suppl.): 105–110.

Hedberg, O. (1975). Studies of adaptation and speciation in the Afroalpine flora of Ethiopia. *S Boissiera* 24: 71–74.

Kerbs, C.J. (1989). *Ecological Methodology.* Harper and Row, New York.

Miehe, S. and Miehe, G. (1994). *Ericaceous Forests and Heathlands in Bale Mountains of South Ethiopia. Ecology and Man's Impact.* Traute Warnke Verlag, Hamburg, Germany.

Mueller-Dombois, D. and Ellenberg, H. (1974). *Aims and Methods of Vegetation Ecology.* John Wiley & Sons, New York.

Negatu, L. and Tadesse, M. (1989) . An ecological study of the vegetation of the HarenNa forest, Bale, Ethiopia, *Sinet: Ethiop J Sci* 12(1): 63–93.

Podani, L. (2000). Syntax — 2000: A New Version for PC. Department of Plant Taxonomy and Ecology, Leotvos University, Budapest, Hungary.

ter Braak, C.J.F. and Smilauer, P. (1998). *CANOCO Reference Manual and User's Guide to CANOCO for Windows: Software for Canonical Community Ordination (Version 4).* Microcomputer Power, Ithaca, New York.

Hillman, J.C. (1986). Bale Mountains Park. Managment Plan. Ethiopia. Wildlife Conservation Organization, Addis Ababa.

Uhlig, S.K. (1988). Mountain forests and the upper tree limit on the southeastern plateau of Ethiopia. *Mount Res Dev* 8: 227–234.

Van der Maarel (1979). Transformation of cover-abundance values in phytosociology and its effect on community similarity. *Vegetatio* 39: 97–114.

Weinert, E. and Mazurek, A. (1984). Notes on vegetation and soil in Bale province of Ethiopia. *Feddes Repertorium* 95: 373–380.

Weinert, E. (1981). Vegetation in Bale mountains near Goba S. Ethiopia. *Wiss Z Univ Halle* 32: 41–67.

Wesche, K. (2002). The high-altitude environment of Mt. Elgon (Uganda, Kenya) — climate, vegetation and the impact of fire. *Ecotropical Monogr* 2: 1–253.

Woldu, Z., Feuli, E., and Negatu, L. (1989). Partitioning an elevation gradient of vegetation from Southeastern Ethiopia by probabilistic methods. *Vegetatio* 81: 189–198.

3 Is Afroalpine Plant Biodiversity Negatively Affected by High-Altitude Fires?

Karsten Wesche

INTRODUCTION

Large-scale fires are especially common in semihumid to semiarid environments all over the globe. Not surprisingly, fires are equally important in tropical and subtropical mountain areas wherever climatic conditions are sufficiently moist to support continuous vegetation as fuel and where, at least occasionally, dry conditions make the vegetation flammable. Most mountains in tropical South America experience at least one dry season a year (Rundel, 1994; Diaz et al., 1997), and one or two dry seasons are also typical for mountains in tropical Africa (Wesche et al., 2000).

The literature on tropical-alpine fires in South America has grown tremendously over the last few decades, and by now fires are widely accepted as being one of the principal ecological factors in the páramos (see Hofstede and Rossenaar, 1995; Hofstede, 1995; Ramsay and Oxley, 1996; Luteyn, 1999, and references therein). The impact of fires on Afroalpine environments has been studied less extensively, but they clearly play an important role in defining the ecosystem. Afromontane vegetation has been influenced by natural fires for thousands of years (i.e., before the first permanent human settlements; Meadows and Linder, 1993), but they have also long been known to occur in Afroalpine environments (Dale, 1940; Hedberg, 1951). Their geographical distribution in tropical Africa is strongly related to the associated climatic conditions. Mountains at the eastern branch of the Rift Valley system generally experience semihumid conditions, and fires have been described from all high mountains in Tanzania, Kenya, and Ethiopia (Hedberg, 1964; Nievergelt et al., 1998; Wesche et al., 2000; Hemp and Beck, 2001). Similar conditions are found on Mt. Cameroon, perhaps the mountain in tropical Africa that has been most strongly affected by wildfires (Hall, 1973). In contrast, mountain ranges at the western part of the Rift Valley system in western Uganda, eastern Congo, and Rwanda receive precipitation from the Congo basin and have humid to perhumid conditions throughout the year. Hence, fires occur only under extreme conditions in the Rwenzori and the neighboring ranges (Langdale-Brown et al., 1964).

Although fires are common in the eastern African mountains, they do not occur every year. The region is well known for its unreliable precipitation (Nieuwolt, 1978), and dry seasons might be entirely absent in some years, making fire impact negligible. On the other hand, periods of severe drought are encountered every 2 to 10 years, and under such pronounced dry conditions, fires become widespread throughout the highlands. On Mt. Elgon, pronounced droughts occurred in 1997 and 1999, subsequently triggering extensive fires (Figure 3.1). The direct physiological impact of drought stress on tropical-alpine vegetation is still being assessed (Beck, 1994; Körner, 1999; Leuschner, 2000), but the indirect impact of drought conditions through fires is obvious (Wesche, 2003; Hemp, this volume).

Thus, the size and intensity of fires are controlled by climatic conditions, but the present fire frequency probably comes as a consequence of human land use. Natural ignition is

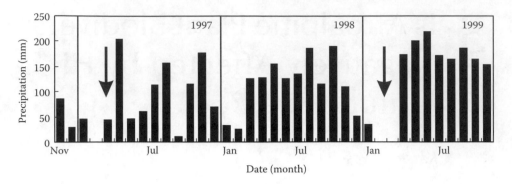

FIGURE 3.1 Total monthly precipitation at the study site from November 1996 to November 1999. Arrows indicate the timing of the extensive fires in 1997 and 1999.

often assumed (Beck et al., 1986), but very few authors have actually seen thunderstorms igniting fires in tropical mountains, and most report man-made fires instead (e.g. Luteyn, 1999; Kessler, 2000). I know of no published account of natural fires for the tropical mountains in Africa, and personal experience during 15 months on Mt. Elgon suggests that lightning is exceedingly rare during the critical dry season conditions. Mountain people frequently light fires for various reasons including improvement of hunting conditions, honey hunting and, at least in Ethiopia, pasture management (promotion of grasses). Thus, although natural fires most likely do occur from time to time (Beck et al., 1986), general fire frequency would clearly be lower than that of the actual fire regime.

The overall drought-triggered fire frequency is of little interest for Afroalpine ecology unless the spatial extent of the fire events is known, because what matters is the frequency with which a given vegetation stand is hit by fires. Comprehensive spatial surveys are rare, but fires in 1996 and 1997 affected large areas on Mt. Kilimanjaro (Hemp and Beck, 2001). On Mt. Elgon, a single fire in March 1997 devastated 70% of the central Afroalpine zone, and the proportion was estimated to be equally high in the ericaceous belt and the lower Afroalpine grasslands (Wesche, 2002a). The severe fires of the next extreme dry season in 1999 consumed most of the remaining intact high-altitude vegetation. Clearly, fires are spatially extensive enough to affect the vegetation structure and

biodiversity of most Afroalpine sites. Some available studies assess this impact by means of comparing burned and unburned sites (e.g. Miehe and Miehe, 1994), but true postfire succession has rarely been monitored. Among the few exceptions are studies on Mt. Elgon (Ekkens, 1988) and Mt. Kilimanjaro (Beck et al., 1986). Only the latter is based on observation plots; hence, I will present some new data from Mt. Elgon, which will serve as a base for a discussion on the impact of fire on Afroalpine biodiversity.

STUDY AREA

Mt. Elgon is an ancient volcano at the interstate boundary of Uganda and Kenya. The summit region collapsed in the tertiary so that the highest peak reaches only 4321 m asl. The upper limit of closed forest is presently found at 3000 to 3200 m and is mainly formed by broad-leaved trees. Remnant *Erica* trees occur up to 3800 m and bear evidence of the potential position of the upper treeline, which was clearly depressed in recurrent fires (Wesche, 2002a). At present, the so-called ericaceous belt (Hedberg, 1951) extends from the upper limit of broad-leaved forests to some 3900 m. It is characterized by patchy mosaics of ericaceous thickets and Afroalpine grasslands. Proper single-stemmed trees are rare in the ericaceous belt. Permanent observation plots were established in 1996 at various elevations and exposures on Mt. Elgon as part of a general study on plant phenology (described in detail in Wesche, 2002a). Here, only data for lower Afroalpine grasslands on the western slopes of Mt. Elgon are reported, in which postfire succession was monitored in detail.

SAMPLING DESIGN, DATA COLLECTION, AND ANALYSIS

SAMPLING DESIGN

The severe dry season of 1997 (Figure 3.1) triggered the occurrence of widespread fires in the lower Afroalpine grasslands and the ericaceous belt. Conditions were suitable for experimental burning, which was carried out on two sites (0.25 ha each; F1 and F2) in February 1997. Fires were moderately hot, i.e. they left some partly charred biomass standing. The plots were located near two permanent plots for phenological observations, which served as unburned controls (C1 and C2). Ten subplots of 10 m × 10 m were selected randomly within plots for phenological observations and within each of the burnt plots. The floristic differences within these subplots, as well as among plots and their respective controls, were tested with a multivariate procedure (multiple response permutation procedure based on Euclidean distance, McCune and Mefford, 1997), suggesting insignificant differences between F1 and C1 and between F2 and C2, respectively. The latter pair was situated at 3750 m asl and was covered by Afroalpine tussock grasslands. The other pair was at 3650 m asl; here, tussock grasslands had a somewhat higher cover of shrubs. Ericaceous scrub grows abundantly on the surrounding rocky outcrops, but the national park administration granted no permission to burn ericaceous vegetation. I monitored recovery only in ericaceous vegetation that burned "naturally." In these cases, there is a lack of data on the prefire conditions, making results incomparable, so they are reported elsewhere (Wesche, 2002a).

DATA COLLECTION

Species inventory and cover (LONDO scale; Londo, 1976) were monitored monthly on each of the 40 subplots in 1997; the sites were revisited 2 years later. However, the year 1999 was once again extraordinarily dry, and almost all of the Afroalpine grasslands were on fire, including the control plots C1 and C2 that were spared in 1997. Plots F1 and F2 remained unaffected. The fires of 1999 were very hot and combusted all standing biomass. Thus, it was impossible to take records of the presence or cover of plant species, and the respective samples are absent in the following data. The area was revisited in August 2002, i.e. more than 5 years after the experimental burning and 3 years after the severe fire in 1999. This time, the survey included only five subplots per plot, because analysis of the 1997 data revealed that differences among subplots on a given site were negligible (Wesche, 2002a). Therefore, this case study is based on the 4 × 5 subplots that were followed over a period of 5 years. Subplots within a plot are spatially autocorrelated; therefore subplots are pseudoreplicates (Hurlbert, 1984) nested within the "true" replicates — i.e. the two fire plots and their controls. True replication of fire events involves a much higher technical effort and has not been carried out in Afroalpine environments so far. Thus, I present these data as the best available to date and caution the reader to keep the weakness of the design in mind (cf. Oksanen, 2001).

DATA ANALYSIS

For analysis purposes, I simply calculated species richness and evenness (based on the Shannon approach; Kovach, 1995), which gives an idea of the relative contribution of the species to the (Shannon) diversity index. Bray–Curtis similarity was expressed in reference to the first sample taken before burning. When calculated only for qualitative data (presence or absence of species), Bray–Curtis similarity is identical to Sörensen similarity. In its quantitative version, it is also known as Steinhaus similarity (Legendre and Legendre, 1998). This asymmetrical coefficient was used because it gives special weight to double occurrences. In addition, I performed a detrended correspondence analysis of the raw data (DCA) to assess overall floristic heterogeneity. This ordination method has the great advantage that distances between samples are rescaled such that they correspond to species turnovers. In DCA, four multivariate standard deviations roughly correspond to one species turnover (cf. Jongman et al., 1995). Calculations were performed with PCORD 3.15 (McCune and Mefford, 1997).

RESULTS

Figure 3.2 summarizes the simple univariate characteristics for the four plots. Vascular plant species richness in the prefire samples comprised between 30 and 35 species on all plots. F1 and F2 were burned in February 1997, and 4 weeks later species richness was below 25. Recovery, however, was fast. Five months after burning, mean numbers had reached the same levels as before. Somewhat surprisingly, richness continued to rise until the first part of the experiment stopped after 9 months (in November 1997). This increase was mainly a consequence of the availability of open sites, which facilitated the establishment of weakly competitive species. This effect was transitory, because 2 years after experimental burning, species numbers were at the same level as before the experiment, and remained so until 2002.

Control plots C1 and C2 were followed for 10 fire-free months in 1997. The slight fluctuation from month to month indicates a weak phenology (Figure 3.2). When the sites burned in 1999, fires were so hot that all standing biomass was consumed, and no meaningful samples could be taken. Three years later, in 2002, stands had largely recovered and species richness achieved the same range as before the fire (Figure 3.2). Thus, the impacts of fire on plant species richness were transitory on all sites.

The evenness was used to assess changes in the dominance structure. Thin lines in the diagrams of Figure 3.2 indicate that the mean evenness was similar on all plots at around 0.70 to 0.75. Neither phenological nor successional changes after burning were observed, suggesting that dominance structures did not change much during the 5-year observation period.

Floristic composition was slightly more affected than mere species richness. In the right part of Figure 3.3, both qualitative and quantitative Bray–Curtis similarity was calculated in reference to the prefire sample. If based on the presence or absence data alone, similarity on the plots of F1 and F2 dropped to below 0.8

immediately after burning, but reached between 0.8 and 0.9 just 3 months after the burning. Again, there were some fluctuations but the similarity remained in this range in 1997 and was similar when sites were resampled in 1999 and 2002. For controls, differences between repeated samples and initial records were almost negligible (right part of Figure 3.3), suggesting that fluctuations were minor before accidental burning. In 2002, i.e. 3 years after burning, the similarity to the initial records was also around 0.85. This indicates some, albeit relatively small change in the floristic composition caused by the severe fires of 1999.

The picture was somewhat different when species-cover values were included in the calculation of Bray–Curtis similarity (Figure 3.3, bold lines). Before burning, controls remained stable, including relative abundances of the species, whereas burned plots did not fully recover from fires in 1999. In 2002, quantitative similarity was at around 0.7. In the experimental plots, burning induced a change in species abundance. Quantitative similarity to prefire conditions dropped below 0.7 in both plots F1 and F2, and remained at this level during all following records. Thus, recovery from fire impact was almost complete in terms of species composition, but abundances of species were permanently altered by fire (at least during the 5-year study period).

As ordination analysis indicates, however, these changes were still relatively small. Simple DCAs of the data (Table 3.1) revealed very short multivariate gradients, and samples within a succession on a given plot were always less than 0.6 s.d. distant from each other. In other words, species turnover rates between the most dissimilar samples of a given subplot were below 15% during the entire study period. This finding corresponds to the results given earlier and to visual inspection of the data, which suggested that plant communities were largely comparable before and after the fires.

Despite being relatively minor, overall fire effects differed somewhat among major plant life-forms (Table 3.2). For most life-forms, numbers of species were depressed after experimental burning of F1 and F2, but usually recovered quickly to preburning levels. Nonetheless, the number of grass species on both

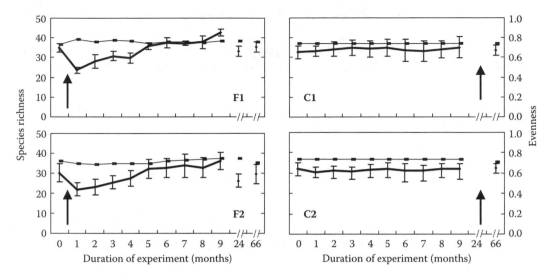

FIGURE 3.2 Mean richness of vascular plant species (bold lines with bars indicating standard deviations) and mean evenness (thin lines without error bars) for richly structured tussock grasslands (F1 and its control C1) and an entirely grassland-dominated plot (F2 and its control C2, "pseudo-n" = 5). Arows indicate timing of fires.

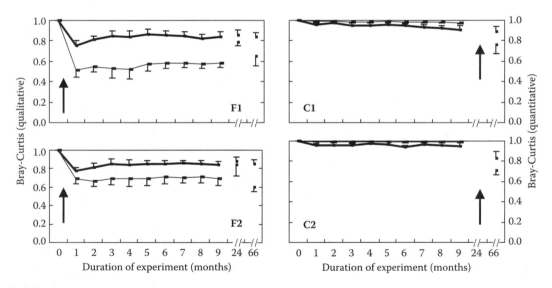

FIGURE 3.3 Bray–Curtis similarity among the initial sample taken before experimental burning and the following stages. Similarity was calculated for presence or absence data alone (i.e. Sörensen similarity: bold line with error bars for standard deviation directed upwards) and for species abundance data (thin lines with error bars directed downwards).

plots increased in the first 10 months and remained above preburning levels during the course of the study. At 24 and 66 months after burning, mean species richness of rosettes was still lower than before the experiment started; trends for erect herbs differed among plots and were weak overall. All other life-forms showed no clear trend in their recovery rate after burning and were hardly affected in terms of richness. The pattern on the controls was similar (Table 3.2). Before incidental burning in 1999, life-form composition fluctuated somewhat due

TABLE 3.1
Results of Detrended Correspondence Analyses for the four plots. Listed are the lengths of multivariate gradients along the first three axes after detrending (4 s.d. = 1 complete species turnover)

	F2	F3	Co2	Co3
Axis 1	1.67 s.d.	1.50	0.99	1.02
Axis 2	1.07	1.18	0.69	0.57
Axis 3	0.83	1.00	0.52	0.49

TABLE 3.2
Mean number of species of different life-forms of two burnt plots (Fire Plot 1, 2) and their respective controls, at 5, 10, 24, and 66 months after experimental fire

Month after fire	Mikro phanerophyte/ shrub	Chamaephyte/ dwarf shrub	Caespitose hemicryptophyte/ perennial grass	Scapose hemicryptophyte/ erect forb	Rosette hemi-cryptophyte/ rosette forb	Geophyte	Therophyte/ annual
Fire plot 1							
0	6.6	2.8	8.0	8.0	7.8	0.6	1.2
5	6.4	3.2	6.4	6.6	6.6	0.0	0.6
10	7.2	3.2	13.0	9.2	9.4	0.8	1.2
24	6.8	3.4	8.6	8.0	6.6	0.0	0.8
66	6.6	3.4	9.6	9.2	6.4	0.6	1.0
Control 1							
0	6.6	3.0	8.0	7.2	6.6	1.2	0.0
5	6.4	3.0	8.2	8.0	7.8	1.0	0.0
10	6.0	3.0	8.8	8.8	7.2	1.2	0.0
24	fire	—	—	—	—	—	—
66	7.0	3.0	9.0	7.8	6.2	0.6	0.0
Fire plot 2							
0	6.4	2.4	6.8	6.6	7.6	0.4	0.0
5	6.0	2.4	6.6	5.0	7.2	0.2	0.0
10	6.2	2.6	10.6	7.4	8.4	1.0	0.0
24	6.2	2.0	6.6	4.2	7.2	0.0	0.0
66	6.2	2.6	8.4	5.0	6.8	0.4	0.0
Control 2							
0	7.6	2.2	6.6	7.2	8.0	0.4	0.0
5	7.6	2.2	6.8	6.6	8.0	0.4	0.0
10	7.6	2.4	7.0	6.6	8.0	0.6	0.0
24	fire	-	—	—	—	—	—
66	7.4	2.2	7.2	7.8	7.4	0.6	0.0

to phenological changes. There were no records directly after the hot fire, but 3 years later, the numbers had reached preburning levels. Trends were small in most life-forms though, again, richness of grasses remained higher in post-burning vegetation, and numbers of rosettes were lower than before burning. In contrast to the fire plots, numbers of erect herbs were higher on both controls after fire; all other life-forms remained more or less constant.

DISCUSSION

The data support the idea that lower Afroalpine grasslands on Mt. Elgon are mostly replacement communities, which have benefited from recurrent fires. The impacts of fire on their diversity are generally minor, and effects quickly disappear if fires are only moderately intense. Even after relatively hot fires, as on C1 and C2, prefire conditions were largely recovered after some years. After fires, open soil was abundantly available for the establishment of new species, which explains some of the changes observed in species richness. However, even changes on intensively burnt plots C1 and C2 remained small, as seen by the relatively small floristic gradients in the data set (Table 3.1). Species turnover values among subsequent samples on all subplots were less than 20% and possibly in the same range of magnitude as long-term successional trends induced by stochastic processes or trends in the climate. The exception is a general, albeit small, increase in the number of grass species, which was found on all plots after fires. Nonetheless, species set and abundances remained similar enough to allow for the grouping of all plots and successional stages in the same plant community (*Festuca pilgeri–Euphorbia wellbyi* community; Wesche, 2002a), thereby corresponding to a typical autosuccession.

This result corresponds to findings on Mt. Kilimanjaro, where Afroalpine fires did not induce a change of grassland communities, and regeneration stages were within the prefire community (Beck et al., 1986). All life-forms were capable of survival, although giant groundsels (*Dendrosenecio* spp.) had often died. This sensitivity was also reported from Mt. Elgon (Ekkens, 1988). More recent studies in 1997 and 1999 suggest that *Dendrosenecio elgonensis* survives fire if the flames do not reach the green leaves (Wesche, 2002a). Once again, the question appears to be one of fire intensity.

Overall plant species richness of Afroalpine grasslands was not adversely affected by fires on Mt. Kilimanjaro, and tussock grasslands had recovered completely when visited 4 years after the fire (Beck et al., 1986). On Mt. Kilimanjaro, however, shrub recovery was clearly slower

than for herbaceous species and was not completed within the same timespan. The high regenerative capability of tussock grasses was also described by Ekkens (1988) and was confirmed in the present study. Species richness of *Erica trimera* scrub on Mt. Kilimanjaro was slightly lower 4 years after the fire (Beck et al., 1986), but the effect was less apparent than the change in the structure of the plant community that had not nearly recovered. In contrast to grasslands, differences in the succession were sufficiently clear to designate separate stages, if not truly distinct communities.

These studies show that fires have only small and mostly transitory effects on the biodiversity and structure of lower Afroalpine grasslands. This finding is not surprising, given that fires are generally common in these environments as described earlier. Any presently widespread vegetation type has to be relatively fire-tolerant or else it would have been replaced during previous fire events. In other words, the vegetation we find on most eastern African mountains today is not changed under a moderate fire frequency because its present structure is the result of a moderate fire frequency. This observation leads to the question of whether less-fire-tolerant communities have been replaced. For the ericaceous belt, the possibly replaced vegetation types include communities dominated by woody perennials. Such thickets are commonly found on rocky outcrops, on boulder streams, or in shady and moist valleys. This ericaceous vegetation is fire-tolerant in the sense that most species, including the *Erica* spp. (Beck et al., 1986; Hemp and Beck, 2001), survive even hot fires and are capable of resprouting from burls or other parts. However, recovery is slower than for many grassland species because shoots of *E. trimera* and *E. excelsa* will not grow more than 60 cm within 2 years (Wesche et al., 2000; Wesche, 2002b). Thus, the present fire frequency often does not allow for complete recovery and has presumably led to a large-scale replacement of ericaceous communities.

Grasses (e.g. *Deschampsia flexuosa*) have been observed invading burnt woody vegetation on Mt. Elgon and on Mt. Kilimanjaro (Beck et al., 1986). Dominance of *D. flexuosa* is temporary in Afroalpine environments as it is in

European *Calluna* heathlands (Mallik and Gimingham, 1983). However, once other large tussock grasses (*Festuca* spp., *Koeleria capensis*) have invaded the stands at a later stage, the establishment of tree seedlings becomes more difficult (Klötzli, 1975) because of the grasses' competitive power and the possibly increased fire frequency. A similar positive feedback switch (Wilson and Agnew, 1992) might well operate in the interplay of ericaceous vegetation and burning-intolerant Afromontane broad-leaved forest below the timberline, where *Erica excelsa* forests proliferate on burnt sites (Hemp and Beck, 2001).

Notwithstanding the secondary character of *Erica* stands below the present timberline, it is clear that the uppermost forest stands in the African mountains would be constituted by ericaceous woody perennials under natural conditions. Most authors agree that the present timberline has been depressed by some 300 to 800 meters vertically and that fires have induced a large-scale replacement of dense ericaceous vegetation by grasslands and scrub (Hall, 1973; Schmitt, 1991; Miehe and Miehe, 1994; Menassie and Masresha, 1996; Miehe, 2000; Wesche, 2002a). This corresponds to observations in Madagascar, where *Erica* species immediately invaded Afromontane grasslands once burning ceased (Bloesch et al., 2002). The situation is analogous to South America, where *Polylepis* forests were widely replaced by open páramo vegetation (e.g. Ellenberg, 1979; Balslev and Luteyn, 1992; Kessler, 2000). Similar replacements have occurred at the treeline in tropical Southeast Asia (e.g. Corlett, 1984).

Thus, we face a different question when assessing the impact of fire on Afroalpine diversity: Is the present vegetation less rich in species than a hypothetical, undisturbed vegetation? This poses a fundamental problem for all further studies on the impact of fire on Afroalpine diversity, because it is difficult to find any truly undisturbed sites. Fires sweep through grasslands and thickets, and I even witnessed them reaching outcrops that were mostly rocky. The few sites that are possibly safe from fire are edaphically very different from the surroundings. *Hypericum revolutum* scrub on boulder streams was always spared from fires on Mt. Elgon, but soils among the rocks are

very different from those on other sites (higher moisture and nutrient levels). However, most plants growing on boulder streams are also found in the burnt grassland and scrub; the only exception includes nitrophytic species such as *Anthriscus sylvestris* and *Parietaria debilis* (Wesche, 2002a). The uppermost *Erica* forests are also exclusively restricted to special site conditions, largely because rocky outcrops and boulder streams offer some fire protection. Even these stands are partly opened and disturbed and can hardly be compared to Afroalpine grasslands on deep soils.

In principle, it is possible to extract data on α-diversity from the various phytosociological surveys that have been conducted for all major mountains; an example is provided in Table 3.3. The data suggest that the *Erica excelsa* forest is generally richer in species than tussock grasslands at similar altitudes. This estimation refers to disturbed forests, as the only unburned stand encountered on Mt. Elgon had some 25% fewer species than the disturbed forest and had an almost equivalent richness as the grasslands (Table 3.3). A single record is not sufficient to draw conclusions, but it supports the idea that an unburned forest is at least not exceedingly more diverse than frequently burned stands. Most of the additional species found in the disturbed forest come from the surrounding grasslands, whereas hardly any species are truly restricted to *Erica* forests (Wesche, 2002b). Except for *E. excelsa* itself, only the small erect herb *Wahlenbergia krebsii* and the shrub *Helichrysum nandense* were common in the upper-montane *E. excelsa* forest on Mt. Elgon and were not found at all in the surrounding vegetation (Wesche, 2002a).

Thus, diversity of the entire mountain is not affected by whether we have grasslands or ericaceous vegetation. Landscape heterogeneity is increased due to the patchy nature of the fires. This development should lead to a higher β-diversity, providing there are no extinctions. This idea is supported by the observation that, for Mt. Elgon, so far none of the high-altitude endemics has become extinct, although fire frequency has increased during the 20th century (Wesche, 2002a).

Clearly, it is hard to draw definite conclusions from this type of data. Apart from the

TABLE 3.3
Mean vascular plant species richness along a transect on the western slopes of Mt. Elgon based on phytosociological relevés

Vegetation	Montane broad-leaved forest	Bamboo forest	*Hagenia* forest	*Erica* forest, intact	*Erica* forest, disturbed	Pure tussock grassland	Shrubby tussock grassland	*Alchemilla* thickets
Mean richness	**53**	**22**	**21**	**35**	**44**	**27**	**28**	**24**
Elevation (masl)	2875	3050	3250	3300	3460	3600	3870	3850
Evidence for fires	No	No	No	No	Recent	Recent	Old	Recent
Samples	4	3	3	1	6	9	19	13

Source: From Wesche (2002a).

problem that site conditions are not strictly comparable between the various communities and, therefore, a space-for-time substitution seems inappropriate, another fundamental problem arises with such an approach: Phytosociological relevés are not sampled randomly, and this introduces a potentially severe bias when biodiversity is analyzed (Chytry, 2001). Therefore, large-scale comparisons of intermittently burned and unburned sites are less promising than long-term experiments with controlled fire (and no-fire) treatments. Unfortunately, to my knowledge these are not yet available for any of the Afroalpine environments.

SUMMARY

Grasslands and ericaceous thickets are the most common vegetation types of the frequently burned ericaceous belt in the African mountains. This study presents data on the impact of fire on species composition in lower Afroalpine grasslands on Mt. Elgon, Uganda. Two plots were burned experimentally, and recovery rates and extent were monitored over a total of 66 months. Additional records were kept for two control sites, which were both incidentally burned by naturally caused fires in the second year of the experiment. Changes in richness and species composition were generally small after burning, and full recovery was quickly achieved. Evenness was also hardly affected by burning, though relative abundance of species remained slightly changed after plots had been burned. The only life-forms that increased in richness after burning were perennial grasses. Nonetheless, the overall impact of fire on these grasslands was relatively minor, which supports the idea that they originated from earlier fires. The grasslands replaced denser woody vegetation that had been suppressed by recurrent burning. Moreover, fire-sensitive plants were still found on the mountain, and overall diversity in the high-altitude environment of Mt. Elgon appears to be largely unaffected by extensive fires.

ACKNOWLEDGMENTS

The fieldwork was financially supported by the Deutsche Forschungsgemeinschaft, Deutscher Akademischer Austauschdienst, and by the Studienstiftung des Deutschen Volkes. The Uganda Wildlife Authority and the Ugandan Council of Science and Technology kindly granted access to the study sites. On Mt. Elgon, Z. Gibaba, V. Clausnitzer, and R. Neumann assisted with data collection. Suggestions by V. Clausnitzer and C. Ohl helped to clarify the ideas presented here. H. Burbank and D. McCluskey corrected my English.

References

Balslev, H. and Luteyn, J.L. (1992). *Páramo. An Andean Ecosystem under Human Influence*. Academic Press, London.

Beck, E. (1994). Cold tolerance in tropical alpine plants. In Rundel, P.W., Smith, A.P., and Meinzer, F.C., (Eds.), *Tropical Alpine Environments*. Cambridge University Press, Cambridge, pp. 77–110.

Beck, E., Scheibe, R., and Schulze, E.D. (1986). Recovery from fire: observations in the alpine vegetation of western Mt. Kilimanjaro (Tanzania). *Phytocoenologia,* 14: 55–77.

Bloesch, U., Bosshard, A., Schachenmann, P., Rabetaliana, H., and Klötzli, F. (2002). Biodiversity of the subalpine forest-grassland ecotone of the Andringitra Massif, Madagascar. In Körner, C. and Spehn, E.M. (Eds.), *Mountain Biodiversity. A Global Assessment*. Parthenon Publishing, London, New York, pp. 165–175.

Chytry, M. (2001). Phytosociological data give biased estimates of species richness. *Journal of Vegetation Science,* 12: 439–444.

Corlett, R.T. (1984). Human impact on the subalpine vegetation of Mt. Wilhelm, Papua New Guinea. *Journal of Ecology,* 72: 841–854.

Dale, I.R. (1940). The forest types of Mt. Elgon. *East African Ugandan Nat Hist Society,* 9: 74–82.

Diaz, A., Pefaur, J.E., and Durant, P. (1997). Ecology of southern American páramos with emphasis on the fauna of the Venezuelan páramos. In Wiegolaski, F.E. (Ed.), *Ecosystems of the World 3. Polar and Alpine Tundra*. Elsevier, Amsterdam, pp. 263–310.

Ekkens, D. (1988). Fire and regrowth on Mount Elgon. *Swara,* 11: 30–31.

Ellenberg, H. (1979). Man's influence on tropical mountain ecosystems in South America. *Journal of Ecology,* 67: 401–416.

Hall, J.B. (1973). Vegetational zones of the southern slopes of Mount Cameroon. *Vegetatio,* 27: 19–69.

Hedberg, O. (1951). Vegetation belts of the East-African mountains. *Svensk Botanisk Tidskrift,* 45: 141–196.

Hedberg, O. (1964). Features of afro-alpine plant ecology. *Acta Phytogeographica Suecica,* 49: 1–144.

Hemp, A. and Beck, E. (2001). *Erica excelsa* as a fire-tolerating component of Mt. Kilimanjaro's forests. *Phytocoenologia,* 31: 449–475.

Hofstede, R.G. and Rossenaar, A.J.G.A. (1995). Biomass of grazed, burned, and undisturbed Páramo grasslands, Colombia. II. Root mass and aboveground:belowground ratio. *Arctic and Alpine Research,* 27: 13–18.

Hofstede, R.G.M. (1995). The effects of grazing and burning on soil and plant nutrient concentrations in Colombian páramo grasslands. *Plant and Soil,* 173: 111–132.

Hurlbert, S.H. (1984). Pseudoreplication and the design of ecological field experiments. *Ecological Monographs,* 54: 187–211.

Jongman, R.H.G., ter Braak, C.J.F., and van Tongeren, O.F.R. (1995). *Data Analysis in Community and Landscape Ecology*. Cambridge University Press, Cambridge.

Kessler, M. (2000). Observations on a human-induced fire event at a humid timberline in the Bolivian Andes. *Ecotropica,* 6: 89–94.

Klötzli, F. (1975). Zur Waldfähigkeit der Gebirgssteppen Hoch-Semiens. *Beitr. naturk. Forsch. Südw. Dtl.,* Band 34: 131–147.

Körner, C. (1999). *Alpine Plant Life*. Springer-Verlag, Berlin.

Kovach, W.L. (1995). *MVSP Plus*. Kovach Computing Services, Pentraeth, Wales.

Langdale-Brown, I., Osmaston, H.A., and Wilson, J.G. (1964). *The Vegetation of Uganda*. The Government Printer, Entebbe.

Legendre, P. and Legendre, L. (1998). *Numerical Ecology*. Elsevier, Amsterdam.

Leuschner, C. (2000). Are high elevations in tropical mountains arid environments for plants? *Ecology,* 81: 1425–1436.

Londo, G. (1976). The decimal scale for releves of permanent quadrates. *Vegetatio,* 33: 61–64.

Luteyn, L. (1999). *Páramos: A Checklist of Plant Diversity, Geographical Distribution and Botanical Literature*. New York Botanical Garden Press, New York.

Mallik, A.U. and Gimingham, C.H. (1983). Regeneration of heathlands following burning. *Vegetatio,* 53: 45–58.

McCune, B. and Mefford, M.J. (1997). *PC-ORD: Multivariate Analysis of Ecological Data*. MJM Software, Gleneden Beach, OR.

Meadows, M.E. and Linder, H.P. (1993). A paleoecological perspective on the origin of Afromontane grasslands. *Journal of Biogeography,* 20: 345–355.

Menassie, G. and Masresha, F. (1996). Plant communities of the Afroalpine vegetation of Sanetti Plateau, Bale Mountains, Ethiopia. *Sinet,* 19: 65–86.

Miehe, G. (2000). Comparative high mountains research on the treeline ecotone under human impact. *Erdkunde* 54: 34–50.

Miehe, S. and Miehe, G. (1994). *Ericaceous Forests and Heathlands in the Bale Mountains of South Ethiopia — Ecology and Man's Impact*. Stiftung Walderhaltung in Afrika, Hamburg, Germany.

Nieuwolt, S. (1978). Rainfall variability and drought frequencies in East Africa. *Erdkunde*, 32: 81–88.

Nievergelt, B., Good, T., and Güttinger, R. (1998). A survey on the flora and fauna of the Simen Mountains National Park. *Walia (special issue)*: 1–109.

Oksanen, L. (2001). Logic of experiments in ecology: is pseudoreplication a pseudoissue? *Oikos*, 94: 27–38.

Ramsay, P.M. and Oxley, E.R.B. (1996). Fire temperatures and postfire plant community dynamics in Ecuadorian grass páramo. *Vegetatio*, 124: 129–144.

Rundel, P.W. (1994). Tropical alpine climates. In Rundel, P.W., Smith, A.P., and Meinzer, F.C. (Eds.), *Tropical Alpine Environments*. Cambridge University Press, Cambridge, pp. 21–44.

Schmitt, K. (1991). *The Vegetation of the Aberdare National Park Kenya*. Wagner, Innsbruck, Austria.

Wesche, K. (2002a). The high-altitude environment of Mt. Elgon (Uganda/Kenya) — climate, vegetation and the impact of fire. *Ecotropical Monographs*, 3: 1–253.

Wesche, K. (2002b). Structure and dynamics of *Erica* forest at tropical-African treelines. *Verhandlungen der Reinhold-Tüxen Gesellschaft* 14: 145–159.

Wesche, K. (2003). The importance of occasional droughts for afroalpine landscape ecology. *Journal of Tropical Ecology*, 19: 197–208.

Wesche, K., Miehe, G., and Kaeppelli, M. (2000). The significance of fire for Afroalpine ericaceous vegetation. *Mountain Research and Development*, 20: 340–347.

Wilson, J.B. and Agnew, A.D.Q. (1992). Positive feedback switches in plant communities. *Advances in Ecological Research*, 23: 264–336.

4 The Impact of Fire on Diversity, Structure, and Composition of the Vegetation on Mt. Kilimanjaro

Andreas Hemp

INTRODUCTION

Fire is a driving ecological force on East African mountains, as thoroughly documented by Beck et al. (1986), Schmitt (1991), Miehe and Miehe (1994), Lange et al. (1997), Wesche et al. (2000), and Hemp and Beck (2001). The impact of fire on vegetation depends largely on the type of vegetation. Savannas have been influenced by fire for millennia while at the same time remaining stable ecosystems (Eva and Lambin 2000) containing many fire-adapted species. By contrast, fire acts as an ecological transformation factor in forests.

On Mt. Kilimanjaro, the number and intensity of wild fires is increasing, which is — similar to the shrinking of the glaciers — linked to regional climatic changes such as warming and decline in precipitation (Hemp, 2005). However, as fires on Mt. Kilimanjaro are, in most cases, lit by man, the increasing number of fires is linked to the increase in human activity. This holds true in particular for the forest fires in the submontane and lower montane zone, where human alterations of the landscape play a very important role. Similar to other moist tropical forests (cf. Eva and Lambin 2000), the increasing number of fires in the lower forest belt is most likely due to a forest cover change in these areas. During intensive vegetation studies since 1996, only very few undisturbed forest plots were found below 1900 m due to selective logging on Mt. Kilimanjaro. This creates a mosaic of clearings and differ-ently closed forest regeneration stages, in which the herb layer is exposed to sunshine and a drier microclimate, which increases the frequency of forest fires.

This differs from the situation in the upper regions of the mountain. In contrast to other high-altitude regions in East Africa, such as Ethiopia or Mt. Elgon in Kenya, there is no grazing or agriculture above the forest belt, and logging in the upper forest zone is very rare until now. However, even these remote areas are not free from human influence, as population on the foothills has increased enormously. Since 1895, population has multiplied 20 times (Hemp et al. in press). This means an increase in human activity in all altitudinal zones and areas, promoted in particular by tourism: Since the establishment of Kilimanjaro National Park in 1972, the number of visitors has multiplied about five times. About 100,000 people visit the upper regions of Mt. Kilimanjaro every year, as porters, guides, and tourists. Such increasing numbers of visitors have, of course, effects on the environment. Therefore, the biological effects of the changing climatic conditions are boosted by human influence.

In this study, the following questions are addressed: Which are the most fire-affected vegetation zones on this mountain? How does biodiversity (species diversity, community diversity, and functional-type diversity) change? What is the consequence for the position and structure of the treeline ecotone? These questions will be explored by the presentation of data based on vegetation relevés representing all the major vegetation types and the complete range of ecological zones of Mt. Kilimanjaro.

STUDY AREA

LOCATION AND TOPOGRAPHY

Mt. Kilimanjaro is located between 2°45 and 3°25 south and 37°00 and 37°43 east, about 300 km south of the equator, in Tanzania on the border with Kenya. The mountain, with a diameter of roughly 90 km, is a more or less eroded remnant of an ancient volcano. The remaining three peaks (Shira, Mawenzi, and Kibo) rise from the savanna plains at 700-m elevation to a snow-clad summit of 5895-m altitude.

CLIMATE

Mt. Kilimanjaro is characterized by a typical equatorial daytime climate. Due to its proximity to the equator, there are two distinct rainy seasons: a long rainy period lasting from March to May, and a short period around November–December. However, rainfall and temperature vary with altitude and exposure to the dominant wind blowing from the Indian Ocean. Annual rainfall reaches a maximum of around 3000 mm at 2100 m on the central southern slope in the lower part of the montane forest belt, clearly exceeding the total precipitation of other East African high mountains (Hemp 2001a, in press). Higher up, at 2400, 2700, and 3000 m, some 80, 70, and 50%, respectively, of this maximum rainfall was observed. In the alpine zone above 4000 m, the mean annual precipitation decreases to 500 mm and less. The northern slopes, on the lee side of the mountain, receive much less rainfall.

The mean annual temperature in Moshi at an altitude of 813 masl is 23.4°C (Walter et al. 1975). It decreases to 18°C at 1500 m, 9.2°C at 3100 m, 4.5°C at 4000 m (author's measurements), and, on top of Kibo at about 5800 m, to −7.1°C (Thompson et al. 2002) with a lapse rate of about 0.56°C per 100 m (Hemp, in press). The climate in the alpine belt above 3500 to 4000 m is characterized by nightly frosts all year round and with intensive sunshine

during daytime — summer every day and winter every night (Hedberg, 1964).

LAND USE AND VEGETATION

On the slopes of Mt. Kilimanjaro, several vegetation zones can be identified (Figure 4.1; cf. Hedberg 1951; Pócs 1994; Hemp et al. 1999, 2002a, in press; vegetation map of the whole mountain in Lambrechts et al. 2002). Between 700 and 1000 m, a dry and hot savanna zone stretches around the mountain base. Most of this area is used for crop production (maize, bean, and sunflower) or for grazing. Remnants of the former savanna vegetation can be found only in the surroundings of Lake Chala in the eastern foothills as well as in the northern and western foothills.

Due to the favorable climate, the main zone of agriculture and horticulture with banana and coffee plantations extends from 1000 to 1800 m. This zone has been inhabited by the Chagga, a Bantu tribe, for several centuries. At present, the population density is more than 500 persons/km².

Today, the intensively cultivated zone borders the montane forest belt between 1600 and 1800 m in the southern and eastern parts of the study area and at 1300 m in the western parts. Closed tall forests extend to about 3100 m, surrounding the whole mountain. In the western and northern parts of the mountain, the comparatively dry submontane forest below 1600 m is dominated by *Calodendrum capense*, *Olea europaea* ssp. *africana*, *Croton megalocarpus*, and *Diospyros abyssinica*. Above 1600 m, *Cassipourea* forests prevail in this area, which are substituted above 2400 m by *Podocarpus* and *Juniperus* forests. On the southern slope, the most important tree is the camphor tree, *Ocotea usambarensis*. In the lower part, it occurs syntopic with *Agauria salicifolia* and *Macaranga kilimanjarica*. In the upper part, it is associated with *Podocarpus latifolius*. Above 2800 m, *Podocarpus latifolius*, *Hagenia abysinica*, and *Erica excelsa* are the dominant tree species.

Between 3100 and 3500 m, the tall (tree height 6 to 15 m) *Erica excelsa* forests are replaced by lower-stature bushlands (height, 1.5 to 2 m) dominated by *Erica arborea*, *Erica trimera*, *Protea caffra*, and *Euryops*

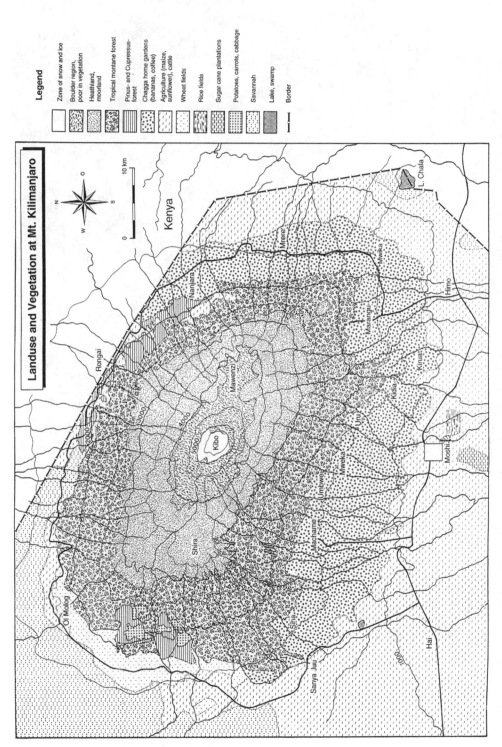

FIGURE 4.1 Land use and vegetation at Mt. Kilimanjaro, based on the interpretation of a SPOT3 satellite image from July 2, 1994. Dotted lines: location of transects.

FIGURE 4.2 Burnt areas and fire regeneration stages since 1996. White areas: fires of the year 2000 on cultivated ground and savanna. Black areas: fire regeneration stages since 1996 in the ericaceous belt. Black dots: fires of restricted extent in the montane forests. Background to the map: two combined Landsat ETM 7 images taken on January 29 and February 21, 2000. Source: USGS/UNEP-GRID-Sioux Falls.

dacrydioides (Figure 4.3a). In the southeastern parts, moorland vegetation, formed by tussock grasses and characterized by giant lobelias, lobes down into the forest belt (Figure 4.3b). At an altitude of about 3900 m, the *Erica* bush grades into an open *Helichrysum* cushion belt (Figure 4.3c), which extends up to 4500 m.

There is very little plant life above 4500 m, and the top of Kibo is covered by bare rock fields and glaciers (for details on the vegetation of Mt. Kilimanjaro, see Hedberg 1951; Klötzli 1958; Beck et al. 1983; and Hemp 2001a, in press).

FIGURE 4.3A *Erica trimera* and *E. arborea* (flowering on the right-hand side) bush at 3200 m substituting upper montane forest after recurrent fires; (b) Mosaic-type interface between *Podocarpus* forests, *Erica excelsa* forests, and different fire-regeneration stages (*Erica* and *Stoebe* bush; *Festuca obturbans* tussock grassland) in the moorland zone in southeast Kilimanjaro at 2600 to 2900 masl (November 2003). Recurrent fires enhance habitat diversity and, with this species richness, enable alpine plants and plant communities to migrate downslope into formerly forested areas; (c) Transition between *Helichrysum newii* cushion vegetation and remnants of *Erica* bush (in the center of the picture with *E. trimera* and yellow flowering *Euryops dacrydioides*) at the potential treeline at 4000 m; Mweka route, March 2003. Recurrent fires have lowered this borderline by several hundred meters in many areas in the last three decades. In the background, Mawenzi, the second highest peak of Kilimanjaro; (d) Burning subalpine cricaccous shrubland between Great and Little Barranco at the northeastern slope of Mawenzi (February 2000), lit by poachers; (e) Burnt lower-montane *Agauria–Ocotea* forest, southwest Mt. Kilimanjaro above Nrwaa at 1800 masl, July 2001. The clearings (above the burnt forest) are densely covered by *Pteridium aquilinum*, impeding forest regeneration for decades. Photo: Christian Lambrechts; (f) Mass flower of the giant groundsel *Senecio johnstonii* var. *cottonii* after fire. These giant rosette plants flower mainly after fire and can survive even with "naked" bare stems without getting damaged by frost; inlets left: unburnt giant groundsel, with dead leaves still coating the stems; below: inflorescence; (g) Regeneration of *Erica excelsa* in a burnt forest at 3400 m asl from seeds and stem bases 5 years after a fire leading to a monodominant *Erica* stand of uniform age; (h) Destroyed (by the fires of the years 1996–1997) *Erica* forests at 3400 masl. Such cloud forests had a high potential of collecting fog water. Forest above Kibosho, Weru-Weru River system, November 2003.

FIGURE 4.3B

FIGURE 4.3C

FIGURE 4.3D

FIGURE 4.3E

FIGURE 4.3F

FIGURE 4.3G

FIGURE 4.3H

METHODS

Data were collected since 1996 by means of about 1400 relevés that were established at 100-m intervals along 30 transects (Figure 4.1), 5 of which were tourist trails (Marangu, Mweka, Umbwe, and Machame routes, and the Maua rescue route), whereas the others were poacher trails or were simply following ridges or valleys without any path.

Using the method of Braun-Blanquet (1964), these relevés cover all important vegetation types (cf. Hemp 2001a). Plot size was chosen with respect to the minimum areas of the different vegetation formations. In forests, it was mainly 1000 m²; in forest clearings, grasslands, and heathlands, 100 m²; in salt marshes, swamps, and ruderal vegetation, 25 m²; and in rock habitats, 5 m². Based on a classification of satellite images, vegetation maps were produced for various years since 1976 to quantify land cover changes.

The low-stature and open *Erica* woodland of the subalpine zone in this chapter is due to its structure and floristic composition, which differs significantly from the tall, closed forests of Kilimanjaro (cf. the vegetation tables in Hemp and Beck 2001, Hemp 2001a), not regarded as forest but as bush. The term *subalpine* is applied to the transition zone between the broad-leaf montane forest and the alpine *Helichrysum* scrub vegetation (the treeline–ecotone in the sense of Körner 2003). It corresponds closely to the "ericaceous belt" of Hedberg (1951; for the definition of the different altitudinal zones, compare Hemp, in press).

RESULTS

RECENT FREQUENCY OF FIRES

The analysis of satellite images clearly shows a bimodal distribution of fire, following the precipitation regime (Figure 4.2): regular fires every year in the colline savanna zone and in the subalpine zone (Figure 4.3d), and — to a lesser degree — in the submontane and lower-montane forest zone. About three quarters of the *Erica* bush and forest of the subalpine zone, which covers over 300 km², were affected by fire since 1976; the fires of the years 1996 and

1997 only destroyed nearly 90 km² of the *Erica* forests and shrubland of the southern slopes of Mt. Kilimanjaro. In contrast, fires are rare in the midmontane forest zone — at least on the southern slope. Most of these fires are lit by man, but natural fires do occur as well, and old charcoal horizons in the soil suggest that fires have been occurring for a very long time, probably back from the ice age (Hemp and Beck, 2001), but must have been far less frequent than today.

ELEVATIONAL GRADIENT IN PLANT SPECIES RICHNESS

As the altitudinal zonation on the wet southern and dry northern slope differs significantly (cp. Hemp, in press), the following explanations focus on the southern slope with its larger altitudinal gradient. Numbers of 2000 vascular plant species (including 143 pteridophytes) at 100-m elevation intervals on the southern slope of Mt. Kilimanjaro are shown in Figure 4.4. Vascular plants reach a maximum of over 900 species at 1000 m in the mosaic-type interface between the colline savanna and the banana–coffee plantations of the submontane zone, supporting the well-known phenomenon that plant species numbers are peaking in moderately cultivated or disturbed areas and not in natural, completely untouched areas. Remarkably, there is a second peak at 2600 m at the upper border of the montane zone. It is at this altitude where fires start to become important on Mt. Kilimanjaro, creating a mosaic of different fire-induced successional stages of forest, shrub, and tussock grassland stands (Figure 4.3b). By numerical evaluation of the relevés, several plant associations were revealed (Hemp 2002b; partly published in Hemp 2001a). The numbers of these plant associations in different altitudes show a clear correlation with the species numbers (Figure 4.4), suggesting that the number of habitats, the beta diversity (as indicated by the number of plant communities) is a major controlling factor of species richness on Mt. Kilimanjaro. This means that high community (beta) diversity in the fire-influenced areas of the upper-montane zone (12 communities) leads to a higher species (alpha) diversity (311 species) as compared to the closed undisturbed forest at lower altitudes with 233 species

in 7 communities and the monotonous *Erica* bush at higher altitudes with about 50 species in 4 to 5 communities.

In the following, pteridophytes will be used to illustrate these patterns. Because of their species richness and continuous distribution, this taxocene is an excellent tool for recognizing the altitudinal zonation of tropical mountains (Hemp 2001a). In Zambia, geophytic and deciduous fern species indicate dry conditions with recurring fires (Kornas', 1978, 1985). Similar observations were made on Mt. Kilimanjaro with such ferns occurring in colline/submontane and subalpine grasslands (Hemp, 2001a). In Figure 4.6, the forests of Mt. Kilimanjaro were divided into three groups (riverine and gorge forests were excluded): low-altitude forests with submontane *Croton–Calodendrum* forests, lower-montane *Cassipourea*, and *Agauria–Ocotea* forests; midaltitude forests with midmontane and upper-montane *Ocotea* and *Cassipourea* forests; high-altitude forests with upper-montane *Podocarpus*, *Hagenia*, *Juniperus*, and *Erica* forests. Figure 4.6 shows that geophytes and deciduous ferns are of lower importance in the midaltitude forests than in the fire-affected lower and higher forest areas. This is reflected by the bimodal discontinuous distribution pattern of deciduous, fire-tolerating fern species such as the geophyte *Pteridium aquilinum* and the hemicryptophyte *Dryopteris pentheri*, which have a distribution gap in the central forest zone (Hemp 2001a, b; Figure 4.5).

INFLUENCE OF FIRE ON LOW-ALTITUDE FORESTS

As in case of the pteridophytes, there is a certain fraction of deciduous trees in the drier submontane *Croton–Calodendrum* forests of the western and northern slopes. About 10% of the tree species were found to be deciduous (Hemp, unpublished data). This may be an adaptation to dry seasons but also to recurring fires in these forests. According to Hall and Swaine (1976), the highest amount of deciduous trees in Ghanaian forests was found in often-burnt areas.

In the same forest type, very distinct fire-induced *Olea europaea* ssp. *africana* dominance stages are quite common. This olive tree species is mainly distributed in the submontane

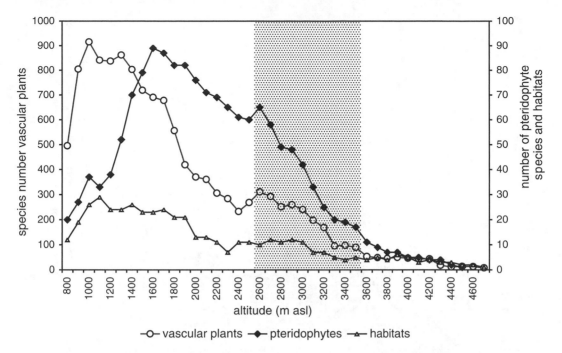

FIGURE 4.4 Altitudinal changes of species numbers of pteridophytes and of all occurring vascular plants on the southern slope of Mt. Kilimanjaro in relation to the number of plant associations (habitats), based on the evaluation of 1270 plots with about 2000 species. Dotted field: Fire-influenced altitudes in the higher zones.

and lower-montane zones and rarely in upper montane forests. The same holds for the fire-resistant trees *Agauria salicifolia* and *Myrica salicifolia* (both with a thick, corky bark), which occur mainly in the lower-montane camphor forests of the southern slope.

In the years 1996 and 1997, considerable areas of *Agauria–Ocotea* forests were destroyed by fire (Figure 4.3e), and in 1998, *Pteridium aquilinum* occupied more than 80% of the forest floor, whereas it covered less than 1% prior to the fire (Hemp 2002a). Gliessman (1978) observed a similar rapid spreading of the bracken in the mountains of Costa Rica. He explained the species' proliferation by the fact that its spores encountered optimal pH conditions in the ash in the burnt soil. However, 1 year is more likely a short period for the development of both gametophytes and sporophytes. Therefore, this phenomenon is probably better explained by the potential of fire to break the dormancy of the rhizomes. The occurrence of pteridium in the forests is of major importance for forest regeneration, as the dense cover of bracken impedes the sprouting of trees. Many clearings covered by *Pteridium* seem to

exist for several decades (Figure 4.3e). Similar observations were made, e.g. by Hartig and Beck (2003), in the Andes.

INFLUENCE OF FIRE ON HIGH-ALTITUDE FORESTS

Induced by a drop of precipitation above the major cloud zone, fire causes sharp discontinuities in composition and structure of the tall (20 to 30 m) upper montane *Hagenia–Podocarpus* forests at 2800 to 3000 masl. The giant heather, *Erica excelsa*, becomes dominant at this altitude, forming dense monospecific stands of about 10-m height (Hemp and Beck 2001). The occurrence of an *Erica* forest is an obvious sign of fire. During longer periods of dry climate with recurrent fires, *Erica* forest moves downslope and advances upslope during wet periods. The presence of *Erica* enhances the risk of fire, because even fresh *Erica* wood burns well, which in turn prevents the *Podocarpus* forest from reestablishing (Figure 4.7). At high fire frequency, the closed *Erica excelsa* forest degrades into open shrubland (height, 1.5 m) dominated by *E. trimera* and *E. arborea* (Figure

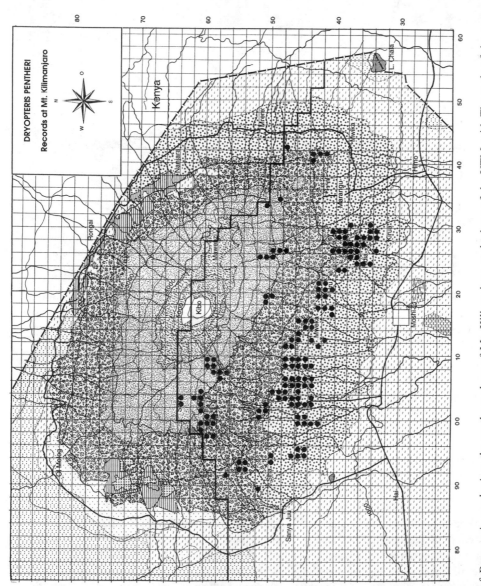

FIGURE 4.5 Records of *Dryopteris pentheri* on the southern slope of Mt. Kilimanjaro, at the base of the UTM grid. The scale of the squares is 4 km². The dark line represents the border of the surveyed area. Legend: see Figure 4.1. *Dryopteris pentheri*, a deciduous terrestrial fern, has a disjunct distribution with a gap in the central forest. It occurs, on the one hand, in fire-affected lower montane *Erica* forests and, on the other hand, in the lower-montane *Agauria–Ocotea* forest and the adjacent disturbed bush zone, as well as on shaded roadsides in the banana plantations.

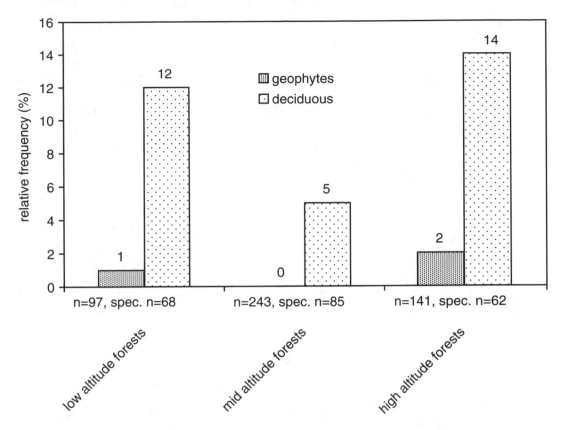

FIGURE 4.6 Relative frequency of fire-indicative ferns (geophytes and deciduous species) in vegetation plots of different forest types. Such ferns have their maximum distribution in the fire-affected low- and high-altitude forests. Low-altitude forests (97 plots, 68 fern species): submontane *Croton–Calodendrum* forests, lower-montane *Cassipourea* and *Agauria–Ocotea* forests; midaltitude forests (243 plots, 85 fern species): middle- and upper-montane *Ocotea* and *Cassipourea* forests; high-altitude forests (141 plots, 62 fern species): upper-montane and subalpine *Podocarpus, Hagenia, Juniperus,* and *Erica* forests. Riverine and gorge forests were excluded.

4.3a) between 3200 and 4000 masl (the potential treeline). A continuously high frequency of fires even destroys this bush, resulting in *Helichrysum* cushion vegetation, which is the climatic climax vegetation at altitudes above 4000 m (Figure 4.3c).

One year after a fire in an *Erica excelsa* forest, the forest starts to "burn" again from flowering *Kniphofia thomsonii*, a fire-resistant geophyte, which dominates the herb layer, together with *Dryopteris pentheri* and *Fumaria abyssinica*. *Senecio johnstonii* var. *cottonii* flowers mainly after fire (Figure 4.3f). New twigs sprouting from the stem bases of *Erica excelsa* were seen on about 80% of the individuals, and regeneration of *Erica* from seeds is — in contrast to the observations of Hemp and Beck (2001) — of additional importance as

well (Figure 4.3g). Therefore, these *Erica excelsa* forests consist of multi- and single-stemmed trees of apparently similar age, suggesting simultaneous sprouting after a fire. Without disturbance by fire, *Erica excelsa* grows as a single-stemmed tree up to 28 m high and with a DBH of 70 cm. Such trees have been observed in *Podocarpus* forests (Hemp and Beck, 2001).

Due to the open canopy of the *Erica* forest, the microclimate changes, as is obvious from the composition of the herb layer: Montane forest species disappear, and light-demanding species of the alpine flora, such as representatives of the genera *Helichrysum* or *Pentaschistis*, become dominant, which cannot successfully compete in shady forests.

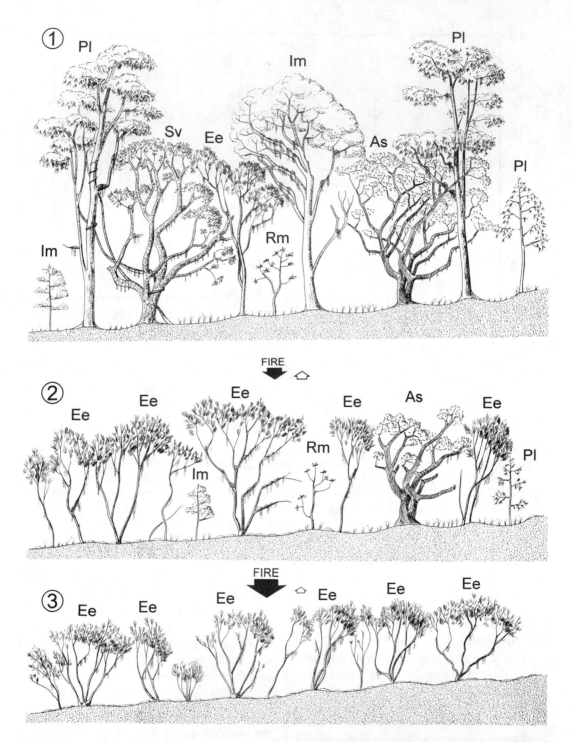

FIGURE 4.7 Regeneration of the canopy in upper-montane forests. (1) *Podocarpus* forest; (2) *Erica excelsa* forest with montane forest species; (3) pure *Erica excelsa* forest. Whether the *Podocarpus* forest is replaced by an *Erica* forest is a question of intensity and frequency of fire (black arrow). The higher the percentage of *Erica* is, the higher the risk is of another fire and the more difficult the reestablishment of a *Podocarpus* forest (white arrows). As: *Agauria salicifolia*; Ee: *Erica excelsa*; Im: *Ilex mitis*; Pl: *Podocarpus latifolius*; Rm: *Rapanea melanophloeos*; Sv: *Schefflera volkensii* (from Hemp and Beck 2001).

IMPACT OF FIRE ON THE UPPER TREELINE

The fires of 1996 and 1997 declined the upper closed forest line by about 400 m; burnt tall (tree height, 10 to 15 m) forests were encountered during fieldwork up to 3600 m (Figure 4.3h). From field observations and historical descriptions (Jaeger 1909; Klute 1920), it can be assumed that different forest types extended up to 4000 m in many areas of Mt. Kilimanjaro at the beginning of the 20th century. Thus, today's treeline is 800 m lower than where it would be without recurrent fires.

Erica excelsa stands on the southeastern slopes at an altitude of 2800 m, and the "moorland" tussock vegetation produces very abrupt boundaries (Figure 4.3b). Tree islands consisting of a core of *Podocarpus* forest are surrounded by a fringe of *Erica* trees and various shrubs such as *Conyza vernonioides* and *Hypericum revolutum*. In this area, *Podocarpus* forest, *Erica* forest, and subalpine grassland occur at the same altitude. These forest islands must be interpreted as remnants of the former forest, rather than as outposts of the recent ones (Hemp and Beck, 2001). Therefore, it can be concluded that the original vegetation of the present subalpine tussock grassland, with its undulating borderline and the tree islands, must have consisted of a mixed-type forest in which *Erica* was a component, but not the dominant, tree species.

The comparison of 1976 and 2000 Landsat images reveals enormous changes in the upper vegetation zones of Mt. Kilimanjaro during the last 24 years. In 1976, the *Erica trimera* bush — today depressed in most areas below 3800 m — reached 4100 m, in part forming a continuous belt in areas covered by *Helichrysum* cushion vegetation today (Figure 4.3c). This vegetation is not threatened by fire as its biomass provides too little fuel for fire to spread; causing it to increase significantly between 1976 and 2000. Closed *Erica* forests covered nearly six times the area of today (187 and 32 km², respectively) in 1976, extending in many places up to 3800 m. This means that nearly 15% of Mt. Kilimanjaro's forest cover was destroyed by fire since 1976.

DISCUSSION

On Mt. Kilimanjaro, especially, the subalpine zone is strongly affected by fires. During the end of the dry season in February and March and in September–October, the whole ericaceous belt is usually so dry that it may easily be set on fire by people (especially by careless honey collectors or poachers; cp. Hemp 1999, Figure 4.3d) or by lightening. Big fires, devastating large areas of the high-altitude forests, have been repeatedly reported (e.g. Meyer 1890, Volkens 1897, Jaeger 1909, Beck et al. 1986, Blot 1999, and Hemp and Beck 2001).

The altitudinal distribution of species richness found in the present study (Figure 4.4) confirmed the respective results presented in Hemp (2001a): In the present study with about 300 additional relevés and more transects, the same main peak of species numbers at 1000 to 1300 m was detected, and the second peak at 2600 m is even more distinct.

The observed phenomenon of monodominant forest stands and discontinuous bimodal distribution patterns is in the same line of *Erica excelsa*. On Mt. Kilimanjaro, *Erica excelsa* forms monotonous stands in the subalpine zone but is also a component of the forests of the lower-montane zone. However, it is rare in the central-montane zone (Hemp and Beck 2001). This distribution pattern is mainly due to both the precipitation and the fire regime. Some trees of the submontane and lower-montane forests such as *Agauria salicifolia*, *Myrica salicifolia*, and *Olea europaea* ssp. *Africana*, as well as some ferns, such as the fire-resistant *Pteridium aquilinum* and *Dryopteris pentheri* (Hemp 2001a, (b), show a similar bimodal altitudinal distribution as *Erica excelsa*, with a gap in the central forest belt, and can therefore be regarded as fire indicators. In some cases, frequently burned areas in the lower-montane forests result in a pure *Erica* bush whose vegetation structure is similar to the subalpine *Erica excelsa* communities. Volkens (1897) has already mentioned the bimodal occurrence of *Agauria*, *Myrica*, and *Erica* on Mt. Kilimanjaro, albeit not referring it to the influence of fire.

Pollen diagrams of East African mountains show considerable treeline dynamics. They indicate a dry climate between 25,000 to

12,500 B. P. (Lind and Morrison 1974) with charcoal horizons in today's lower-montane forests on Mt. Kilimanjaro, suggesting an extent of ericaceous forest 1000 m below its current lower distribution limit (Hemp and Beck 2001). Theoretically, the rising temperatures and the enrichment of atmospheric CO_2 and nitrogen, associated with the increased burning of fossil fuels and pollution since the Industrial Revolution, could be consistent with an increase in the elevation of mountain vegetation zones. An upslope movement of alpine plant species was observed, e.g. in the Alps by Grabherr and Pauli (1994). Late successional closed vegetation, however, will change very slowly (Körner 2003). On Mt. Kilimanjaro, this effect is superimposed by the increasing fire activity, resulting from a typical biological feature of high mountains in Africa: the vast ericaceous belt. This plant formation is very inflammable and more sensitive to fire in a warmer climate. Therefore, alpine vegetation on Mt. Kilimanjaro migrates downslope, replacing high-altitude forests.

These findings are in contrast with observations of Shugart et al. (2001), who reported an upslope movement of vegetation zones on Mt. Kilimanjaro, stressing the above-mentioned atmospheric and climatic factors and comparing two satellite images from 1984 and 1993. However, first, their observations were based on only two small sites and, second, they were obviously not aware that the vegetation in the alpine zone of Mt. Kilimanjaro is a mosaic of fire-induced regeneration stages. Certainly, there are places in such a mosaic where vegetation is recovering and increasing. Nevertheless, in general, there is a strong decrease caused by fires. Evaluation of satellite imagery has to be based always on an extensive ground survey.

In addition to the function of filtering and storing water, the upper-montane and subalpine forests have a high potential of collecting cloud water (cp., e.g. Cavelier and Goldstein 1989, Juvik and Nullet 1995, Cavelier et al. 1996, Bruijnzeel 2001), especially the ericoid type of forest (Kerfoot 1968). It can be estimated that the 1000 km² of forest on Mt. Kilimanjaro annually receive 5% of their 1600-million-m³ precipitation by fog interception, mainly from the upper-montane and subalpine forests (Hemp, 2005; Figure 4.3h). Thus, the loss of such cloud forests due to fires — as well as the loss of montane forest due to clearing — means a considerable reduction and increase in variability of water yields of the Mt. Kilimanjaro catchments, affecting over one million people living on the mountain.

CONCLUSIONS

Fire influences vegetation on Mt. Kilimanjaro to a great extent and in different ways. It promotes, to some extent, biodiversity and species with specific physiological and morphological adaptations, driving the distribution of both plants and plant communities. However, in the forests and bushlands of the treeline ecotone, where large areas are covered by different fire-tolerating *Erica* species, fire plays an important role as an ecological transformation factor. In recent years, a drier climate in combination with increasing human activities (poaching, honey collecting, and tourism), have facilitated the spreading of fires dramatically, leading to an almost complete destruction of the former closed *Podocarpus*, *Juniperus*, *Hagenia*, and *Erica* forest between 3000 m and the treeline at 4000 m. In contrast to common predictions, this has opened land for a downward migration of the alpine flora, which leads to a distinct peak in species richness at 2600 m. On the other hand, fires have affected biodiversity in such a way that a key functional group of tall woody nanophyllous plants, largely representatives of the genus *Erica*, have been eliminated or diminished in the upper forest zone, which has led to a massive reduction of fog trapping and, thus, of water yielding.

SUMMARY

Fire is a major ecological factor on Mt. Kilimanjaro that influences vegetation in different ways. In addition to wildfires, humans have always used bush and forest fires for hunting and land clearing, and some fires were lit accidentally. In recent years, a drier climate has facilitated the spreading of these fires dramatically, leading to an almost complete destruction of the former closed forest between 3000 m and the treeline at 4000-m altitude. Therefore, the current upper forest limit is suppressed by nearly 800 m in many areas. Simultaneously, the easily inflammable subalpine heathland areas became much larger, increasing the risk of fire. In contrast to common predictions, this has opened land for a downward migration of the alpine flora. Fires have affected biodiversity in such a way that a key functional group of tall woody nanophyllous plants, largely representatives of the genus *Erica*, have been eliminated or diminished, which has led to a massive reduction of fog trapping and, thus, of water yielding. Although negatively influencing ecosystem services such as water yielding, soil protection, and provision of fuel wood and timber, fire has increased the patchiness of vegetation (beta-diversity) and with this, the overall alpha-diversity of vascular plants in the treeline ecotone.

ACKNOWLEDGMENTS

I gratefully acknowledge support by the Deutsche Forschungsgemeinschaft for funds and the Tanzanian Commission for Science and Technology for permitting research. For support of fieldwork, I owe gratitude to Chief Park Wardens Moirana and Mafuru, of Kilimanjaro National Park, and to my research counterpart Mr. Mushi, Moshi.

References

Beck, E., Scheibe, R., and Schulze, E.D. (1986). Recovery from fire: observations in the alpine vegetation of western Kilimanjaro (Tanzania). *Phytocoenologia*, 14(1): 55–77.

Beck, E., Scheibe, R., and Senser, M. (1983). The vegetation of the Shira plateau and the western slopes of Kibo (Mount Kilimanjaro, Tanzania). *Phytocoenologia* 11(1): 1–30.

Blot, J. (1999). The incidence of forest fire in Kilimanjaro. In *Mount Kilimanjaro: Land Use and Environmental Management*. French Institute for Research in Africa. IFRA Les Cahiers 16: 85–86.

Braun-Blanquet, J. (1964). *Pflanzensoziologie*. Wien 865 pp.

Bruijnzeel, L.A. (2001). Hydrology of tropical montane cloud forests: a reassessment. *Land Use Water Resour Res* 1: 1.1–1.18 (www.luwrr.com).

Cavelier, J. and Goldstein, G. (1989). Mist and fog interception in elfin cloud forests in Colombia and Venezuela. *J Trop Ecol*, 5: 309–322.

Cavelier, J., Solis, D., and Jaramillo, M.A. (1996). Fog interception in montane forests across the Central Cordillera of Panamá. *J Trop Ecol*, 12: 357–369.

Gliessman, S.R. (1978). The establishment of bracken following fire in tropical habitats. *Am Fern J*, 68: 41–44.

Grabherr, G. and Pauli, H. (1994). Climate effects on mountain plants. *Nature*, 369: 448.

Hall, J.B. and Swaine, M.D. (1976). Classification and ecology of closed-canopy forest in Ghana. *J Ecol*, 64: 913–951.

Hartig, K. and Beck, E. (2003). The bracken fern (Pteridium arachnoideum (Kaulf.) Maxon) dilemma in the Andes of southern Ecuador. *Ecotropica*, 9: 3–13.

Hedberg, O. (1951). Vegetation belts of the East African mountains. *Svensk Bot Tidskrift*, 45: 140–202.

Hedberg, O. (1964). Features of Afroalpine plant ecology. *Acta Phytogeographica Suecica*, 49: 1–144.

Hemp, A. (1999). An ethnobotanical study of Mt. Kilimanjaro. *Ecotropica*, 5: 147–165.

Hemp, A. (2001a). Ecology of the pteridophytes on the southern slopes of Mt. Kilimanjaro. Part II: Habitat selection. *Plant Biol*, 3: 493–523.

Hemp, A. (2001b). Life form and strategies of forest ferns on Mt. Kilimanjaro. In Gottsberger, G. and Liede, S. (Eds.), *Life Forms and Dynamics in Tropical Forests*. Disserationes Botanicae, 346: 95–130.

Hemp, A. (2002a). Ecology of the pteridophytes on the southern slopes of Mt. Kilimanjaro. Part I: Altitudinal distribution. *Plant Ecol*, 159: 211–239.

Hemp, A. (2002b). *Fränkische Steppenheide, tropische Bergregenwälder — zwei gegensätzliche Florenregionen, eine gemeinsame Klassifikationsmethode?* Habilitationsschrift, Universität Bayreuth.

Hemp, A. (in press). Continuum or zonation? Altitudinal gradients in the forests on Mt. Kilimanjaro. *Plant Ecology*.

Hemp, A. (2005). Climate change driven forest fires marginalize the impact of ice cap wasting on Kilimanjaro. *Global Change Biology* 11(7): 1013–1023.

Hemp, A. and Beck, E. (2001). *Erica excelsa* as a fire-tolerating component of Mt. Kilimanjaro's forests. *Phytocoenologia*, 31(4): 449–475.

Hemp, A., Hemp, C., and Winter, J.C. (1999). Der Kilimanjaro — Lebensräume zwischen tropischer Hitze und Gletschereis. *Natur und Mensch*, 1998: 5–28.

Hemp, A., Lambrechts, C., and Hemp, C. (in press). Global Trends and Africa. The Case of Mt. Kilimanjaro. UNEP, Nairobi.

Jaeger, F. (1909). Forschungen in den Hochregionen des Kilimandscharo. *Mitteilungen aus den Deutschen Schutzgebieten*, 22: 113–146; 161–196.

Juvik, J.O. and Nullet, D. (1995). Relationships between rainfall, cloud-water interception, and canopy throughfall in a Hawaiian montane forest. In Hamilton, L.S., Juvik, J.O., and Scatena, F.N. (Eds.), *Tropical Montane Cloud Forests*. Ecological Studies 110, Springer-Verlag, Berlin, pp. 165–182.

Kerfoot, O. (1968). Mist precipitation on vegetation. *Forestry Abstracts*, 29: 8–20.

Klötzli, F. (1958). Zur Pflanzensoziologie des Südhanges der alpinen Stufe des Kilimanjaro. *Ber Geobot Inst Rübel*, 1957: 33–59.

Klute, F. (1920). *Ergebnisse der Forschungen am Kilimandscharo 1912*. Reimer, Berlin. 136 pp.

Kornaś, J. (1978). Fire-resistance in the Pteridophytes of Zambia. *Fern Gaz*, 11(6): 373–384.

Kornaś, J. (1985). Adaptive strategies of African pteridophytes to extreme environments. *Proc Roy Soc Edinb*, 86B: 391–396.

Körner, C. (2003). *Alpine Plant Life. Functional Plant Ecology of High Mountain Ecosystems*. Springer-Verlag, Berlin, Heidelberg. 344 pp.

Lambrechts, C., Woodley, B., Hemp, A., Hemp, C., and Nnyiti, P. (2002). Aerial Survey of the Threats to Mt. Kilimanjaro Forests. UNDP, Dar es Salaam, 33 pp.

Lange, S., Bussmann, R.W., and Beck, E. (1997). Stand structure and regeneration of the subalpine *Hagenia abyssinica* forests of Mt. Kenya. *Bot Acta* 110: 473–480.

Lind, E.M. and Morrison, M.E.S. (1974). *East African Vegetation*. London, 257 pp.

Meyer, H. (1890). *Ostafrikanische Gletscherfahrten*. Forschungsreisen im Kilimandscharo-Gebiet. Leipzig, 376 pp.

Miehe, S. and Miehe, S. (1994). *Ericaceous Forests and Heathlands in the Bale Mountains of South Ethiopia. Ecology and Man's Impact*. Hamburg, 206 pp.

Pócs, T. (1994). The altitudinal distribution of Kilimanjaro bryophytes. In Seyani, J.H. and Chikuni, A.C. (Eds.), *Proc XIII Plenary Meeting AETFAT*, Malawi, 797–812.

Schmitt, K. (1991). *The Vegetation of the Aberdare National Park Kenya*. Universitätsverlag Wagner, Innsbruck, Austria. 259 pp.

Shugart, H.H., French, N.H.F., Kasischke, E.S., Slawski, J.J., Dull, C.W., Shuchman, R.A., and Mwangi, J. (2001). Detection of vegetation change using reconnaissance imagery. *Global Change Biology*, 7: 247–252.

Thompson, L.G., Mosley-Thompson, E., Davis, M.E., Henderson, K.A., Brecher, H.H., Zagorodnov, V.S., Mashiotta, T.A., Lin, P.N., Mikhalenko, V.N., Hardy, D.R., and Beer, J. (2002). Kilimanjaro ice core records: evidence of holocene climate change in tropical Africa. *Science*, 298: 589–593.

Volkens, G. (1897). *Der Kilimandscharo. Darstellung der allgemeineren Ergebnisse eines fünfzehnmonatigen Aufenthalts im Dschaggalande*. Berlin, 389 pp.

Walther, H., Harnickell, E., and Mueller-Dombois, D. (1975). *Climate-Diagram Maps of the Individual Continents and the Ecological Climatic Regions of the Earth*. Springer, Berlin.

Wesche, K., Miehe, G., and Kaeppeli, M. (2000). The significance of fire for Afroalpine ericaceous vegetation. *Mount Res Dev*, 20: 340–347.

5 Effects of Fire on the Diversity of Geometrid Moths on Mt. Kilimanjaro

Jan C. Axmacher, Ludger Scheuermann, Marion Schrumpf, Herbert V.M. Lyaruu, Konrad Fiedler, and Klaus Müller-Hohenstein

INTRODUCTION

Fires often occur along the upper margin of the montane rain forest on Mt. Kilimanjaro (Beck et al. 1986; Blot 1999; Meyer 1890; Salehe 1997). This is especially true for exceptionally dry periods which regularly appear during the El Niño southern oscillation (ENSO) phases. As a consequence, the forest boundary has been profoundly lowered, and the forest has been replaced by heathland dominated by *Erica excelsa*, *E. trimera*, and *E. arborea* (Hemp and Beck 2001), with some small isolated patches of forest remaining above the current closed forest boundary.

This change in the vegetation is accompanied by various changes in ecological conditions. The temperature drops more rapidly at nightfall outside the forest, and the microclimate is drier within the open heath vegetation, leading to the disappearance of many ferns and epiphytic plants outside the forest.

The objective of our study was to explore how these fire-induced changes affect herbivorous insects along the upper forest margins and in forest remnants. We investigated changes in species composition and diversity of nocturnal geometrid moth communities along an altitudinal transect covering intact closed forests below 3000 m, a mosaic of forest remnants and heathland at 3100 m, and heathland vegetation at 3300 m. The sites at 3100 m, therefore, enable a direct comparison of geometrid moth communities in heathland and forest fragments, whereas the lower and higher portions of the transect, albeit the altitudinal differences between the sites, provide comparison of community structure and diversity of geometrid moths between large closed forest and heathland habitats. Geometrids were selected for four reasons: (1) They are one of the three most diverse families of Lepidoptera. (2) They are taxonomically rather well known (Scoble 1999). (3) They proved to be suitable indicators of environmental change in a number of studies in tropical forests (e.g. Beck et al. 2002; Brehm 2002; Holloway 1984; Intachat et al. 1999; Kitching et al. 2000). (4) They regularly occur in quite large numbers at high altitudes at which most other insects are not very common (Brehm 2002).

TEST SITES AND METHODS

STUDY AREA

Mt. Kilimanjaro, the highest mountain in Africa (5892 m), is situated 300 km south of the equator in Tanzania. It is a volcanic complex composed of three volcanic centers, covering an area of roughly 80 km × 40 km.

Within the Mt. Kilimanjaro area, the study sites are located in the southwest part of the mountain along the Machame tourist route below the Shira plateau, along an altitudinal gradient from 2300 to 3300 m. The upper portions of this area were heavily affected by forest and heath fires during the last El Niño event (1996 to 1997), when large areas of the ericaceous belt on the Shira plateau burned (Salehe 1997). These fires also affected the heath sites included in this study, whereas the forest remnants and forest sites remained intact.

METHODS

Geometrid moths were investigated at 13 sites. Eight of these sites were located within the intact forest belt in the ranges 2300 m, 2580 m, 2700 m, and 2900 m; two plots consisted of isolated patches of remnant forest at 3100 m; and three sites of heathland ranged from 3100 to 3300 m.

Moths were attracted by an accumulator-powered weak UV-containing light source (Sylvania blacklight-blue, F 15 W/BLB-TB) placed within a white reflective gauze cylinder (diameter 0.8 m, height 1.6 m). From the gauze, insects were sampled manually using plastic jars prepared with diethylether. The samples were taken from 7 to 10 P.M. local time between October 2001 and January 2002. Catches were restricted to periods without strong moonlight, avoiding the period from 5 d before to 5 d after full moon. For each site, two to four night catches were pooled for analysis. Subsequently, specimens were spread, sorted to morphospecies level, and taxonomically identified as far as possible at the Zoologische Staatssammlung in Munich.

Data handling and statistical analysis were carried out using Microsoft Access, SPSS, EstimateS (Colwell 2000), and an Excel macro provided by Meßner (1996) for calculating the normalized expected species shared (NESS) index and Sørensen index of similarity.

RESULTS

A total of 2158 individuals representing 92 morphospecies were sampled (Table 5.1); 41 (44%) of these morphospecies representing 1563 individuals (72%) could be identified to species level.

PROPORTIONS OF SUBFAMILIES WITHIN GEOMETRIDAE

In the study area, three subfamilies of geometrid moths, Larentiinae, Ennominae, and Geometrinae, are largely represented and can be found along the whole altitudinal gradient. Two other subfamilies, Sterrhinae and Desmobathrinae, also occur but account for less than 3% of individuals as well as species at any site and disappear at sites above 2900 m. Concerning the number of individuals caught, Larentiinae was the dominant subfamily throughout the whole study area, always accounting for more than 50% of the total catch. Within the closed forest, Ennominae proved to be the second most abundant subfamily, whereas they were much rarer in the heath, where the proportion of Geometrinae was very high. In the forest fragments, Geometrinae and Ennominae were caught in roughly equal proportions (Figure 5.1).

When considering the proportions of subfamilies according to the occurrence of morphospecies on the plots, Larentiinae are dominant on all plots, followed by Ennominae. Geometrinae accounted for a smaller share of species richness, and the proportions seen in heathland are not different from those obtained at some lower forest sites (Figure 5.2).

TABLE 5.1
Number of catches performed and number of geometrid moth individuals and morphospecies caught at each site

Site	2300 F1	2300 F2	2580 F1	2580 F2	2700 F1	2700 F2	2900 F1	2900 F2	3100 Ff1	3100 Ff2	3100 E	3300 E1	3300 E2
Catches	4	3	3	3	2	2	3	3	4	4	2	2	2
Individuals	224	138	148	177	215	178	181	147	138	75	196	184	157
Morphospecies	32	31	28	33	31	37	38	32	34	24	26	19	21

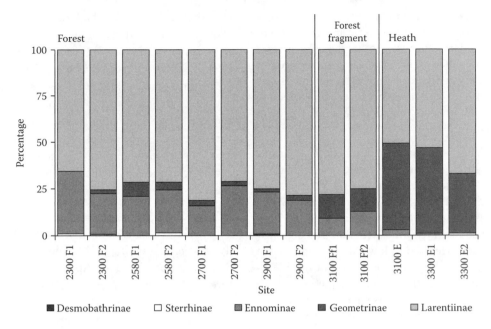

FIGURE 5.1 Proportions of subfamilies, based on numbers of individuals. Sites are sorted by altitude. (Site codes: altitude in meters, F = forest, Ff = forest fragment, E = heath vegetation).

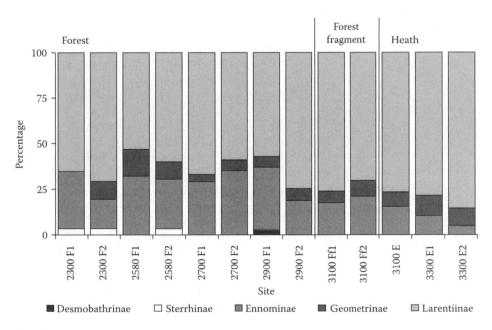

FIGURE 5.2 Proportions of subfamilies, based on numbers of morphospecies. Sites are sorted by altitude. (Site codes: altitude in meters, F = forest, Ff = forest fragment, E = heath vegetation).

ALPHA-DIVERSITY

Fisher's alpha proved to be a reliable measure of the diversity of moth communities in a num- ber of studies (e.g. Brehm 2002; Chey et al. 1997; Schulze 2000), as long as the samples fit a log-series distribution. This is true for all the sites sampled in this study. Once sample size exceeds 120 individuals per site in species-poor communities as in the case of Mt. Kilimanjaro, Fisher's alpha shows a greater independence of sample size than most other alpha diversity measures (Hayek and Buzas 1997). This number was exceeded in our study at all but one site, namely, the forest fragment 3100 Ff 2.

When comparing geometrid diversity from different sites, two main results emerge. First, geometrid communities at Mt. Kilimanjaro are not very diverse overall (values ranging from 9 to 15 in closed forests above 2250 m). Second, the values from the forest below the upper treeline below 3000 m and in the forest fragments at 3100 m do not differ significantly, whereas diversity in the heath both at 3100 m and 3300 m is significantly reduced (Figure 5.3) compared to all forested sites.

BETA-DIVERSITY

To study the beta-diversity of the moth communities in the study area, two similarity measures,

the Sørensen index and the NESS index, were chosen. As an ordination method, nonlinear multidimensional scaling was applied (see Beck et al. 2002 and Schulze and Fiedler, in press, for methods). Figure 5.4 shows the ordinations of the samples for the Sørensen index and the NESS index for a low sample size parameter $m = 1$ to emphasize the similarity of moth ensembles with regard to the most abundant species and a higher parameter $m = 37$ with stress on the occurrence of rare species (Grassle and Smith 1976)

In all three ordinations, the sites are primarily arranged according to their altitudinal position along the first axis. Apart from this, the forest plots are always positioned closely together and separated from the remaining sites by the first dimension. In all three ordinations, communities from forest fragments are positioned closer to the heathland than to the forest sites. A clear distinction between heath and forest fragments is only discernible with the NESS index, particularly with $m = 1$. According to the

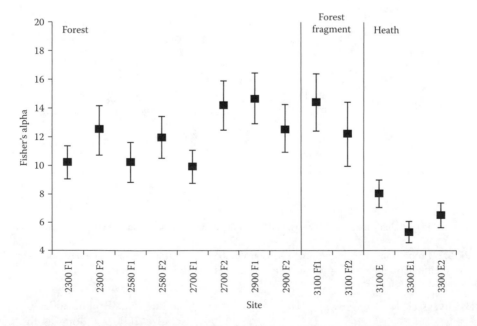

FIGURE 5.3 Diversity of Geometridae: Fisher's alpha. The bars indicate the standard deviation, calculated using EstimateS. (From Colwell, R.K. [2000]. Sites are sorted by altitude. (Site codes: altitude in meters, F = forest, Ff = forest fragment, E = heath vegetation.)

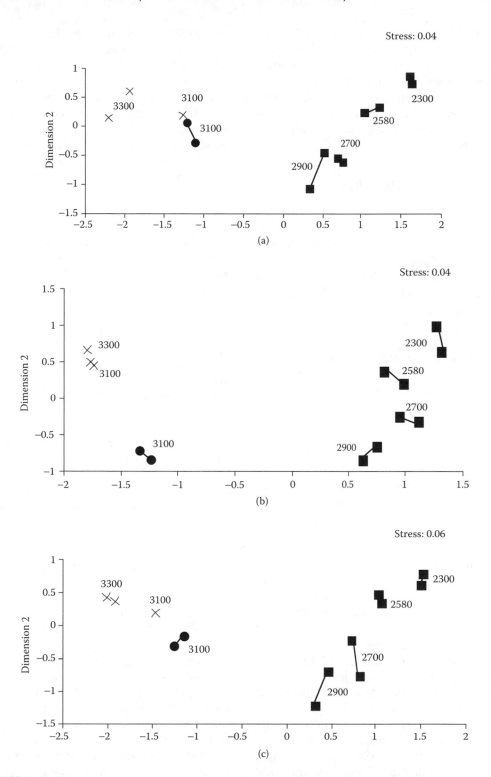

FIGURE 5.4A Two-dimensional nonlinear scaling of samples of Geometridae, based on matrices calculated with (A) Sørensen and NESS similarity indices, (B) small sample-size parameter $m = 1$, and (C) large sample-size parameter $m = 37$. Numbers correspond to the altitude in meters; ■ = forest, ● = forest fragment, X = heath. Low stress values indicate an excellent goodness-of-fit of the ordinations to the initial similarity data.

Sørensen index, forest fragments are indistinguishable from the heath plot at the same altitude (3100 m). Therefore, differentiation between heathland and forest fragments is more a matter of altered abundance relationships, whereas species composition in forest fragments cannot be distinguished clearly from heathlands at identical altitudes.

DISCUSSION

Our data reveal strong effects of fire-induced spread of heathland on geometrid moth communities at the three levels of subfamily relationships, alpha-diversity, and species composition. On the level of subfamilies, in addition to the complementary altitudinal increase of Larentiinae and decrease of Ennominae, Geometrinae appear to benefit from fire, as they are much more numerous at the heathland as compared to forest fragments situated at similar altitudes. However, this applies only to two species of the genus *Comostolopsis* (Warren), which has become very abundant in this habitat. The altitudinal decrease of Ennominae is even more accentuated in open heathland, probably as a consequence of the rather strong link of most Ennominae to forested habitats. The overall predominance of Larentiinae has generally been observed at both high altitudes and latitudes that might reflect particular thermal adaptations of this moth subfamily (Brehm 2002).

Within the intact montane rain forest occurring in the lower parts of the study area below 3000 m, there was no observed decrease in overall geometrid moth diversity with increasing altitude. The highest forest sites at 2700 and 2900 m had even slightly more diverse geometrid moth communities. Overall diversity in this tropical montane forest was nowhere higher than in temperate zone woodlands and much lower than in other tropical mountain regions (Schulze 2000; Brehm 2002). Yet, species composition within the forest belt revealed a continuous change from 2300 to 2900 m, regardless of the analytical methods chosen.

Above 3000 m, the marked drop in geometrid alpha-diversity of the heath in comparison to forest fragments in the direct vicinity (Plot 3100 Ff 1 and 2 in comparison to 3100 E) shows that the destruction of the forest along the upper forest line can trigger distinct effects on the diversity of these insects. Forest remnants do not show a similar impoverishment. However, with regard to species composition, forest remnants are more similar to the surrounding heathland than to intact forest. It is, therefore, not yet clear whether such forest remnants could serve as a permanent refuge for forest species, offering the possibility of recolonization if the forest could encroach into the heathland again with time. Alternatively, these forest fragments might actually be sinks rather than sources, if the forest moth species are only present due to permanent immigration from the intact forest.

CONCLUSION

Fire-induced conversion of montane forest into heath scrubland severely reduces diversity of geometrid moths. If ENSO-triggered drought periods and related forest fires increase in frequency and severity, as is suggested by recent climate models, our data suggest massive consequences for large proportions of insects that are specifically dependent on the more humid and thermally less extreme and less variable conditions prevailing in the high-altitude closed forest.

SUMMARY

Due to reoccurring fires, the upper forest boundary at Mt. Kilimanjaro has become distinctly lowered and fragmented. These large-scale and long-term vegetation changes are reflected in the changes to geometrid moth communities. These changes were studied on 13 sampling sites located within the intact forest below 3000 m, as well as in forest fragments and in the heath vegetation replacing the forest above 3000 m. The overall diversity of geometrid moths decreases when the forest is replaced by heath. The species composition, especially the abundance pattern of the species, is dramatically altered. Forest fragments remaining within heathland areas might serve as refuge areas for some of the moth species, but only to a limited extent, as is reflected by strong faunal dissimilarities between intact forest and forest fragments.

ACKNOWLEDGMENTS

We would like to thank the German Research Foundation (DFG, Mu 364 14-1, 2 & 3), Forestry and Beekeeping Division (FBD), Kilimanjaro National Park (KINAPA), Tanzania National Parks (TANAPA), Tanzania Commission for Science and Technology (COSTECH), the CITES-Office, Dr. Axel Hausmann at the Zoologische Staatssammlung, Munich, and numerous other helping hands and minds for their kind help and cooperation.

References

Beck, E., Scheibe, R., and Schulze, E.D. (1986). Recovery from fire: observations in the alpine vegetation of western Mt. Kilimanjaro (Tanzania). *Phytocoenologia*, 14: 55–77.

Beck, J., Schulze, C.H., Linsenmair, K.E., and Fiedler, K. (2002). From forest to farmland: diversity of geometrid moths along two habitat gradients on Borneo. *Journal of Tropical Ecology*, 17: 33–51.

Blot, J. (1999). The incidence of forest fire in Kilimanjaro. French Institute for Research in Africa, *IFRA Les Cahier* 16: 85–86.

Brehm, G. (2002). Diversity of Geometrid Moths in a Montane Rainforest in Ecuador. Ph.D. dissertation, University of Bayreuth.

Chey, V.K., Holloway, J.D., and Speight, M.R. (1997). Diversity of moths in forest plantations and natural forests in Sabah. *Bulletin of Entomological Research*, 87: 371–385.

Colwell, R.K. (2000). EstimateS: Statistical Estimation of Species Richness and Shared Species from Samples, Version 6.0 b1. http://viceroy.eeb.uconn.edu/estimates.

Grassle, J.F. and Smith, W. (1976). A similarity measure sensitive to the contribution of rare species and its use in investigation of variation in marine benthic communities. *Oecologia*, 25: 13–22.

Hayek, L.A. and Buzas, M.A. (1997). *Surveying Natural Populations*. Columbia University Press, New York.

Hemp, A. and Beck, E. (2001). *Erica excelsa* as a fire-tolerating component of Mt. Kilimanjaro's forests. *Phytocoenologia*, 31: 449–475.

Holloway, J.D. (1984). Moths as indicator organisms for categorizing rain-forest and monitoring changes and regeneration processes. In Chadwick, A.C. and Sutton, S.L. (Eds.). *Tropical Rain-Forests: The Leeds Symposium*. Leeds Philosophical and Literary Society, Leeds, 235–242.

Intachat, J., Holloway, J.D., and Speight, M.R. (1999). The effects of logging on geometroid moth populations and their diversity in lowland forests of Peninsular Malaysia. *Journal of Tropical Forest Science*, 11: 61–78.

Kitching, R.L., Orr, A.G., Thalib, L., Mitchell, H., Hopkins, M.S., and Graham, A.W. (2000). Moth assemblages as indicators of environmental quality in remnants of upland Australian rain forest. *Appl Ecol*, 37: 284–297.

Meßner, S. (1996). *Untersuchungen zur Biodiversität der Myrmecofauna (Formicidae) im Parc National de la Comoe (Elfenbeinküste)*. Diploma thesis, University of Würzburg.

Meyer, H. (1890). *Ostafrikanische Gletscherfahrten. Forschungsreisen im Kilimandscharo-Gebiet*. Duncker und Humblot, Leipzig.

Salehe, J. (1997). Preliminary Assessment of the Mount Kilimanjaro Forest Fires of February and March 1997. Consultant's Report, IUCN, Nairobi.

Schulze, C.H. (2000). Auswirkungen anthropogener Störungen auf die Diversität von Herbivoren. Ph.D. dissertation, University of Bayreuth.

Schulze, C.H. and Fiedler, K. (2003). Vertical and temporal diversity of a species-rich moth taxon in Borneo. In Basset, Y., Novotny, V., Miller, S., and Kitching, R. (Eds.). *Arthropods of Tropical Forests: Spatio-Temporal Dynamics and Resource Use in the Canopy*. Cambridge University Press, Cambridge, pp. 68–85.

Scoble, M. (1999). *Geometrid Moths of the World: A Catalogue (Lepidoptera, Geometridae)*. 2 Volumes. The Natural History Museum, London.

6 The Influence of Fire on Mountain Sclerophyllous Forests and Their Small-Mammal Communities in Madagascar

Bernardin P.N. Rasolonandrasana and Steven M. Goodman

INTRODUCTION

One of the classical explanations for the Holocene extinction of nearly 20 species of large mammals, birds, and reptiles on Madagascar is, in part, human-induced fires and associated modifications of the natural environment. On the basis of several different pollen profiles cored from the Malagasy central highlands, it is clear that natural wildfires occurred well before the first human occupation of the island about 2300 years ago (Burney, 1987; Burney et al., 2004). In certain regions of the island, particularly the high-mountain zones, fire is a natural aspect of the environment. However, because humans have been colonizing portions of the central highlands during the course of the past 1000 years, they have modified its effects on the environment and influenced its frequency.

A number of researchers have investigated the effects of fire on populations of small mammals (e.g. Howard et al., 1959, Chew et al., 1959; Ahlgren, 1966; Krefting and Ahlgren, 1974; Christian, 1977), but most of these studies were based on sites in savanna habitats or temperate forests. The effects of fire on local biota and ecological processes in savanna or in forests are generically different (Daubenmire, 1968). The implications of these perturbations on ecological processes associated with small mammals living within tropical latitudes are still poorly understood. Fire can directly kill certain animals during its passage (Chew et al., 1959; Tevis, 1956). As the passage of fire in a natural area affects plant succession, the specific composition of animals in a burned area can also be considerably modified (Beck and Vogl, 1972; Cook, 1959; Layne, 1974). Alteration of the microclimate, changes in litter accumulation, burning of soil organic material, and availability of different types of food have considerable implications for demographic parameters of small mammals that may have limited dispersal ability.

Variation in soil litter depth influences the distribution and abundance of small mammals (Kaufman et al., 1989). For example, the majority of Tenrecidae, a family of small mammals (order Lipotyphla) endemic to Madagascar and largely forest dwelling, are terrestrial and feed primarily on soil invertebrates (Soarimalala, 1998). Effectively, a study conducted in the central highlands of Madagascar showed the close relation between the diversity and the distribution of *Microgale* spp. (a speciose genus of Tenrecidae) and the diversity of soil invertebrates (Goodman et al., 1996). Thus, in burned forest areas, litter recovery and the recolonization by different groups of macroinvertebrates are presumably determining factors in the immigration rates of these mammals.

Three points can be advanced to explain how small mammals are affected and how they repopulate after the passage of a forest fire associated with their type of locomotion, ecological niche, and the intensity and speed of the fire: (1) some individuals succumb to the fire, (2) some find refuge in natural shelters (in holes or tree trunks) if the fire is not too intense, and (3) others escape and find refuge in nearby forest habitats that are more humid or not touched by the fire.

With the preceding points in mind, the purpose of this current study was to investigate the speed of recolonization by small mammals in a fire-disturbed habitat. An attempt was made to correlate certain parameters of small-mammal relative density and diversity with the regeneration of the local plant communities, based on monitored survey plots.

GEOGRAPHICAL SETTING AND METHODS

The study area is located in Andringitra National Park (22°10 S, 46°56 E) within the Fianarantsoa Province, Madagascar, about 140 km north of the Tropic of Capricorn, at an altitude between 1,920 and 2,000 m. The park has a surface area of 31,160 ha with elevations ranging from 650 to 2,658 m, culminating at Pic Boby, the second highest peak on the island. A portion of the massif is situated within a forested zone, which ranges from lowland forest to montane sclerophyllous forest. Between 1,900 m and 2,000 m, the forest grades into a mosaic of ericoid bush (mountain thicket) and open grassland savanna. This ecotone is presumed to be natural, although the historical use of fire in the region may have modified the plant community and elevation where this transitional zone occurs. The climatic regime varies from warm lowland areas to the seasonally cold summit zone. During the months of June to August (generally the coldest months), the higher elevations often experience temperatures below 0°C, and snow has been recorded (Saboureau, 1962; Paulian et al., 1971). The upper slopes of the massif were glaciated during the late Pleistocene–Holocene epoch (Vidal Romani et al., 2002).

The natural vegetation of our study site, with a western exposure, is upper-montane sclerophyllous forest with year-round leaf cover. Lichens and bryophytes are abundant on the ground or as epiphytes. These characteristics are evident at the sites that were not burned by an intensive October 1995 human-set fire that passed through the upper portions of the massif. The burned area, nearly 2340 ha in total, contained trees that lacked foliage but were identifiable to genus or species either by their physical aspect or by bark characteristics. Further, the subsequently regenerated trees, almost exclusively by root–shoots, are easily recognizable.

To monitor woody plant regeneration, a permanent 50 m × 50 m plot was delimited in the burned forest immediately after the fire. The method used was to survey, count, and mark trees with a diameter at breast height (DBH) equal to or greater than 10 cm. Monitoring sessions were conducted once per year during the rainy season. During each session, both the DBH of trees and aspects of their phenology and vitality were recorded. To monitor regeneration, three separate permanent plots measuring 10 m × 10 m were established to count the emergence of new woody plants.

The sampling method used to estimate the sequence and rate of recolonization by nonflying mammals was based on inventories of small mammals in forests burned by the fire in October 1995 and, following the same protocol, in nearby nonburned forests that served as controls. Two different standardized trapping regimes were used: pitfall traps and live traps (Sherman and Nationals). For precise details on these techniques, see Goodman and Carleton (1996) and Goodman et al. (1996). Trap lines were established at the control and study sites within the elevational range of 1960 m ± 40 m. Sampling began in July 1997, 22 months after the fire, and continued until April 2000. During the study, nine sampling periods were completed (three 10-d sessions/year). The animals studied include the order Lipotyphla (formerly placed in the order Insectivora), which on Madagascar is composed of the family Soricidae, and the endemic family Tenrecidae, as well as the order Rodentia (Muridae), which includes the endemic subfamily Nesomyinae and two introduced genera of the subfamily Murinae (*Rattus* and *Mus*).

Very little precise information is available on the dietary regime of Malagasy Lipotyphla and Rodentia. In 2001, a study was conducted on central highland species, using the available literature (Goodman and Rasolonandrasana, 2001; Table 6.1). During the course of the field-work reported herein, feeding studies were conducted for three species of captive rodents (*Rattus rattus, Monticolomys koopmani*, and *Eliurus majori*), which were presented seeds collected at the site.

RESULTS

RICHNESS OF SMALL MAMMALS

On the basis of previous surveys and other sources of information, 16 species of Lipotyphla and 11 species of Muridae rodents have been documented in the park across an elevation range from 720 to 2450 m (Goodman and Rasolonandrasana, 2001; Table 6.1). Most small mammals are found in broad elevational ranges. The highest species diversity occurs in the middle to upper portion of the forest zone, with Lipotyphla peaking at 1990 m and Rodentia between 1625 and 1960 m. In natural forest formations on the same massif between 1960 and 1990 m, 11 Lipotyphla and 8 Rodentia (2 introduced) have been documented (Table 6.1).

At least 17 species of small mammals (11 Lipotyphla and 6 Rodentia) were recorded during the 3 years of the study. These included 15 species (11 Lipotyphla and 4 Rodentia) in the burned forest and 16 species (10 Lipotyphla and 6 Rodentia) in the unburned forest (Table 6.2). *Setifer setosus* was not found in the control site, and only one individual was found in the burned forest.

Among the Rodentia, no individual of *Eliurus minor* or *Nesomys rufus* was trapped in the burned forest. When considering all trapping sessions, the lipotyphlans *Microgale longicaudata* and *M. fotsifotsy* were better represented in terms of numbers of individuals captured in burned forests than in the unburned forests. The two introduced rodents, *Rattus rattus* and *Mus musculus*, showed a clear affinity for burned forests. The abundance of *Microgale cowani*, *M. dob-*

soni, and *Monticolomys koopmani* was nearly the same for the burned and unburned sites.

IMPACT OF FIRE ON VEGETATION

Woody forest understory plants were notably modified, and forest regeneration was not common (Figure 6.1a). Although most *Agauria* trees recuperated after the fire, they retained evidence of physical shock (burned branches and bark damage) with noticeable damage to the cambium layer and other aspects of their physiognomy and physiology (loss of reserves and reduced growth). The regeneration of *Agauria* by seeds rarely happens naturally (Berner, 1996). On the other hand, 50% of the *Agauria* marked before the fire produced new shoots during the first season after the fire, at which time they also produced abundant fruit. The regeneration of species other than *Agauria* was low, particularly in areas where the fire was intense. In general, burned thicket (shrubs) did not produce shoots with the exception of *Vaccinium* sp. and *Vernonia* sp. (Figure 6.1b). Based on these data, gathered up to 4 year after the passage of the fire, forest regeneration will be a long process, given the almost complete absence of new trees in the burned area.

In 2001, of the 156 originally monitored marked trees (*Agauria polyphylla* [134 individuals], *Ilex mitis*, *Pittosporum* sp., *Polyscias* sp., and *Weinmannia* sp.), only 121 *Agauria* trees survived, and all of the other forest species died. In comparison, thicket (shrub) species regenerated during the first year after the fire. They constitute pioneer species of vegetation succession. The average growth of living trees was rather limited, about 0.3 mm increase in DBH for the 4-year study. The regeneration of vegetation after the fire, in terms of species (10) and abundance (30 to 300 new plants), was largely dominated by thicket species.

FOOD AVAILABILITY (RODENTIA)

One year after the fire, the thicket species such as *Acacia farnesiana*, *Mimosa* sp., and *Indigofera lyallii* (Fabaceae), *Buddleja fusca* (Loganiaceae), and *Vaccinium emirnense* and *V. secondiflorum* (Ericaceae) were abundant in the burned forest and produced seeds in the

TABLE 6.1
Elevation distribution of lipotyphlans and rodents in the Andringitra National Park

Species	Guild[a]	Elevation Zone (m)							
		720	810	1210	1625	1960	1990	2050	2450
Order Lipotyphla									
Microgale cowani	TI		+	+	+	+	+	+	+
Microgale taiva	TI	+	+	+	+	+	+		
Microgale drouhardi	TI	+	+						
Microgale longicaudata	TI	+	[+]	[+]	+	+	+		
Microgale parvula	TI	+	+	+	+	+	+		
Microgale dobsoni	TI/C			+	[+]	+	+	+	
Microgale talazaci	TI/C						+		
Microgale soricoides	TI/C			+	+	+	+		
Microgale gracilis	TI			+	[+]	+	+		
Microgale gymnorhyncha	TI			+	+	+	+		
Microgale fotsifotsy	TI			+	[+]	+	+		
Oryzorictes hova	TI					+	+		
Oryzorictes tetradactylus	TI							+	+
Hemicentetes nigriceps	TI						+		
Setifer setosus	TI		+						
Tenrec ecaudatus	TI	+	+						
Total number of Lipotyphla		5	7	9	9	10	11	4	2
Total number of *Microgale* spp.		4	5	9	9	9	10	2	1
Order Rodentia									
Mus musculus[b]	TGr					+	[+]	+	
Rattus rattus[b]	TAGr		+	+	+	+	+	+	+
Brachyuromys betsileoensis	Tfol/Gr					+	+	+	+
Brachyuromys ramirohitra	Tfol/Gr			+	+	+			
Eliurus majori	ATGr			+	+	+	+		
Eliurus minor	ATGr	+	+	+	+	+			
Eliurus tanala	ATGr		+	+	+				
Eliurus webbi	TAGr	+	+						
Gymnuromys roberti	TGr	+	+	+	+				
Monticolomys koopmani	TAGr					+	+	+	
Nesomys rufus	TGr	S	+c	+	+	+	S		
Total number of rodent species		4	6	7	8	8	6	3	2
Total number of native rodent species		4	5	6	7	6	4	1	1

Note: Key: + = specimens examined for rodents and lipotyphlans; S = sighting for rodents; [+] = local altitude occurrence derived based on records above and below this zone.

[a]Guild codes according to Goodman and Rasolonandrasana (2001). TI = terrestrial invertebrates, TI/C = terrestrial invertebrates and vertebrates, TGr = terrestrial graminivore, TAGr = largely terrestrial and secondarily arboreal graminivores, TFol/Gr = terrestrial folivore and graminivore, ATGr = largely arboreal and secondarily terrestrial graminivores.
[b]Introduced to Madagascar.
[c]*Nesomys audeberti* was previously reported from this transect zone (Goodman and Carleton, 1996). The specimen identified as this taxon is *N. rufus.*

subsequent year. The surviving *Agauria* bore fruit the first year after the fire. Dietary tests on captive wild animals showed that the fruits of these species were consumed by *Rattus rattus,* *Eliurus majori,* and *Monticolomys koopmani.* At the burned site 2 years after the fire, the fruits of these plants constituted, among others, the primary food of an abundant rodent population.

TABLE 6.2
Number of individuals of all species captured during the study per type of habitat (not including recaptures)

	Unburned Forest	Burned Forest
Lipotyphla (2838 pitfall bucket days)		
Microgale longicaudata	39	66
M. cowani	147	137
M. fotsifotsy	8	19
M. dobsoni	29	27
M. taiva	18	4
M. gymnorhyncha	17	6
M. soricoides	21	9
M. gracilis	3	2
M. parvula	5	1
Oryrorictes hova	1	2
Setifer setosus	—	1
Total number of individuals	288	274
Rodentia (4400 trap days)		
Rattus rattus	35	152
Mus musculus	1	12
Monticolomys koopmani	244	218
Eliurus majori	34	4
Eliurus minor	4	—
Nesomys rufus	4	—
Total number of individuals	322	386

Note: All Lipotyphla species were trapped (with pitfall buckets) in both burned and unburned forests for 2838 pitfall bucket days; the Rodentia species were trapped for 4400 trap days, respectively.

ORDER OF RECOLONIZATION OF SMALL MAMMALS IN THE BURNED FOREST

Based on the lipotyphlans and rodents captured in the control site, it is presumed that the burned area contained ten and four species, respectively, before the fire. The species accumulation curves show that all the lipotyphlan species likely to be present in the unburned forest were found after 660 pitfall bucket days, with the exception of *Setifer setosus* that was found only once in the burned forest (Figure 6.2a). With the same trapping protocol, the accumulation of lipotyphlan species in the burned forest was notably slower. In the burned habitat, *Microgale gymnorhyncha* was captured after 927 pitfall bucket days, *M. soricoides* after 1254 pitfall bucket days, *Oryzorictes hova* and *M. gracilis* after 1320 pitfall bucket days, *M. parvula* after 2277 pitfall bucket days, and *S. setosus* (the

single individual captured during this study) after 2376 pitfall bucket days. We presume these data reflect the approximate period that these species recolonized the burned area.

The rodents, *Rattus rattus*, *Monticolomys koopmani*, and *Eliurus majori* appeared during the first trapping sessions in the burned and control sites (Figure 6.2b). *Mus musculus* was first captured during the first session in the burned forest (after 200 trap days), but was only trapped in the control site after 1200 trap days. This indicates a particular preference of *Mus* for the burned forest, as has been documented for this species in other areas of the world (Newsome et al., 1975). After the trapping sessions, there was nearly a complete overlap in the species of small mammals documented in the burned and unburned forests. In regard to lipotyphlans, the only exception was *Setifer setosus*, which was captured

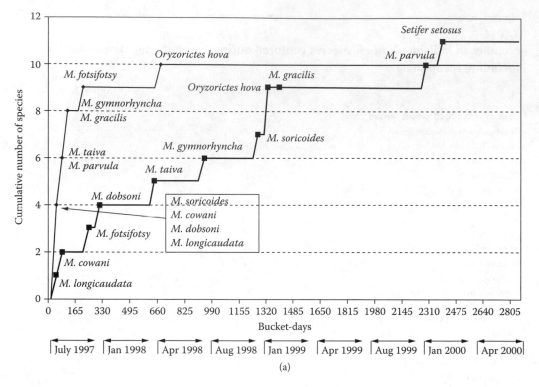

(a)

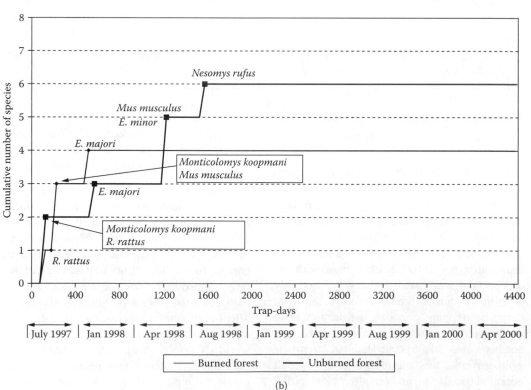

(b)

FIGURE 6.1 Forest regeneration 24 months (September 1995) to 65 months (February 2001) after the passage of a fire, including measures of abundance of individual plants and species richness for (A) forest (woody) species and (B) thicket (shrub) species. The seven forest species at the start of the study included *Ilex mitis* (Aquifoliaceae), *Polyscias* sp. (Araliaceae), *Weinmannia* sp. (Cunoniaceae), *Agauria polyphylla* (Ericaceae), an unidentified Euphorbiaceae, *Pittosporum* sp. (Pittosporaceae), and an undetermined tree. It was only *Agauria* that remained at the site in stable numbers after the passage of the fire. The thicket species censused after the passage of the fire included: April 1996, April 1997, and June 1998 — *Indigofera lyallii* (Fabaceae) and *Vernonia delapsa* (Asteraceae); February 1999 and June 2000 — *I. lyallii* and *Mundulea barclayi* (Fabaceae); *Philippia floribunda, P. gracilis, P. trichoclada, P.* sp., and *Vaccinium secundiflorum* (Ericaceae); and *Vernonia delapsa*; February 2001 — *I. lyallii, M. barclayi,* and *Mimosa descarpentriense* (Fabaceae); *P. floribunda, P. gracilis, P. trichoclada, P.* sp., *Vaccinium secundiflorum, Vernonia delapsag* and *V.* sp.

exclusively at the burned site. *Nesomys rufus* and *Eliurus minor* were the rodents captured at the control site. The upper elevational limit of these two species is near 1950 m, and their late documentation at the control site presumably reflects their low density. Given this point, it is difficult to interpret their absence in the burned area.

DISCUSSION

INSIGHT INTO SMALL-MAMMAL COMMUNITIES OF THE CONTROL AND BURNED AREAS

The community of small mammals documented at the control site is largely typical of the upper-montane sclerophyllous forest of the central highlands in regard to species representation and diversity (Goodman and Rasolonandrasana, 2001; Goodman et al., 2003). No site in the upper-montane sclerophyllous forests of Madagascar has been as extensively sampled as the Andringitra Massif. Perhaps the two best comparisons are the Andohahela Massif, 300 km to the south of Andringitra, and the Marojejy Massif, about 900 km to the north of Andringitra. In unburned montane forests of Marojejy, 10 of the 11 species of Tenrecinae (genera *Microgale* and *Oryzorictes*, the nonspiny members of the family Tenrecidae) found at Andringitra are shared in common between these two sites, as are 3 of the 5 species of native and introduced rodents (Carleton and Goodman, 2000; Goodman and Jenkins, 2000), and for largely unburned areas of Andohahela, these figures are 9 out of 11 for Tenrecinae and 3 out of 5 for rodents (Goodman, Jenkins et al., 1999; Goodman, Carleton et al., 1999). Thus, based on

broad-scale inventories, it can be concluded that, in general, these three massifs share a considerable proportion of their small-mammal fauna.

The relative abundance of small mammals at the burned site was different from that in the control area. This is most notable for rodents, which had a greater trap success rate in the burned area than the control area, and a preponderance of introduced species (*Rattus rattus* and *Mus musculus*) — 164 (42%) captures as compared to 36 (11%) of the individuals trapped in the control site. Excluding the introduced species, the burned area had lower densities of native rodents than that of the control site (286 [88%] of captures vs. 222 [58%] of captures).

This study showed a preference between the two species of Tenrecidae, *Microgale longicaudata* and *M. fotsifotsy*, for burned forest areas. Both of these species consume terrestrial invertebrates (Table 6.1), and this would indicate that sufficient numbers of soil invertebrates had recolonized the site to support these two species. In contrast, three other species, *M. taiva, M. gymnorhyncha,* and *M. soricoides*, were distinctly more common in the unburned site than the burned site. The first two of these animals feed on terrestrial invertebrates, whereas the third also consumes small vertebrates (Table 6.1). Thus, it would appear that some parameters of life history traits for these different *Microgale* spp. are not in parallel between the unburned and burned area. We presume that this is associated with their different dietary regimes.

Among rodents, *Rattus rattus* and *Mus musculus*, either terrestrial or largely terrestrial graminivores, have broad dietary regimes and

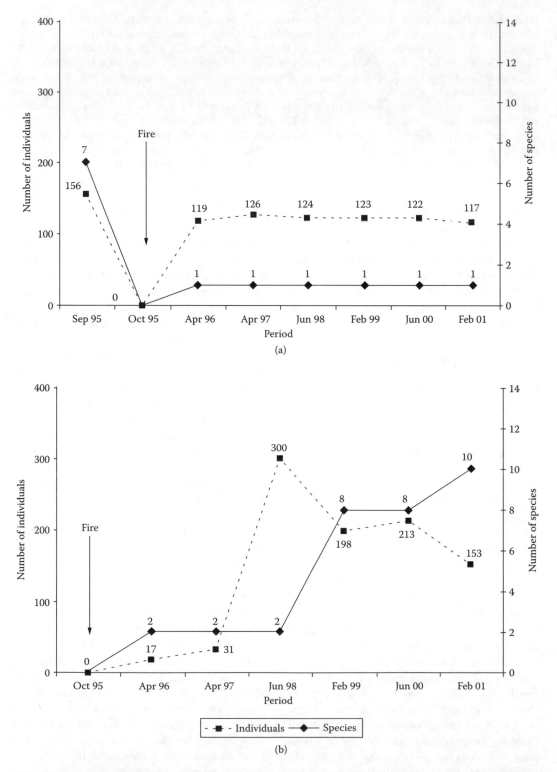

FIGURE 6.2 Small-mammal species accumulation curves for both burned and nonburned sites in relation to the nine sampling periods: the total number of (A) bucket days for Lipotyphla and (B) trap days for Rodentia.

presumably consume a wide variety of foods found in the burned zones. At least *R. rattus* has successfully invaded the sclerophyllous forest. On the basis of specimens collected from 1970 to 1971 (Paulian et al., 1971), *R. rattus* has been known to occur on the mountain for at least 30 years. The western side of the Andringitra Massif has been settled and cleared for a considerable period of time and contains numerous agricultural areas, particularly rice paddies. Also found in the environment of the study site are open areas used by local residents for cattle pasture. Because *R. rattus* is often a synanthropic species and prospers around dwellings, grazing areas, and agricultural clearings, its abundance at the study site is not surprising. The presence of *Eliurus minor* and *Nesomys rufus* was never established in the burned forest. We speculate that *N. rufus* may be a rodent with a restricted diet and habitat preferences. *E. majori* and *E. minor*, which are primarily arboreal graminivores, preferred an intact habitat rather than the burned forest in which all trees except *Agauria* were dead. Presumably, an intact lower story of the forest, with its complete array of lianas, bushes, and other sorts of arboreal connections, is important for the locomotion of these two species. In the unburned forest, the number of trapped individuals of *R. rattus* and *E. majori* were similar. On the basis of feeding trials with captive individuals on the Andringitra Massif, it was found that *R. rattus* and several rodent species of the family Nesomyinae feed on similar types of forest fruits. Thus, the potential for direct competition between these rodents exists if indeed these resources are at times in short supply; however, this needs to be more rigorously tested with free-living animals.

CHANGES IN SMALL-MAMMAL COMMUNITY RESOURCES

As mentioned earlier, most members of the family Tenrecidae consume terrestrial and soil-dwelling invertebrates, and the density and diversity of these prey items are presumably closely tied to soil productivity, particularly organic content, and degree of water saturation. Forest soils are covered with a mosaic of litter at various depths. After the passage of the fire, there was a reduction in the surface layer of decomposing litter. The invertebrates living in the soil could have been affected directly by the fire or by the change in the surface litter associated, for example, in the depletion in organic content or to the increase in pH levels. A fire does not necessarily raze a region uniformly — unburned zones exist in the forest and constitute the refuge areas that provide food for small mammals that feed on soil invertebrates. Over time, reconstitution of the soil litter and associated fauna occurs, presumably with the decomposition of vegetation. Further, after the fire, the local forest was transformed into a mosaic of thicket species with simplified forest diversity dominated by *Agauria*. We presume that after the passage of the fire, the diversity of fruits and nuts, the principal foods consumed by nesomyine rodents, was more restricted in general abundance or seasonally in the burned area. Hence, this would reduce the augmentation of local populations either by means of successful breeding of any remaining individuals or dispersal from nearby intact areas, at least until regeneration advanced.

CONCLUSION

The present study suggests that on a long-term scale, the passage of fire in a sclerophyllous forest is not harmful to the small-mammal population when the fire frequency remains moderate. However, the speed of recolonization by small mammals is variable among species. The most important factor influencing recolonization seems to be the availability of food fitting with the dietary niche of each species. The appearance of small mammals in the burned forest habitat can be considered as a secondary stage in the resilience of the forest to fire. Thus, the frequent passage of fire in the same forest has negative effects, as it increases the period of recovery for natural vegetation. In effect, the forests studied are situated in areas where the abiotic conditions are relatively severe, such as large fluctuations during a 24-h period in climate conditions and extreme annual differences in rainfall (Goodman and Andrianarimisa, 1996), making the forest less resistant to repeated disturbances. Finally, given that fires occur naturally in the region, there is presumably some adaptation to the associated changing conditions among the indigenous biota.

SUMMARY

Occasional natural fires have been a part of the environment in the central highlands of Madagascar throughout recent geological time, but since the arrival of man about a millennia ago to this portion of the island, the frequency and intensity have increased. As the occurrence of fire in a natural forest habitat affects plant composition and succession, we investigate how animal composition, distribution, and population dynamics are linked to such types of disturbance. We examine the impact of an extensive fire on nonflying small mammals (Rodentia and Lipotyphla) in a montane sclerophyllous forest at 1960-m elevation and compare these results to an unburned control site at the same elevation. The study commenced 22 months after the fire and continued over 3 years. At the burned site, two introduced rodent species represented 42% of the rodents captured as compared to 11% of those at the control site. An attempt was made to correlate small-mammal numbers, species composition, and order of recolonization with the regeneration of plant communities and food availability. The recolonization pattern of the burned area by different small-mammal species was not uniform and appears to be a function of food availability and understory structure.

References

Ahlgren, C.E. (1966). Small mammals and reforestation following prescribed burning. *Journal of Forestry*, 64: 614–618.

Beck, A.M. and Vogl, R.J. (1972). The effects of spring burning on rodent populations in brush prairie savanna. *Journal of Mammalogy*, 53: 336–346.

Berner, P. (1996). La conservation des forêts de montagne de la partie nord-est des massifs rocheux de l'Andringitra, Madagascar. Rapport de consultation. Document interne du PCDI Andringitra, Pic d'Ivohibe.

Burney, D.A. (1987). Late quaternary stratigraphic charcoal records from Madagascar. *Quaternary Research*, 28: 274–280.

Burney, D.A., Burney, L.P., Godfrey, L.R., Jungers, W.L., Goodman, S.M., Wright, H.T., and Jull, A.J.T. (2004). A chronology for late prehistoric Madagascar. *Journal of Human Evolution*, 47: 25–63.

Carleton, M.D. and Goodman, S.M. (2000). Rodents of the Parc National de Marojejy, Madagascar. In Goodman, S.M. (Ed.), *A Floral and Faunal Inventory of the Parc National de Marojejy, Madagascar: With Reference to Elevational Variation. Fieldiana: Zoology*, new series, 97: 231–263.

Chew, R.M., Butterworth, B.B., and Grechman, R. (1959). The effects of fire on the small mammal populations of a chaparral. *Journal of Mammalogy*, 40: 253.

Christian, D.P. (1977). Effects of fire on small mammal populations of a chaparral. *Journal of Mammalogy*, 58: 423–427.

Cook, S.F. (1959). The effects of fire on a population of small rodents. *Ecology*, 40: 102–108.

Daubenmire, R. (1968). Ecology of fire in grasslands. *Advances in Ecological Research*, 5: 209–266.

Goodman, S.M. and Andrianarimisa, A. (1996). Meterology. In Goodman, S.M. (Ed.), *A Floral and Faunal Inventory of the Réserve Naturelle Intégrale d'Andringitra, Madagascar: With Reference to Elevational Variation. Fieldiana, Zoology*, new series, 85: 20–23.

Goodman, S.M. and Carleton, M.D. (1996). The rodents of the Réserve Naturelle Intégrale d'Andringitra, Madagascar. In Goodman, S.M. (Ed.), *A Floral and Faunal Inventory of the Réserve Naturelle Intégrale d'Andringitra, Madagascar: With Reference to Elevational Variation. Fieldiana: Zoology*, new series, 85: 257–283.

Goodman, S.M., Ganzhorn, J.U., and Rakotondravony, D. (2003). Introduction to the mammals. In Goodman, S.M. and Benstead, J.P. (Eds.), *The Natural History of Madagascar*. The University of Chicago Press, Chicago, pp. 1159–1186.

Goodman, S.M. and Jenkins, P.D. (2000). Tenrecs (Lipotyphla: Tenrecidae) of the Parc National de Marojejy, Madagascar. In Goodman, S.M. (Ed.), *A Floral and Faunal Inventory of the Parc National de Marojejy, Madagascar: With Reference to Elevational Variation. Fieldiana: Zoology*, new series, 97: 201–229.

Goodman, S.M., Jenkins, P.D., and Pidgeon, M. (1999). The Lipotyphla (Tenrecidae and Soricidae) of the Réserve Naturelle Intégrale d'Andohahela, Madagascar. In Goodman, S.M. (Ed.), *A Floral and Faunal Inventory of the Réserve Naturelle Intégrale d'Andohahela, Madagascar: With Reference to Elevational Variation.* Fieldiana: Zoology, new series, 94: 187–216.

Goodman, S.M., Carleton, M.C., and Pidgeon, M. (1999). Rodents of the Réserve Naturelle Intégrale d'Andohahela, Madagascar. In Goodman, S.M. (Ed.), *A Floral and Faunal Inventory of the Réserve Naturelle Intégrale d'Andohahela, Madagascar: With Reference to Elevational Variation.* Fieldiana: Zoology, new series, 94: 217–249.

Goodman, S.M. and Rasolonandrasana, B.P.N. (2001). Elevational zonation of birds, insectivores, rodents and primates on the slopes of Andringitra Massif, Madagascar. *Journal of Natural History*, 35: 285–305.

Goodman, S.M., Raxworthy, C.J. and Jenkins, P.D. (1996). Insectivore Ecology in the Réserve Naturelle Intégrale d'Andringitra, Madagascar. In Goodman, S.M. (Ed.), *A Floral and Faunal Inventory of the Réserve Naturelle Intégrale d'Andringitra, Madagascar: With Reference to Elevational Variation.* Fieldiana: Zoology, new series, 85: 191–217.

Howard, W.E., Fenner, R.L., and Childs, H.E., Jr. (1959). Wildlife survival in bush burns. *Journal of Range Management*, 12: 230–234.

Kaufman, D.W., Kaufman, G.A., and Finck, E.J. (1989). Rodents and shrews in ungrazed tallgrass prairie manipulated by fire. In Bragg, T.B. and Stubbendieck, J. (Eds.), *Prairie Pioneers: Ecology, History and Culture. Proceedings of the Eleventh North American Prairie Conference*, Nebraska, pp. 173–177.

Krefting, L.W. and Ahlgren, C.E. (1974). Small mammals and vegetation changes after fire in a mixed conifer-hardwood forest. *Ecology*, 55: 1391–1398.

Layne, J.N. (1974). Ecology of small mammals in a flatwoods habitat in north-central Florida, with emphasis on the cotton rat (*Sigmodon hispidus*). *Am Mus Novitates* 2544: 1–48.

Newsome, A.E., McIlroy, J., and Catling, P. (1975). The effects of an extensive wildlife fire on populations of twenty ground vertebrates in south-east Australia. *Proceedings of the Ecological Society of Australia*, 9: 107–123.

Paulian, R., Betsch, J.M., Guillaumet, J.L., Blanc, C., and Griveaud, P. (1971). RCP 225. Etudes des écosystèmes montagnards dans la région malgache. I. Le massif de l'Andringitra. 1970–1971. Géomorphologie, climatologie et groupements végétaux. *Bulletin de la Société d'Ecologie*, II(2–3): 198–226.

Saboureau, M. (1962). Note sur quelques températures relevées dans les réserves naturelles. *Bulletin de l'Académie Malgache*, nouvelle série, 40: 12–22.

Soarimalala, R.A.L. (1998). Contribution à l'étude du régime alimentaire des insectivores du Parc National de Ranomafana. Mémoire pour l'obtention d'un diplôme d'études approfondies. Département de Biologie Animale, Faculté des Sciences, Université d'Antananarivo.

Tevis, L. (1956). Effect of slash burn on forest mice. *Journal of Wildlife Management*, 20: 405–409.

Vidal Romani, J.R., Mosquera, D.F., and Campos, M.L. (2002). A 12,000 year BP record from Andringitra Massif, (southern Madagascar): post-glacial environmental evolution from geomorphological and sedimentary evidence. *Quaternary International*, 93: 45–51.

7 Fire, Plant Species Richness, and Aerial Biomass Distribution in Mountain Grasslands of Northwest Argentina

Roxana Aragón, Julietta Carilla, and Luciana Cristóbal

INTRODUCTION

Grazing and fire are the most common disturbances in many grassland ecosystems around the world (McNaughton et al. 1993; Vogl 1974 in Oesterheld et al. 1999, De Baro et al. 1998), and they both affect biodiversity and plant community dynamics. Grazing and fire influence species composition and richness, determine dominant life-forms and therefore the general structure of the community (Belsky 1992; Diaz et al. 1992; Milchunas and Lauenroth 1993; Collins et al. 1998). They can also regulate ecosystem processes such as nutrient cycling (Hobbs et al. 1991) and plant productivity (McNaughton 1985; Rusch and Oesterheld 1997). Importantly, grazing and fire often occur together, and they interact deeply.

Grazing and fire are both consumers of plant production. Herbivores feeding on forage can determine the fuel load. Fire, in turn, consumes accumulated biomass that could be used by herbivores (Oesterheld et al. 1999). Grazing can influence fire frequency and intensity, and fire determines what is left for herbivores, not only in terms of quantity but also in terms of forage quality (Hobbs et al. 1991). In addition, these disturbances provide open space for colonization that, in turn, can modify species diversity, promote the establishment of certain species, and change the general structure of the community (Collins 1987; Pucheta et al. 1998;

Valone and Kelt 1999). Grazing and fire occur naturally in many grasslands and savannas, and they also are part of many management practices. In addition, burning in grasslands and savannas has an important worldwide effect because it is one of the major sources of atmospheric methane and CO_2, especially in tropical areas (Crutzen et al. 1985 in Hobbs et al. 1991).

Livestock raising is one of the most important land uses in many montane grasslands (Eckholm 1975). Particularly in Andean grasslands, extensive cattle grazing is often combined with burning of the natural vegetation (Schmidt and Verweij 1992 in Hofstede et al. 1995, Grau and Brown 2000). Fire promotes resprouting and is believed to encourage the development of more palatable life-forms (Grau and Brown 2000). However, grazing and fire can also increase soil susceptibility to erosion, reduce species or functional richness (Lloret and Vila 2003), and modify community composition (Pucheta et al. 1998; Diaz et al. 1992). Eventually, their positive or negative effects depend on an array of factors such as grazing intensity, fire frequency, and climate.

Mountain grasslands are one of the most species-rich habitats of northwest Argentina. They are important in regulating the hydric regime and in providing economic resources (e.g. cattle ranching and scenic values). In spite of their ecological and economical importance, mountain grasslands are scarcely represented in the protected areas of Argentina, and little is known about their functioning. The study site of this work, the valley of Los Toldos, is located in the upper Bermejo River basin and is considered an area of high conservation priority at a national level (Brown et al. 2001). The dominant land use is for grazing by cattle, and this is combined with periodic fires. As was observed in other neotropical mountains, recent works suggest a decrease in land use intensity in this area (Grau et al. *submitted*). The reduction in the density of animals may produce changes in fire frequency and intensity that can, in turn, affect plant communities in different ways. In this chapter, we describe a study on how fires affect vegetation structure in the mountain grasslands of northwest Argentina that are used for grazing. More specifically, this study intends to investigate the effect that the time since the last fire event may have on plant species richness, vegetation structure, and biomass dynamics.

METHODS

STUDY AREA

The study was performed at the valley of Los Toldos (22°30 S, 64°50 W), Santa Victoria, Salta, Argentina. The study area consists of a mosaic of mountain grasslands and *Alnus acuminata* forest patches at an altitude of about 1700 masl. This area lies in the upper altitudinal level of the phytogeographic province of the Argentinean Yungas (subtropical montane forest) (Cabrera 1976). The original vegetation seems to have been dominated by forest patches, but a long history of grazing in the valley probably shaped the current vegetation physiognomy (Malizia 2003). The mean annual temperature is 15°C, and the average precipitation is 1300 mm (Ramadori 1995). The precipitation is highly seasonal, with most of the rain falling during the summer months (Bianchi 1981).

The main disturbances at this altitudinal range are grazing, fire, and landslides (Grau, 2005), and livestock raising is the most com-

mon land use. Cattle grazing is extensive with no fences limiting individual properties. There are no data on grazing intensity in Los Toldos, but information provided by national agricultural censuses for Santa Victoria shows a decrease in the population of domestic animals during the 20th century (Grau et al. submitted). These data, together with information provided by local people, suggest that grazing intensity in Los Toldos is currently low (between 0.5 and 1 cow per 10 hectares). The pastoral system involves transhumance, a seasonal movement of cattle from the highlands to midaltitude and piedmont forests (Grau and Brown 2000). Cattle are driven up to the highland grasslands at the beginning of the summer period and, in March, they are brought back to lower ranges (piedmont forest). During the summer period (from November to March), the animals feed mainly in grassland patches, but also browse in the *Alnus* forest understory. Summer grazing by cattle is usually prepared for by burning the vegetation in spring. The extent and frequency of burning seem to depend on the proximity to settlements and on the weather conditions (wind, temperature, and soil humidity) when the fire is started. As a result of these management practices, the landscape consists of a mosaic of vegetation patches, differing in the time since the last burning event occurred.

SAMPLING DESIGN AND DATA ANALYSIS

In November 2000, we conducted a survey in the study site, looking for evidence of previous fire events. Based on this survey and on information provided by local inhabitants, we identified three types of vegetation patches that differed in the time since the last fire event occurred:

1. Areas burned during the ongoing growing season (the last fire event probably occurred during spring 2000). These areas showed evident signs of fire, such as abundant charcoal, ashes, and burned vegetation.

2. Areas burned during the previous growing season (spring 1999), with

some evidence of fire (mainly the remains of charcoal).

3. Areas not burned recently. In this case, the last fire event apparently took place at least 5 years ago (spring 1995 or earlier). This information was checked with local residents.

We selected three patches of each vegetation type (nine in total). The time since the last fire event was regarded as "treatment." Hereafter, we will refer to the different treatments as: <1 year (areas burned during the ongoing growing season); >1 year (areas burned during the previous season), and >5 years (areas not burned for at least 5 years). Unfortunately, since burning is a common practice in this area, we did not have any plots that had no fire and could have served as a control. All the vegetation patches included in our sample were no more than 3 km apart from each other, had areas of less than 500 m², and were in similar topographic positions. Because of the absence of fences, vegetation patches had potentially similar grazing pressure.

In December 2000, we conducted plant relevés in 1 m × 1 m plots with five plots per patch. The plots were placed every 10 m in a 50-m transect. The transects were placed at random in each patch (i.e. 1 transect in each patch). Each plot was divided into four 0.5 m × 0.5 m quadrats, and all the plant species present were recorded. In addition, we collected all the aerial biomass in ten 0.2 m × 0.2 m plots in each patch. Whenever possible, the plots were placed in two 50-m transects that were separated by 10 m. If the vegetation patches were not big enough, we placed plots in shorter transects, but always used the same number of plots. We collected biomass in December 2000 and in January, February, March, and August 2001 (before the next burning event). The biomass was classified into live biomass, standing dead, and litter. Live biomass was further classified for different life-forms (i.e. graminoids, tussock grasses, erect species, rosettes, prostrate species, ferns, and woody species). All the material was classified, dried to constant weight at 70°C, and weighted.

We used ANOVA tests for the comparisons between treatments (both for total biomass and proportions). In the case of species richness and total number of species, we used Kruskal–Wallis tests, a nonparametric technique, because the assumptions required for parametric tests were not met. Differences in the biomass collected throughout the year were tested through repeated measures ANOVA. Species frequency was computed as the number of plots per treatment in which each species was recorded. Equativity was computed as:

$$E_m = -\Sigma \ (p_i \log p_i)/\log N$$

where p_i is the proportion of the species recorded in transect m that belong to life-form i, and N is the total number of different life-forms in that transect. Small values of E imply that one or a few life-forms are dominant in the community; in other words, this index is an indication of evenness (O'Neill et al. 1988). To measure compositional similarity among plots, we performed a detrended correspondence analysis (DCA) with downweighting of rare species. Only the species that were recorded in at least two plots were considered. We computed a nonparametric Kendall's tau correlation between plot scores in the ordination space and the time since the last fire event. The analyses were performed in Statistica (StatSoft 1993) and PCORD (McCune and Mefford 1997).

RESULTS

We recorded a total of 149 species in the study area. In Table 7.1, we have included only the species that had a frequency 0.4 in at least one of the treatments. Among these 45 species, 32 were common to all the treatments. The number of species per square meter did not differ between the treatments (Kruskal-Wallis test, KW 3.29, $p = .19$) (Table 7.1), and the most common species were present in all the patches independently of their fire history. The most common grasses were *Elionurus muticus* and *Paspalum notatum; Stevia yaconensis* was the most frequent woody species, and the *Cuphea* sp. was the dominant prostrate species.

The ordination of plots in the DCA was not clearly linked to the treatments. The first axis of the ordination explained approximately 30% of the overall variance in species data $\lambda_1 =$ 0.259, total inertia = 0.859), and plot scores were not significantly correlated with the time since the last fire event (Kendall's tau = 0.48 p = .07). However, there seemed to be some minor changes in species composition in response to the treatments because we found that some species were differentially recorded in certain plots. *Anemone decapetala* and *Tessaria fastigiata* were recorded only in plots that were recently burned. *Eupatorium bupleurifolium* and *Ophioglossum* sp. were more abundant in the >5-year treatment, whereas *Baccharis tridentata* and *Setaria* sp. were predominantly recorded in <1-year treatment (Table 7.1). In addition, the equativity of life-forms showed a slight tendency to decrease in the patches that were not burned for 5 years (KW 5.42, p = .06) (Table 7.1). The decrease in equativity in areas that were not burned for 5 years was related to the increasing dominance of woody species in comparison to other life-forms that were less frequently recorded, such as erect species and rosettes.

The total aerial biomass was significantly higher in the patches that were not recently burned (F = 69.59, p < .001). The biomass in the >1-year treatment was almost twice as high as the biomass in the <1-year treatment, and the biomass in the >5-year treatment was more than 3 times the biomass in <1-year treatment (431.48 ± 27.95, 738.09 ± 63.64, and 1303.25 ± 58.79 g m^{-2} for <1-year, >1-year, and >5-year treatments, respectively, Figure 7.1). There was no difference between <1-year and >1-year treatments with respect to total live biomass, but live biomass was highest in the >5-year plots (F = 22.02, p < .01). There was no difference in total standing dead material (F = 3.33, p = .10), but the total amount of litter differed between the treatments (F = 71.17, p <.001). Patches that were burned in the ongoing growing season (<1-year) had considerably less litter than both >1-year and >5-year patches (Figure 7.1).

The relative contribution of the different biomass categories differed between the treatments. Live biomass had a high contribution to the total biomass in the patches that were burned during the ongoing growing season (<1-

yr) (60 ± 0.5%), whereas the proportion of litter was minimum in this treatment (15, 27, and 41% in <1-year, >1-year, and > 5-year treatments, respectively) (F = 5.14, p = .04 for ANOVA on live biomass and F = 73.33, p < 0.001 for ANOVA on litter) (Figure 7.2). The contribution of standing dead material was reduced in the patches that were not burned for 5 years (22, 29, and 10%, respectively) (F = 9.77, p < .01).

The proportion of live biomass differed between <1-year and >1-year treatments, but there was no difference between these two treatments and the >5-year treatment. But importantly, although the overall proportion of live biomass was similar between the <1-year and >5-year treatments, the relative contribution of the different life-forms to the total of live biomass was quite distinct. Patches that were recently burned (<1 year) had a high proportion of erect species and ferns compared to the other treatments (Table 7.2). The proportion of tussock grasses plus graminoids did not differ between <1-year and >1-year treatments but their contribution was significantly smaller in the patches that were not burned for 5 years (28 and 33% in <1-year and >1-year and 15% in >5-year treatments) (Table 7.2). This difference was mainly due to tussock grasses that were reduced in >5-year patches (23, 25, and 9%, respectively, in the <1-year, >1-year, and >5-year treatments). The biomass of graminoids was similar in all three treatments. Woody species accounted for 72 ± 9% of the live biomass in the >5-year treatment (Table 7.2).

The seasonal dynamics of the total live biomass, standing dead, and litter showed some similarities between the treatments. Biomass assigned to the standing dead compartment showed a peak in August in all three treatments (Table 7.3). Similarly, litter had its maximum value in August in the <1-year and >5-year treatments, but we did not detect any seasonal trend in the >1-year patches. Live biomass showed a significant decrease in August in the >1-year patches, and a small peak in March and December; however, no similar trend was detected in the other two treatments (Table 7.3). Interestingly, the relative contribution of live biomass throughout the year strongly differed between the treatments. Patches that were

TABLE 7.1
Species with frequencies $\geq$ 0.4 in at least one of the different treatments[a]

	<1 year	>1 year	>5 years	p Values[b]
Mean number of species per m^2	25.20	25.67	20.27	.19
Total number of species	89	96	83	.11
Equativity of life-forms	0.88	0.93	0.77	.06

Species	Life-forms	<1 year	>1 year	>5 years
Achyrocline sp.	Rosette	0	0.40	0.20
Agrostis sp.	Graminoid	0.13	0.40	0.07
Anemone decapetala	Prostrate	0.40	0	0
Baccharis coridifolia	Woody	0.73	0.73	0.53
Baccharis rupestris	Woody	0.60	0.60	0.27
Baccharis tridentata	Woody	0.73	0.40	0.13
Chaptalia modesta	Rosette	0.53	0.20	0.2
Chevreulia acuminata	Rosette	0.20	0.53	0.40
Clitoria cordobensis	Prostrate	0.67	0.07	0.13
Croton sp.	Woody	0	0.33	0.47
Cuphea sp.	Prostrate	1	1	1
Cynodon sp.	Graminoid	0.33	0.67	0.13
Cyperaceae	Graminoid	0.40	0	0.20
Desmodium affine	Prostrate	0.40	0.73	0.67
Desmodium sp.	Prostrate	0.40	0.40	0.80
Elionurus muticus	Tussock grass	1	1	0.67
Eupatorium bupleurifolium	Woody	0.13	0.40	0.93
Euphorbiaceae	Prostrate	0.87	0.13	0.27
Gamochaeta sp.	Rosette	0.40	0	0.27
Hybanthus parviflorus	Erect	0.27	0.67	0.20
Hypericum sp.	Erect	0.40	0.53	0.07
Hyptis mutabilis	Woody	0.20	0.40	0.60
Juncus sp.	Graminoid	0.67	0.40	0.20
Unknown 3 (Myrtaceae)	Woody	0.20	0.40	0.27
Lepechinia vesiculosa	Woody	0	0.80	0.67
Lobelia nana	Prostrate	0.67	0.53	0.67
Malvaceae	Prostrate	0.33	0.40	0.33
Ophioglossum sp.	Fern	0	0.26	0.47
Panicum ovuliferum	Graminoid	0	0.47	0.33
Panicum sp.	Graminoid	0	0.60	0.33
Paspalum notatum	Graminoid	0.53	0.47	0.47
Paspalum sp.	Tussock grass	0.87	0.47	0.67
Plantago sp.	Woody	0.60	0.07	0.20
Polygala pulchella	Erect	0.60	0.20	0
Polygala sp.	Erect	0.20	0.40	0
Pteridium aquilinum	Fern	0.47	0.27	0.07
Ranunculus praemorsus	Erect	0.20	0.40	0.20
Richardia sp.	Erect	0.53	0.13	0.13
Setaria sp.	Graminoid	0.53	0.07	0
Stenandrium dulce	Rosette	0.13	0.53	0.20
Stevia alpina	Woody	0.80	0.67	0.13
Stevia yaconensis	Woody	1	1	0.93
Tagetes filifolia	Prostrate	0.40	0.20	0.07
Tessaria fastigiata	Woody	0.40	0	0
Tibouchina sp.	Woody	0.13	0.20	0.40

[a]<1 year: patches that were burned in the ongoing growing season, >1 year: burned the previous season, or >5 years: not burned for 5 years; mean number of species per m^2. Total number of species and equativity of life-forms per treatment is also shown.

[b] *p* values correspond to Kruskal–Wallis tests.

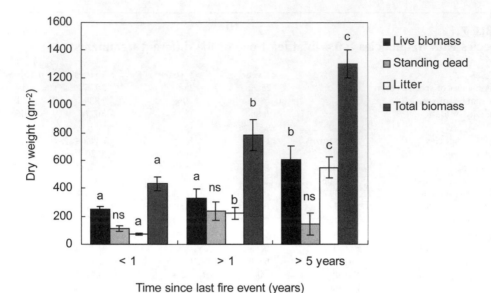

FIGURE 7.1 Variation of biomass according to different treatments. Live biomass, standing dead, litter, and total biomass (g m^{-2}) in patches that were burned in the ongoing growing season (<1 year), burned in the previous season (>1 year), or not burned for 5 years (>5 years). Different letters denote significant differences at $p < .05$ according to an ANOVA test, and ns indicates no significant differences. The comparisons were made between treatments within each biomass category.

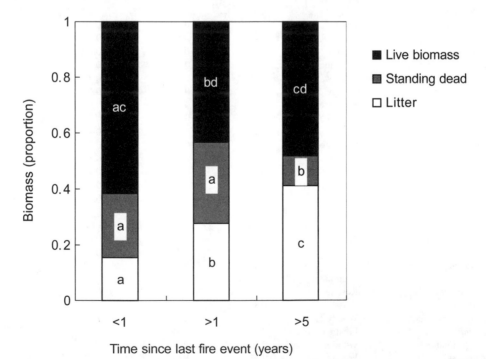

FIGURE 7.2 Relative contribution of live biomass, standing dead, and litter to the total biomass in the different treatments: patches that were burned in the ongoing growing season (<1 year), in the previous season (>1 year), or not burned for 5 years (>5 years). Different letters stand for significant differences at $p < .05$ according to an ANOVA test. The comparisons were made between treatments within each biomass category.

TABLE 7.2
Relative contribution (mean and standard deviation) of the different life-forms to the live biomass

Life-Forms	<1 year[a]	>1 year[a]	>5 year[a]	p-Values[b]
Erect species	0.18 ± 0.02	0.08 ± 0.04	0.07 ± 0.03	.06
Ferns	0.10 ± 0.09	0.03 ± 0.03	0.02 ± 0.01	.67
Graminoids	0.05 ± 0.01	0.09 ± 0.01	0.06 ± 0.06	.30
Prostrate species	0.05 ± 0.02	0.05 ± 0.02	0.03 ± 0.01	.39
Rosettes	0.01 ± 0.01	0.01 ± 0.01	0.01 ± 0.01	.20
Tussock grasses	0.23 ± 0.04	0.26 ± 0.03	0.09 ± 0.09	.06
Woody species	0.37 ± 0.09	0.49 ± 0.05	0.72 ± 0.09	.03

[a] <1 year: patches that were burned in the ongoing growing season, >1: burned the previous season or >5 years: not burned for 5 years.

[b] p values correspond to Kruskal–Wallis tests.

TABLE 7.3
Seasonal dynamics of total live biomass, standing dead, and litter (mean and standard error)[a]

Biomass Compartment	December	January	February	March	August	F[b]	p
<1 year							
Live	207.64 ± 16.18	371.77 ± 95.92	209.05 ± 29.84	262.85 ± 41.31	185.11 ± 34.51	1.81	0.21
Standing dead	23.18 ± 5.22	85.87 ± 43.11	100.80 ± 37.90	98.09 ± 23.25	252.39 ± 37.77	**6.12**	0.01
Litter	32.40 ± 11.27	49.55 ± 19.27	73.15 ± 18.21	38.72 ± 7.03	167.80 ± 19.38	**10.80**	0.002
>1 year							
Live	391.53 ± 66.42	288.44 ± 44.93	327.87 ± 25.26	445.95 ± 30.70	184.76 ± 50.86	**13.22**	0.001
Standing dead	196.02 ± 47.97	159.19 ± 20.78	172.43 ± 32.48	184.92 ± 59.82	462.35 ± 56.28	**20.01**	0.003
Litter	235.88 ± 80.12	175.20 ± 18.92	172.84 ± 20.62	209.41 ± 39.03	313.61 ± 62.52	1.38	0.32
>5 years							
Live	666.15 ± 52.82	542.86 ± 55.39	557.21 ± 94.12	611.96 ± 46.36	662.14 ± 80.68	1.80	0.22
Standing dead	142.35 ± 38.75	84.80 ± 35.89	92.57 ± 29.44	110.06 ± 29.17	298.33 ± 117.10	**3.90**	0.04
Litter	603.98 ± 112.18	409.74 ± 10.33	549.87 ± 65.30	367.83 ± 82.96	816.37 ± 84.52	**5.88**	0.05

[a] In areas that were burned in the ongoing growing season (<1 year), burned the previous season (>1 year), or not burned in 5 years (>5 years).

[b] Boldened cells indicate significant differences at $p < .05$.

burned during the ongoing growing season (<1 year) had 70 to 80% of their biomass as live biomass during December and January (Figure 7.3), whereas in the >1-year and >5-year treatments, this proportion hardly approached 50%. The contribution of live biomass to the total was higher in the <1-year patches almost throughout the year, which may represent substantial changes in the seasonal pattern of forage availability.

DISCUSSION

Time since the last fire event affected the total aerial biomass, the proportion of live biomass, standing dead, and litter, and the contribution of the different life-forms, both in terms of biomass and life-form frequency. Nevertheless, species richness was similar among all treatments, and species composition showed only small variations. Our results differ from those of Collins (1987) and Pucheta et al. (1998), who found that disturbances such as grazing and fire increased both species richness and diversity. In many cases, the increment in the number of species results from the colonization by exotic species or from the predominance of small-sized species, which are tolerant to disturbance (Belsky 1992; Pucheta et al. 1998). In our study site, we did not record exotic species, and because all our patches have a long history of grazing, most of these species may indeed be tolerant to disturbances. Fire and grazing may produce the same kind of selective pressure, and they can both favor fast-growing or small-sized species, especially tussock grasses and annuals. For these reasons, fire suppression may not cause compositional changes in areas such as our study site, where grazing occurs simultaneously.

Although species richness remained similar in the different treatments, we detected changes in the life-form spectrum and in the distribution of aerial biomass. The reduction in functional or species diversity as a consequence of a decrease in disturbance frequency has been observed in many cases (e.g. Pucheta et al.

1998; Valone and Kelt 1999) and is often attributed to a strengthening in species competition. In the grasslands of Los Toldos, fire suppression caused an increase in the dominance of woody species; many of these species were present in burned plots, but they became more abundant and of a bigger size in plots that were not recently burned. Tussock grasses were favored by fire, but their contribution was reduced in >5-year treatment. This change in the dominance of woody species alters site flammability that might reduce fire frequency in these plots in the future.

In addition to the changes in life-form contribution, there was a change in the distribution of aerial biomass. Fire reduced the total biomass by more than two-thirds (1303 gm^{-2} in areas not burned for 5 years compared to 431 gm^{-2} in areas that were recently burned), and the amount of litter was reduced in a similar way. This reduction in aboveground biomass that is associated with changes in the contribution of different life-forms results in changes in vegetation structure that may alter soil cover. Modification of soil cover can, in turn, affect

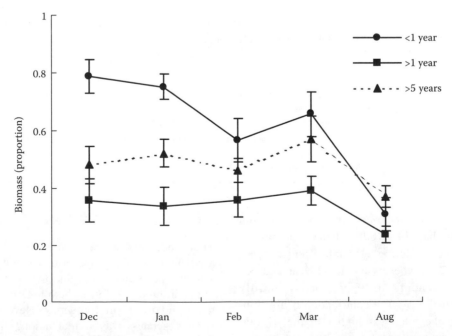

FIGURE 7.3 Relative contribution of live biomass to the total biomass throughout the year in patches that were burned in the ongoing growing season (<1 year), in the previous season (>1 year), or not burned for 5 years (>5 year)

erosion hazards that may have further implications on the hydrology and nutrient dynamics of the system (Hofstede et al. 1995). The differential allocation of biomass to the distinct biomass compartments and, especially, the variation in the amount of dead material that reaches the soil, can alter the decomposition rate and, consequently, the nutrient pools (Hobbs et al. 1991). Unfortunately, due to the lack of sound information, at present we can only hypothesize about these effects in the grasslands of Los Toldos. On the other hand, the short-term effect of fire on forage availability to herbivores in this site is easier to appreciate.

Patches that were burned in springtime had more than 70% of their total biomass as live biomass in the following summer (December and January). Therefore, fire modifies the seasonal dynamics of aerial biomass and changes forage availability at least for the summer period, when livestock is brought up to these mountain grasslands. The availability of green forage, especially in the form of highly palatable grasses, is particularly important for cattle after a period when they have had access only to low-quality winter forage. This means that cattle obtain, in proportion, more green biomass per bite in the patches that were recently burned. This can explain why these patches are often preferred (Coppock and Detling 1986; Hobbs et al. 1991). Consequently, the proportion of live biomass can have important effects on livestock energy budgets and determine their local movements. Importantly, fire promotes more palatable life-forms (grasses instead of woody species), and this makes the effect of fire even more meaningful to herbivores.

Our results indicate that changes in the fire frequency strongly affect vegetation dynamics in the montane grasslands of northwest Argentina. However, it is worth pointing out some limitations of this study. First, we were unable to find areas that were not burned and, therefore, we lacked a true control for our treatments. Our conclusions refer to the effects of a change in the fire frequency from once a year to once in 5 years. We do not know if there is a threshold after which a reduction in fire frequency produces no further changes in plant communities, so we cannot say if our >5-year treatment patches represent a transitional or a steady state. Second, our sample size was rather small, especially with regard to species composition. This is why we gave major emphasis to the results referred to biomass distribution, the variability of which seems to have been sufficiently accounted for by our samples. Third, an assumption of the present study is that our sample patches experience a similar grazing pressure. Even though there are no fences or other obstacles, and livestock have free access to all patches in the study area, which are also very close to one another, animals, as mentioned earlier, may prefer recently burned grasslands. As a consequence, these patches may receive higher grazing pressure. All these limitations have to be taken into account when considering our conclusions. To overcome these inherent difficulties, we are currently carrying out a controlled experimental study with a bigger sample size in the Los Toldos grasslands, which aims at separating the effects of fire and grazing. The preliminary results of this new experimental setup, which has been running for more than 2 years, seem to support the findings reported here.

SUMMARY

Fire and grazing are the most common disturbances in the mountain grasslands of northwest Argentina. They can affect species composition and richness, determine dominant life-form, and the general structure of the community. This work aims to determine the effect of burning on species richness, vegetation structure, and aerial biomass distribution in the grasslands of northwest Argentina that are subjected to grazing. We performed a comparative study at Los Toldos, Salta, Argentina (22°30 S, 64°50 W) at 1700 masl and surveyed patches that differed in the time since the last fire event. We considered three treatments: patches that were burned during the ongoing growing season (in spring 2000), burned the previous season, or not burned for at least 5 years. Treatments did not cause differences in species richness, and caused only small changes in species composition. The equativity of life-forms showed a tendency to decrease with fire suppression, with woody species becoming more dominant in plots that were not recently burned. Total biomass and the proportions of live biomass, standing dead, and litter varied among treatments. Fire caused a reduction in total biomass, but increased the contribution of live biomass and encouraged the development of more palatable growth forms (mainly grasses). Patches that were burned during the ongoing growing season had 80% of their biomass as live biomass in December and January. In these months, livestock are moved from forests at lower altitudinal levels to these highland grasslands. This modification in the seasonal dynamics of aerial biomass may represent a substantial change in the pattern of forage availability, especially at this time of the year.

ACKNOWLEDGMENTS

We are grateful to phytogeography students of Universidad Nacional de Tucumán for the assistance during fieldwork. The manuscript benefited from suggestions from three anonymous reviewers and from colleagues from LIEY. International Foundation for Science and Fundación PROYUNGAS provided financial support for this study.

References

Belsky, A. (1992). Effects of grazing, competition, disturbance and fire on species composition and diversity in grassland communities. *Journal of Vegetation Science*, 3: 187–200.

Bianchi, A. (1981). *Las precipitaciones del Noroeste Argentino*. INTA, Salta, Argentina.

Brown, A.D., Grau, H.R., Malizia, L.R., and Grau, A. (2001). Argentina. In Kappelle, M. and Brown, A.D. (Eds.), *Bosques Nublados del Neotropico*. INBIO, Costa Rica.

Cabrera, A. (1976). *Regiones Fitogeográficas Argentinas*. ACME, Buenos Aires, Argentina.

Collins, S. (1987). Interaction of disturbances in tallgrass prairie: a field experiment. *Ecology*, 68: 1243–1250.

Collins, S., Knapp, A., Briggs, J., Blair, J., and Steinauer, E. (1998). Modulation of diversity by grazing and mowing in native tallgrass prairie. *Science*, 280: 745–747.

Coppock, D.L. and Detling, J.K. (1986). Alteration of bison and black-tailed prairie dog grazing interaction by prescribed burning. *Journal of Wildlife Management* 50:452–455.

Crutzen, P.J., Delany, A.C., Greenberg, J., Haagenson, P., Heidt, L., Lueb, R., Pollock, W., Seiler, W., Wartburg, A., and Zimmerman, P. (1985) Tropospheric chemical composition measurement in Brazil during the dry season. *Journal of Atmospheric Chemistry* 2:233–256.

De Baro, L., Neary, D., and Folliot, P. (1998). *Fire Effects on Ecosystems*. John Wiley & Sons, New York.

Diaz, S., Acosta, A., and Cabido, M. (1992). Morphological analysis of herbaceous communities under different grazing regimes. *Journal of Vegetation Science*, 3: 689–696.

Eckholm, E.P. (1975). The deterioration of mountain environments. *Science*, 189: 764–770.

Grau, H., Gil-Montero, R., Villalba, R., Carilla, J., Araoz, E., Masse, G., and Membiela M. Submitted. Environmental history and forest regeneration dynamics in a degraded valley of NW Argentina cloud forests. In Juvik, J., Bruijnzeel, S., and Scatenna, F. (Eds.). *Forests in the Mists, Ecology and Conservation of Tropical Montane Cloud Forests*. University of Hawaii Press.

Grau, A. and Brown, D. (2000). Development threats to biodiversity and opportunities for conservation in the mountain ranges of the upper Bermejo River Basin, NW Argentina and SW Bolivia. *Ambio*, 29: 445–450.

Grau, H. (2005). Dinámica de bosques en el gradiente altitudinal de las yungas. In Arturi, M.F., Frangi, J., and Goya, J.L. (Eds). Ecología y Manejo de los bosques naturales Argentinos. Univ. EDULP (Ediciones de la Universidad Nacional de la Plata). Argentina.

Hobbs, T., Schimel, D., Owensby, C., and Ojima, D. (1991). Fire and grazing in the tallgrass prairie: contingent effects on nitrogen budgets. *Ecology*, 72: 1374–1382.

Hofstede, R., Mondragón, C., and Rocha, C. (1995). Biomass of grazed, burned and undisturbed Páramo grasslands, Colombia. *Arctic and Alpine Research*, 27: 1–12.

Lloret, F. and Vila, M. (2003). Diversity patterns of plant functional types in relation to fire regime and previous land use in Mediterranean woodlands. *Journal of Vegetation Science*, 14: 387–398.

Malizia, A. (2003). Host tree preference of vascular epiphytes and climbers in a subtropical montane cloud forest of Northwest Argentina. *Selbyana*, 24: 196–205.

McCune, B. and Mefford, M.J. (1997). *PC-ORD. Multivariate Analysis of Ecological Data. Version 3. 0.* MjM Software Design, Gleneden Beach, OR.

McNaughton, S. (1985). Ecology of a grazing ecosystem: the Serengeti. *Ecological Monographs*, 55: 259–294.

Milchunas, D. and Lauenroth, W. (1993). Quantitative effects of grazing on vegetation and soils over a global range of environments. *Ecological Monographs*, 63: 327–366.

Oesterheld, M., Loreti, J., Semmartin, M., and Paruelo, J. (1999). Grazing, fire, and climate effects on primary productivity of grasslands and savannas. In Walker, L. (Ed.), *Ecosystems of disturbed ground*. Elsevier, Amsterdam.

O'Neill, R.V., Krummel, J.R., Gardner, R.H., Sugihara, G., Jackson, D.L., Milne, B.T., Turner, M.G., Zygmunt, B., Christensen, S.W., Dale, V.H., and Graham, R.L. (1988). Indices of landscape pattern. *Landscape Ecology*, 1: 153–162.

Pucheta, E., Vendramini, F., Cabido, M., and Diaz, S. (1998). Estructura y funcionamiento de un pastizal de montaña bajo pastoreo y su respuesta luego de la exclusión. *Revista de la Facultad de Agronomía, La Plata*, 103: 77–92.

Ramadori, D. (1995). Agricultura migratoria en el valle del Rio Baritú, Santa Victoria, Salta. In Brown, A. and Grau, H. (Eds.), *Investigación, Conservación y Desarrollo en Selvas Subtropicales de Montaña*. Proyecto de Desarrollo Agroforestal, LIEY, Tucumán, Argentina.

Rusch, G. and Oesterheld, M. (1997). Relationship between productivity and species and functional group diversity in grazed and nongrazed Pampas grassland. *Oikos*, 78: 519–526.

Schmidt, A.M. and Verweij, P.A. (1992). Forage intake and secondary production in extensive livestock systems in páramo. In Baslev, H. and Luteyn, J.L. (eds.) Páramo: *An Andean EcoSystem Under Human Influence*. Academic Press, London.

Vogl, R.J., (1974). Effects of fire on grasslands. In Kozlowski, T.T. and Ahlgren, C.E. (Eds.). *Fire and Ecosystems*. Academic Press, New York.

Valone, T. and Kelt, D. (1999). Fire and grazing in shrub-invaded arid grassland community: independent or interactive ecological effects? *Journal of Arid Environments*, 42: 15–28.

Part III

Effects of Grazing on Mountain Biodiversity

8 The Biodiversity of the Colombian Páramo and Its Relation to Anthropogenic Impact

Jesus Orlando Rangel Churio

INTRODUCTION

The environments in which páramo vegetation predominates are found above the treeline in the northern Andes (in Colombia, Venezuela, Ecuador, the north of Peru, and recently, Bolivia) and in Central American countries such as Panama and Costa Rica; here, the open type of vegetation (pajonales, prados, and rosette vegetation) predominates. The establishment of the vegetation in a variable climate of sunny days and cold-to-freezing nights clearly depends on the latitudinal and longitudinal location, soil conditions, topo-graphy, and exposure, as well as the human impact and historical biogeographical factors. The physiognomic ensembles are similar, especially between those vegetation types with the greatest distribution; for example, the dense formations dominated by Graminaceae in "macollas" (Andean pajonales), the rosette vegetation or "frailejonales," and the shrub vegetation or "matorrales." There is a high degree of convergence in the use of available environmental resources, as well as in the degree of conversion of the original conditions of the landscape by human disturbance. Despite this convergence, there are marked differences in the expression of alpha-diversity (taxonomical) and beta-diversity (ecological), which highlight the particular conditions of each locality. In the natural region (Colombia), there exists a clear relationship between soil, climate, biota, and human influence. The soils have a dense top layer of organic material, which in some cases extends to more than 1 m in depth. The average annual temperature fluctuates between 4 and 10°C (8°C); in the lower belts (subpáramo), temperatures of between 8 and 10°C, and in the superpáramo, temperatures of 0°C, are reached (Aguilar-P and Rangel-Ch., 1996; Sturm, 1998). The altitudinal gradient allows the subdivision of the páramo into belts or zones: low páramo or subpáramo (from 3200 to 3500 [3600] masl) is characterized by the predominance of matorrales (shrub vegetation) dominated by species of *Diplostephium*, *Monticalia*, and *Gynoxys* (Asteraceae), of *Hypericum* (*H. laricifolium*, *H. ruscoides*, and *H. juniperinum*), and of *Pernettya*, *Vaccinium*, *Bejaria*, and Gaultheria (Ericaceae). The limits of the páramo proper or grass páramo extend from 3500 (3600) to 4100 masl; the diversification of its plant communities is maximal, almost all vegetation types are found in this zone, although frailejonales or rosette vegetation with species of *Espeletia*, the pajonales with species of *Calamagrostis*, and the chuscales with *Chusquea tessellata*, predominate. The superpáramo, the zone situated above 4100 masl, extends as far as the lower limits of perpetual snow and is characterized by the patchiness of the vegetation and an appreciable amount of bare soil (Figure 8.1).

The cover and diversity of the vegetation are visibly reduced, and may result in the growth of few isolated plants, and the rocky substrate predominates. The superpáramo communities are low rosettes, with species of draba:

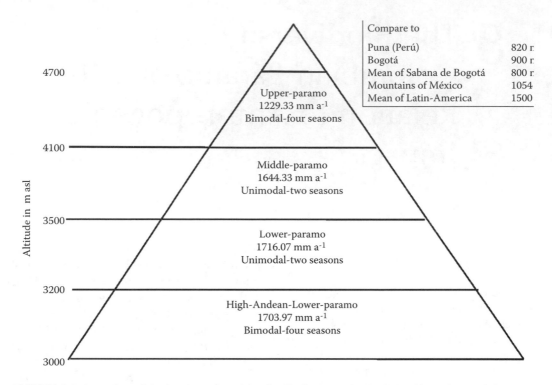

FIGURE 8.1 Annual precipitation (mma⁻¹) and its distribution types in the Colombian páramo region.

D. cheiranthoides, *D. cryophilla*, *D. litamo*, and *D. alyssoides*.

In this chapter, some basic questions about the páramo region, which represents between 1.5 and 2% of the Colombian region, are addressed: How great is the biodiversity (basic inventory of α, β, and γ diversity)? What is the ranking obtained from the inventory of species richness? What has been the principal use of the habitat and the biota from the first settlers up to the present day? And finally, what are the problems associated with the use of natural resources, and what are their effects on the natural function of these environments in Colombia? This use may be direct, influencing the conversion of natural conditions, or indirect, influencing the persistence of biodiversity.

METHODOLOGY

The information on the presence and distribution of plant and animal species was taken from the lists given in Rangel-Ch. (2000a). The flora was complemented by cross-references with the COL, US, MO, and NY herbaria. The data of the páramo flora of countries other than Colombia were taken from the catalogs of Brako and Zarucchi (1993), Jørgensen and León-Yáñez (1999), and Luteyn (1999). The classification of other vegetation types follows Cuatrecasas (1934), Cleef (1981), and Rangel-Ch. et al. (1997) and Rangel-Ch. (2000b). The publications of Sturm and Rangel-Ch. (1985), Witte (1994), and Rangel-Ch. (2000c) were taken into account for data analysis. There are two areas of focus in the analysis of species richness: the taxonomic level (families and genera) and eco-geographical level (richness per altitudinal zone of the páramo and selection of species by precipitation).

EXPRESSION OF THE BIOTA AND THE DISTRIBUTION OF SPECIES RICHNESS OF SPERMATOPHYTES IN THE PÁRAMO BELTS

The floristic diversity of the entire páramo region (Costa Rica to Peru) is represented by

5168 species of 735 genera and 133 families, which makes the vegetation of this upper-montane zone one of the most diverse in its category in the world. This confirms the initial findings of Cleef (1981) and of Sturm and Rangel-Ch. (1985), which also classify the páramo as one of the most diverse vegetation zones in its category. The greatest diversification at the family level is found in the Asteraceae (141 genera/1165 species), Orchidaceae (60/661), and Poaceae (56/292) and at the generic level in *Espeletia* (133), *Epidendrum*, and *Miconia* with 116 species each (Luteyn, 1999; Rangel-Ch., 2000a). The species richness of the Colombian páramo flora consists of around 3173 species of vascular plants, which is almost 60% of the total species richness of the entire páramo region. The families with the greatest relative species richness are the Asteraceae with 100 genera and 598 species, the Orchidaceae (57/578), and the Poaceae (46/153). The genera with the greatest numbers of species are *Espe-*

letia (83), *Epidendrum* (103), and *Pleurothallis* (78). With regard to the habit or growth form, there is an even distribution between families, with woody representatives (Melastomataceae, Rubiaceae, Asteraceae, Rosaceae, and Ericaceae) and families with mostly herbaceous species, such as the Orchidaceae and the Poaceae (Table 8.1).

The belt of the Colombian páramo with the greatest diversity is the transitional zone between the upper-Andean flora and the sub-páramo, with 2385 species from 487 genera and 115 families (Rangel-Ch., 2000a). In general, the mean species richness and diversity decrease with increasing altitude. The Asteraceae is the family with the overall greatest number of species present in the area from the upper-Andean belt to the superpáramo belt, and this family also has the greatest number of species restricted to any one zone, with the exception of the upper-Andean belt, where there are a greater number

TABLE 8.1
Families and genera of the most diversified angiosperms in the geographic páramo region and Colombia

Family	Global Páramo[a]		Colombia		Genus	Global Páramo	Colombia
	Genera	Species	Genera	Species		Species	Species
Asteraceae	141	1165	100	598	*Epidendrum*	116	103
Orchidaceae	60	661	57	578	*Espeletia*	133	83
Poaceae	56	292	46	153	*Pleurothallis*	90	78
Melastomataceae	17	194	13	105	*Diplostephium*	102	73
Scrophulariaceae	20	184	19	77	*Miconia*	116	64
Bromeliaceae	7	146	7	99	*Monticalia*	85	57
Rubiaceae	22	119	16	69	*Hypericum*	70	53
Ericaceae	21	115	18	85	*Lepanthes*	58	52
Gentianaceae	5	113	5	54	*Senecio*	89	51
Rosaceae	17	112	14	62	*Masdevallia*	51	49
Cyperaceae	13	103	10	68	*Stelis*	52	49
Fabaceae	17	103	13	62	*Baccharis*	74	46
Solanaceae	14	103	9	56	*Weinmannia*	44	42
Campanulaceae	8	99	7	33	*Lupinus*	68	41
Apiaceae	20	86	19	58	*Solanum*	75	40
Brassicaceae	16	84	13	44			
Total	735	5168	586	3173	Total	5168	3173

[a]Global páramo: Costa Rica, Panama, Colombia, Venezuela, Ecuador, and Peru.

of Orchidaceae restricted to this zone. No species of the Bromeliaceae or the Melastomataceae are found in the superpáramo (Rangel-Ch., 2000a), and the species richness of the Ericaceae and Scrophulariaceae is greatest in the lower páramo zones (Table 8.2).

The relative richness of species present in a zone vs. the number of species restricted to that zone is greatest in the superpáramo (4443/35 = 12.7), which is the environment with the least area and the one exposed to climatic extremes, where the differentiation of new lineages is probably linked, among other factors, to the low temperatures. The ratio is 6.26 (1958/313) in the subpáramo and 6.97 (1575/226) in the páramo proper.

FLORISTIC COMMUNITIES

In the different zones of the Colombian páramo, 327 plant communities have been recorded matorrales (shrubs) are predominant in all the belts, and the zone with the greatest expression of this vegetation type is the grass páramo or páramo proper. Forests are frequent in the subpáramo, expressing the continuity of the vegetation from the upper-

Andean zone and, with the exception of *Polylepis* forests, they do not extend to the superpáramo; grass communities are the most frequent vegetation type extending up to the higher belts, and the chuscales do not reach the superpáramo (Table 8.3). The greatest diversity of grass communities is found in the páramo proper, and the most frequent physiognomic types are the matorrales and pajonales (grassland). The Colombian páramo mirrors the phytoecological compositions of the whole geographic region. The chuscales of Costa Rica are well represented in the Cordillera Occidental (Macizo del Tatamá) and in the humid páramo of the Cordillera Central and the Cordillera Oriental. The pajonales of the Sierra Nevada de Mérida and other Venezuelan páramos are well represented in the Colombian Cordillera Central. The frailejonales in Ecuador are quite similar to those growing in the páramos on volcanoes in the south of ColombiaThe Colombian frailejonales, similar to the rosette vegetation in Venezuela, are as varied and have ecological spectra as wide as that of the neighboring country (2000b).

TABLE 8.2
Species richness per altitudinal belt in the most diversified families of the Colombian páramo

Family	Upper Andean		Subpáramo		Grass Páramo		Superpáramo	
	A	B	A	B	A	B	A	B
Apiaceae	43	10	49	9	46	10	19	2
Asteraceae	427	127	437	58	427	78	156	16
Bromeliaceae	71	36	55	15	34	4	—	—
Cyperaceae	41	8	49	5	45	6	8	1
Ericaceae	72	19	62	8	46	4	6	0
Melastomataceae	72	36	68	20	41	4	—	—
Orchidaceae	523	357	203	35	105	10	10	2
Poaceae	90	16	103	12	99	10	45	3
Scrophulariaceae	68	5	68	7	51	0	19	1

Note: A = Species present in the belt in question that may also occur in other belts; B = species restricted to the belt in question.

TABLE 8.3
Distribution of vegetation types in the belts of the Colombian páramo

Vegetation Type	Subpáramo	Grass Páramo	Superpáramo	Páramo Total
Sparse forest	12	6	1	19
Matorrales (shrub)	39	46	19	104
Frailejonales (rosette)	15	16	4	35
Pajonales (grassland)	9	22	14	45
Other types (rosettes, pastures, chuscales, or aquatic)	37	56	31	124
Total	112	146	69	327

CLIMATE

Precipitation distribution regimes can be: unimodal–biseasonal, bimodal–tetraseasonal, trimodal–hexaseasonal, and tetramodal–octaseasonal (Rangel-Ch., 2000c). The distribution of the annual total rainfall in each of the páramo belts is shown in Figure 8.1. The most humid zone is the subpáramo, with 1716 mm per year, and the least humid is the superpáramo, with 1229 mm. It is interesting to note that the upper-Andean subpáramo and superpáramo belts (base and apex of the pyramid), which are exposed to extreme climatic variation, express the bimodal–tetraseasonal rainfall distribution pattern, whereas the internal or protected zones of the grass páramo and subpáramo have a unimodal–biseasonal rainfall distribution type. In comparison to the reference values of precipitation (i.e. that of locations in the Peruvian puna, montane sites in Mexico, and the mean of Latin America), the mean rainfall of all páramo belts is greater than the reference means. The Colombian páramo therefore can be regarded as humid to very humid.

ECOLOGICAL VARIABILITY AND FLORISTIC SELECTION

When the floristic species richness of spermatophytes is considered along with the annual rainfall totals, the páramo regions of Colombia can be separated into the following categories.

ARID PÁRAMO

Typical, well-documented locations representative of this climate type are found in Berlín (B) (07°11 N, 72°53 W) and Vetas (V) (Dept. of Santander), where 34 families, 84 genera, and 142 species have been recorded. The annual mean of precipitation for the two zones is 805 mm (B: 623.52; V: 985.88). The páramos of the Nariño volcanoes can also be included in this type [Chiles, Cumbal, Azufral (01°04 N, 77°41 W), and Galeras (01°12 N, 77°20 W)], where 47 families, 127 genera, and 227 species have been recorded, and the mean annual rainfall is 999 mm.

SEMIHUMID PÁRAMO

In the Sumapaz Massif (03°45 N, 74°25 W), 77 families, 251 genera, and 619 species have been recorded; the mean annual rainfall is 1500 mm.

HUMID PÁRAMO

In the Puracé National Park (02°21 N, 76°23 W), 63 families, 175 genera, and 409 species have been recorded; the mean annual rainfall is 2120 mm. In the Chingaza Nature Park (04°31 N, 74°35 W), 76 families, 247 genera, and 534 species have been recorded, and the mean annual precipitation is 2394.3 mm. Although these are only preliminary values, they show a tendency toward increasing floristic species richness with increasing mean annual rainfall. The resulting flora in Chingaza and Chisaca are very similar, although the inventory in Chisaca was carried out in greater detail. In Vetas–Berlín and the Nariño páramos, the floristic species richness is definitely lower, which is directly related to the lower rainfall in these areas.

In the humid and superhumid páramo vegetation in Colombia (mean annual precipitation greater than 2000 mm), the bamboo *Chusquea tessellata* forms highly homogeneous communities; in other areas, it is an important species in community physiognomy. There also exists a relationship between humidity and the ground cover by cushion forms of vascular plants, which show greater cover in the humid páramos and on montane buildings of high altitude, such as Chingaza and Chisaca, than in the arid páramos and low mountains, such as Berlín and El Hato (Cleef, 1981; Sánchez-M and Rangel-Ch., 1990; Rangel-Ch., 2000).

Sturm and Rangel = Ch(1985) identified the species that establish particularly well in the humid páramos of the Cordillera Oriental, where they also reach the greatest ground cover and develop most vigorously: *Chusquea tessellata*, *Calamagrostis bogotensis*, *Calamagrostis effusa*, *Rhynchospora macrochaeta*, *Espeletia grandiflora*, *Blechnum loxense*, *Pernettya prostrata*, *Paepalanthus karstenii*, *Arcytophyllum nitidum*, *Arcytophyllum muticum*, *Aragoa abietina*, *A. corrugatifolia*, *Lycopodium contiguum*, *Castilleja fissifolia*, *Castratella piloselloides*, *Vaccinium floribundum*, *Diplostephium revolutum*, *Disterigma empetrifolium*, *Puya santosii*, *Hypericum goyanesii*, *Halenia asclepiadea*, *Oritrophium peruvianum*, *Monticalia vacciniodes*, *Gentianella corymbosa*, *Festuca dolichophylla*, and *Bartsia santolinaefolia*.

In the arid páramos of the Cordillera Oriental, the frequent species are, among others: *Diplostephium phylicoides*, *Bucquetia glutinosa*, *Brachyotum strigossum*, *Gualtheria cordifolia*, and *Gaylusaccia buxifolia*. In some floristic groups such as in Aragoa (Scro-

phulariaceae), there are also a series of species (section Ciliatae) with preferential distribution in arid páramos such as the core of the páramo in the Sierra La Culata (Cordillera de Mérida, Venezuela) and the Sierra Nevada de Santa Marta (Fernández-Alonso, 1993). Curiously, the majority of species showing a preference for the páramos of arid climates have a woody habit; this is of importance because one of the main concerns of the use of natural resources in the arid páramos is the impoverishment and disappearance of species that are sensitive to fire.

FAUNA

The zonal distribution of páramo fauna is shown in Table 8.4; the number of species in each belt and the number of species restricted to that belt (R) are also given. The transitional upper-Andean–subpáramo zone has the highest values in both categories. The decrease in mean species richness with increasing altitude is apparent in all groups of fauna.

RANKING OF THE PÁRAMO BIOTA IN COLOMBIA

The biota of the páramo is comparatively species rich and varied. It represents 12% of the total flora of Colombia (26,500 species) and 29% of the Cordillera or Andes region (11,000 species). The most highly represented groups of fauna are the mammals (14% of the total in Colombia) and the birds (8%).

Table 8.5 gives the values for the floristic diversity in the geographic páramo region from Costa Rica to the north of Peru and for the Colombian páramo. Values from the yungueño páramos of Bolivia, which are typical representatives of the Andean páramo, are not included. In Colombia, the species richness of the páramo ranges from 60% in angiosperms to 98% in ferns (Table 8.5) and therefore qualifies it as the area with the highest diversity and greatest species richness in the entire páramo region. In Bolivia, matorrales, or shrubby vegetation, can be found in humid montane forest formations that have a border of yungueña scrub above the

TABLE 8.4
Species richness of the fauna in the páramo belts

Fauna	Global Páramo	Upper and Sub-Andean	Subpáramo	Grass Páramo	Superpáramo
Mammals	70	68 (24 R)	45 (No R)	32 (1 R)	1 (No R)
Birds	154	134 (17 R)	117 (2 R)	70 (No R)	46 (No R)
Amphibians	90 (39 R)	77 (36 R)	49 (2 R)	30 (5 R)	5 (No R)
Reptiles	16	12 (5 R)	8 (2 R)	5 (1 R)	2 (No R)
Butterflies	131	117 (92 R)	28 (3 R)	18 (8 R)	1 (No R)

Note: R = Number of restricted species.

Sources: From Ardila-R, M.C. and A. Acosta. 2000. Anfibios. En J.O. Rangel-Ch. (Ed.). *Colombia Diversidad Biótica III. La Región Paramuna.* Instituto de Ciencias Naturales, Universidad Nacional de Colombia. Bogotá. pp. 617–628; Castaño, O., E. Hernández, and G. Cárdenas. 2000. Reptiles. En J.O. Rangel-Ch. (Ed.). *Colombia Diversidad Biótica III. La Región Paramuna.* Instituto de Ciencias Naturales, Universidad Nacional de Colombia. Bogotá. pp. 612–616; Andrade-C., M.G. and J.A. Álvarez. 2000. Mariposas. En J.O. Rangel-Ch. (Ed.). *Colombia Diversidad Biótica III. La Región Paramuna.* Instituto de Ciencias Naturales, Universidad Nacional de Colombia. Bogotá. pp. 645–652; Delgado, A.C. and J.O. Rangel-Ch. 2000. Aves. En J.O. Rangel-Ch. (Ed.). *Colombia Diversidad Biótica III. La región Paramuna.* Instituto de Ciencias Naturales, Universidad Nacional de Colombia. Bogotá. pp. 629–644; Muñoz, Y., A. Cadena, and J.O. Rangel-Ch. 2000. Mamíferos. En J.O. Rangel-Ch. (Ed.). *Colombia Diversidad Biótica III. La región Paramuna.* Instituto de Ciencias Naturales, Universidad Nacional de Colombia. Bogotá. pp. 599–611.

TABLE 8.5
Floristic diversity of Páramos in the biogeographic Páramo region and Colombia

	Group	Families	Genera	Species
Global Páramo[a]	Lichens	51	114	465
	Mosses	57	186	544
	Liverworts	36	102	442
	Ferns	24	56	352
	Angiosperms	140	735	5168
Colombia	Lichens	47	98	361
	Mosses	52	162	459
	Liverworts	34	99	423
	Ferns	24	54	345
	Angiosperms	118	586	3173

[a]Global páramo: Costa Rica, Panama, Colombia, Venezuela, Ecuador, and Peru.

treeline; this is followed by the yungueño páramo with representatives of woody species from the genera *Oreopanax* and *Gynoxis*, as well as *Escallonia myrtilloides* and *Polylepis pepei* (Beck et al., 1993). A recent excursion with other Latin American colleagues allowed us to confirm the initial characterization of these environments as being typically páramo; the species observed, the vegetation types (matorrales, pajonales, cushion plants, and herbaceous vegetation) the soil conditions with a thick black horizon, the humidity of the belt, and the gradient of the vegetation directly related to the geomorphology are the criteria that were taken into account; unfortunately, the detailed data are not available to be included in this discussion.

ENDEMISM AND WIDESPREAD DISTRIBUTION OF THE PÁRAMO FLORA

Table 8.6 shows the values of autochthonal (by country) floristic species richness, following taxonomic categories. The high degree of endemism of the Colombian and Ecuadorian páramo flora is evidental though in the former country the values are minor. Sixteen species with widespread distributions can be identified: *Eryngium humile* (Apiaceae), *Monticalia andicola* (Asteraceae), *Arenaria lanuginosa* (Caryophyllaceae), *Gaultheria erecta* and *Pernettya prostrata* (Ericaceae), *Gentiana sedifolia* (Gentianaceae), *Escallonia myrtilloides* (Grossulariaceae), *Gaiadendron punctatum* (Loranthaceae), *Miconia chionophila* (Melastomataceae), *Myrsine dependens* (Myrsinaceae), *Agrostis tolucensis* and *Cortaderia hapalotricha* (Poaceae), *Hesperomeles obtusifolia* and *Lachemilla aphanoides* (Rosaceae), *Galium hypocarpium* (Rubiaceae), and *Xyris subulata* (Xyridaceae). This result shows a low expression of "cosmopolitanism" or common elements, perhaps due to the inclusion of Panama with small area of páramo vegetation and low species richness. When the comparison is repeated with only the core area consisting of Colombia, Venezuela, Ecuador, and Peru, the number of species with widespread distribution increases to 160. This shows again the fact that available resources are highly particular to each location. On the other hand, the floristic simi-larity between the Andean páramo (which is intrinsically humid) and the puna (which is arid) is very low. The flora of upper-montane Peru includes 1945 species, 432 genera, and 101 families (Brako and Zaruchi, 1993). Of this total, only 30 species are also present in the typical páramo vegetation of the provinces of northern Peru (Luteyn, 1999); 405 species are restricted to the páramo region. In our opinion, these estimates demonstrate clearly the floristic difference between two large regions (páramo and puna) that have very distinct climates.

CONSEQUENCES OF THE EXCESSIVE DEMAND OF NATURAL RESOURCES

The uses of the Colombian páramo given in Table 8.7 allow the conclusion that the available resources exceeded the demand during pre-Colombian times, whereas the demand on natural resources is excessive today.

ENDANGERED SPECIES AND VEGETATION TYPES

The endangered species of the Colombian páramo (344), in terms of family, represent 11% of the páramo flora (3173); the most affected are the Asteraceae (44 spp.), Ericaceae (85), Bromeliaceae (33), and Scrophulariaceae (23). The Bromeliaceae are also possibly affected by changes to swamp habitats (species of Puya).

TABLE 8.6
Autochthonal floristic species richness (endemism) in countries with páramo vegetation

Country	Families	Genera	Species	Proportion of Total Species Richness (%)
Costa Rica	32	63	85	1.64
Panama	4	7	7	0.14
Colombia	**103**	**423**	**2045**	**39.57**
Venezuela	46	122	353	6.83
Ecuador	65	232	716	13.85
Peru	72	232	620	12
Global páramo	133	735	5168	—

TABLE 8.7
Former and present-day uses of the Colombian páramo

Former Use	Present-Day Use
Rocky shelters as protection during hunting trips (indigenous communities)	Logs (firewood and fences)
	Grasses (roofing for rural houses)
Lagoons and lakes for religious and cosmological ceremonies (payment of tributes)	Ornamental native flora
	Use of pajonales for pastures (burning and grazing)
Medicinal plants (Kogui culture)	Drainage of peat bogs — advancing potato crops
Small mammal fauna (guinea pigs, rabbits) as a food source	Hydrological reservoirs (lagoons — generation of electricity)
	Reforestation programs
	Mining, urban settlements
	Tourism (badly managed!)
	Exploitation of ice (Nevado del Cumbal)
Natural resources — abundant	**Demands on the environment — excessive**

Sixty-seven vegetation types in the páramo are endangered; the most endangered are the matorrales (Rangel-Ch., 2000d). The health of the Colombian páramo depends on the predominance of the matorral (shrubby) vegetation, as has been shown in páramo zones without significant human impact (Díaz, 2002). The disappearance of shrubby vegetation (matorral) in the Colombian páramo is therefore indicative of its deficient health (Table 8.8).

ALTERATION OF THE HABITAT AND IMPOVERISHMENT OF THE FLORISTIC SPECIES COMPOSITION

Fundamental work that has been carried out in the Cordillera Central (Verweij, 1995) and in Cordillera Oriental (Hernández, 2002) illustrates the aspects mentioned earlier. Table 8.9 shows the values for species richness of the most diversified families in different geographical regions of the páramo and Cordillera Oriental and the three principal types of formationswith diverse numbers of associations, which correspond to the *Paepalanthus columbiensis* and *Diplostephium phyllicoides* matorrales (PCDP around Bogota), the *Pernettya prostrata* and *Chusquea tessellata* chuscales (PPCT), and the herbaceous vegetation with *Acaena cylindrostachya* and *Orthosanthus*

chimboracensis (ACOC). Comparisons of the α, β, and γ diversity show the impoverishment of formations dominated by *A. cylindrostachya*, in particular, of the families with many woody representatives, such as the Asteraceae, Melastomataceae, Ericaceae, and Rosaceae, and also of the genera *Monticalia*, *Miconia*, and *Baccharis*. Cleef and Rangel-Ch. (1984), Rangel-Ch. and Aguirre-C (1986), and Salamanca et al. (1992) described the predominance of the rosette vegetation of *Acaena cylindrosthachya* in zones with greatly transformed conditions as a result of disturbance by cattle, as cattle are ideal dispersers of the fruits and seeds of *A. cylindrostachya*. These results, and the descriptions given by Verweij (1995), are evidence that the transformation of the habitat and the effects of cattle farming reduce floristic diversity.

TRANSFORMATION OF THE HABITAT

Van der Hammen et al. (2002) documented the loss of extensive areas of original habitat (ground cover) along the borders of the páramo at Laguna Verde (Cundinamarca), and calculations of productivity emphasized those changes. Table 8.10 shows the extent and type of ground cover in two periods; the reasons for the conversions and the incentives for the change as loss (P) and gain (G) are given. In general, an

TABLE 8.8
Endangered vegetation types and their distribution by páramo belt

Belt	Subpáramo	Grass Páramo	Super páramo	Total Number of Endangered Types	Number of Vegetation Types in the Páramo
Forests	3	3	—	6	19
Matorrales (shrub)	9	11	7	26	104
Frailejonales (rosette)	6	3	1	10	35
Pajonales (grassland)	4	3	1	8	45
Rosettes	1	1	—	1	6
Pastures	1	2	5	8	44
Chuscales	1	2	—	3	18
Aquatic or marshland vegetation	2	1	2	5	46

TABLE 8.9
Species richness in the páramo vegetation of the Cordillera Oriental

Family	Global Páramo	Colombia	Cordillera Oriental.	PCDP[a]	ACOC[b]	PPCT[c]
Asteraceae	1165	598	355	50	20	41
Orchidaceae	661	578	330	6	2	3
Poaceae	292	153	79	19	12	11
Scrophulariaceae	184	77	53	5	3	5
Melastomataceae	194	105	45	10	2	7
Ericaceae	115	85	37	11	4	6
Rosaceae	112	62	35	12	4	10
Cyperaceae	103	68	27	12	5	9
Genera						
Espeletia	133	83	57	4	5	3
Epidendrum	116	103	55	1	1	2
Hypericum	70	53	38	5	4	9
Monticalia	85	57	30	6	1	5
Diplostephium	102	73	29	2	3	3
Miconia	116	64	26	6	1	4
Baccharis	74	46	25	6	1	3

[a]PCDP: Matorrales dominated by *Paepalanthus columbiensis* and *Diplostephium phyllicoides*.
[b]ACOC: Herbaceous vegetation dominated by *Acaena cylindrostachya* and *Orthosanthus chimboracensis*.
[c]PPCT: Chuscales dominated by *Pernettya prostrata* and *Chusquea tessellate*.

uncontrolled increase of events leading Delgado, A.C. and J.O. Rangel-Ch. 2000. Aves. En J.O. Rangel-Ch. (Ed.). *Colombia Diversidad Biótica III. La región Paramuna*. Instituto de Ciencias Naturales, Universidad Nacional de Colombia. Bogotá. pp. 629–644.

to increased areas of pastures, crops, and human settlements caused the notable decline in the upper-Andean forests and matorrales.

In the northwest of the Bogotá savanna (Villapinzón, Chocontá, and Umbita), disturbance to the original vegetation has been very large,

TABLE 8.10
Conversion of the original area in the region of Laguna Verde, Cundinamarca, Cordillera Oriental, Colombia

	1970 Area (ha)	1990 Area (ha)	Conversion of original area in 1970–1990 (%)
Bodies of water	1,013.16	886.31	+12,52
Páramo vegetation	12,942.86	14,595.19	−13
Páramo shrubby vegetation	10,078.74	618.58	+93
Upper-Andean forest	14,754.12	12,828.02	+13
Pastures	4,094.03	8,448.28	−106,35
Crops	3,913.62	10,312.68	−163,5
Area for human settlements	22.82	319.75	−1301,18

affecting an area of 50,000 ha, including the natural vegetation of the sub-Andean, Andean, and páramo regions. Sixty percent of the upper-Andean vegetation practically disappeared as agricultural activities increased significantly (Cortés et al., 2003). Activities related to the extension of agricultural boundaries (potato crops) are evident in the majority of the páramo regions of Colombia: indiscriminate burning, extensive cattle farming, small-scale mining and, especially in the Cordillera Oriental, the concentration of human populations in the proximity of these natural areas.

FUTURE OF THE COLOMBIAN PÁRAMO — ALTERNATIVES FOR ITS PRESERVATION

The extension of agricultural boundaries, particularly in relation to potato crops, is the cause behind the use of ever-higher altitudinal zones and the planning and acceptance of the nontraditional utilization of the natural resources in those zones. One example of this is the prime objective of using the water sources of the páramo for water supply to the human populations adjoining the area. In Colombia, the cultivation of potatoes is a critical factor for the continuity of natural conditions in the páramo regions; the extent of the area under potato cultivation is 170,000 ha, and the annual harvest amounts to 3 million t. At present, the cultivation of potatoes extends farther up the altitudi-

nal gradient every day with the development of new varieties, which are not as strongly affected by the prevalent climatic conditions, but which accumulate high concentrations of NO_3 (nitrate) and NO_2 (nitrite) at levels of 0.79 to 1% for nitrate and 0.53 to 1.9% for nitrite; these levels are above the permissible or threshold levels for human health of 216 mg of nitrates and 3.6 mg of nitrites in the daily diet (Uribe, 2003). There are conflicts of interest in this land use, and it would be advisable to examine whether the utilization of páramo regions favors small farmers or landowners with considerable tracts of land.

The distribution of the upper-montane population (>2,700 masl) in Colombia shows that there exist 36 municipalities with 433,676 inhabitants representing 1.10% of the population in the montane gradient. In the páramo region, there are 13 municipalities with 171,686 inhabitants (0.40% of the population of the country); the municipality located at the greatest altitude is Vetas (Dept. of Santander), at 3300 masl. The majority of the population is rural and is present at low densities of 30 inhabitants/km^2 (Falla and Rolón, 2002); furthermore, the land is concentrated in medium-sized (10 to 20 ha) and large (>20 ha) properties (CAR, 1990). The economic situation of the small farmers, for example, those in the Páramo de Villapinzón (Cundinamarca), shows that the exploitation of land does not meet the economic requirements needed to provide appropriate subsistence; small land-owners who own less than 2.2 ha produce 12 t/ha, which generates returns of $1,694,473, or a monthly income of $242.067, less than the minimum wage for the year 2001. The exploitation of the land under the present socioeconomic conditions cannot meet the minimum requirements of the population; if this is true, then we must find another solution, so that the small farmer can remain in the zone in decent living conditions.

The adequate utilization of natural resources in the páramo region is bound, by necessity, to the water resources. The accumulated annual rainfall in the belts of the páramo in the 3 cordilleras amounts to between 3900 and 8300 mm, depending on the mountainside considered, with a marked difference between the external humid and superhumid slopes and the internal humid slopes; these amounts allow the conclusion that the páramo has an excess of water, and it should therefore be used as a source of water supply for the population living in the montane gradient. However, the amount of water taken up by swamps and peat bogs should be added to these direct measures of water volume. In the páramo de Frontino (Parra, 2002), it has been shown that for every centimeter of sediment determined as organic hydrocoloids, 87% is water (almost greater weight in water per volume than water itself) which are the key to the hydrological balance of the watershed; it is obvious that their protection must have first priority.

Faced with the situation of the biological importance of the páramo and the claims for its anthropogenic appropriation, it becomes necessary to reconcile the interests of those who defend the direct utilization of the biological and physical resources of the páramo in the traditional manner and those who defend their conservation with social aims, essentially the water supply, for the entire cordillera system. Based on this approach, it is possible to consider the creation of a foundation based on the rate of water use to subsidize those small farmers who wish to change their occupation. The compliance with norms for the investment made by the municipalities (1% of the budget) must be enforced to carry out environmental rehabil-itation and to buy the land designated for conservation.

CONCLUSIONS

The páramo region that extends from Costa Rica to the north of Peru is an upper-montane zone with one of the greatest floral diversities in the world; this confirms the findings of Cleef (1981) and of Sturm and Rangel-Ch. (1985). The Colombian páramo sites show the highest diversity: spermatophytes (62%), mosses (85%), liverworts (60%), lichens (77%), and ferns (98%). The greatest numbers of species restricted to the páramo are found in Colombia (2045) and in Ecuador (716), and the greatest floristic similarity among countries is between Colombia and Ecuador (15% at the species level), followed by Colombia and Venezuela (10.5%).

The vegetation types of the Colombian páramo (327) represent almost all of the phytoecological combinations of the biogeographic region. The belt with the greatest within-community species richness is the páramo proper. The greatest biodiversity (flora and fauna) and the number of species with restricted areas of distribution are found in the belts with an upper-Andean subpáramo ecotone; the lowest values are found in the superpáramo. The precipitation distribution regimes in the upper-Andean subpáramo and the superpáramo regions are bimodal–tetraseasonal, whereas those in the intermediate or interior zones are unimodal–biseasonal. The páramo regions can be classified as arid (with annual precipitation between 620 and 1196 mm) to wet (with more than 4000 mm of rain annually).

SUMMARY

The páramo region from Costa Rica to the north of Peru has 5168 species from 735 genera and 133 families of vascular plants, which makes it one of the world's most floristically diverse montane regions. The vegetation and flora of the Colombian páramo is the most diverse and varied within this large region. Compared to other geographic regions, the Colombian páramo is humid and, as a consequence, the main vegetation forms have the physiognomy of matorral or shrubby vegetation (semiclosed vegetation), which is converted to open vegetation (pasture and rosette vegetation) by natural causes (climate and volcanic eruptions) or by human disturbance. The different types of Colombian páramo vegetation (327) represent almost all the possible phytoecological compositions of the biogeographic páramo region. The natural limits of the páramo lie between 3200 (3500) and 4600 masl, but there also exist extrazonal páramo formations. The climate in páramo regions can be defined as dry, with 620 to 1196 mm of annual rainfall, to wet, with over 4000 mm of annual rainfall per year.

There is evidence of anthropogenic disturbance (cattle farming, agriculture, inadequate reforestation, and urbanization) in the majority of the páramo regions of Colombia. The transformation of the original floristic conditions and the vegetation through agriculture and cattle farming has been recorded in sectors of the Oriental and Central Cordilleras. Today, the location of natural areas of the Colombian páramo is linked to the control of the extension of potato crops.

References

Aguilar-P, M. and J.O. Rangel-Ch. 1996. Clima de alta montaña en Colombia. *El Páramo Ecosistema a Proteger.*, Fundación de Ecosistemas Andinos ECOAN: Serie montañas tropandinas, Vol. 2. Santafé de Bogotá. pp. 73–130.

Andrade-C., M.G. and J.A. Álvarez. 2000. Mariposas. In J.O. Rangel-Ch. (Ed.). *Colombia Diversidad Biótica III. La Región Paramuna.* Instituto de Ciencias Naturales, Universidad Nacional de Colombia. Bogotá. pp. 645–652.

Ardila-R, M.C. and A. Acosta. 2000. Anfibios. In J.O. Rangel-Ch. (Ed.). *Colombia Diversidad Biótica III. La Región Paramuna.* Instituto de Ciencias Naturales, Universidad Nacional de Colombia. Bogotá. pp. 617–628.

Beck, S.G., T.J. Killeen, and E. García. 1993. Vegetación de Bolivia. En T. Killeen, E. García, and S. Beck (Eds). *Guía de árboles de Bolivia.* Herbario Nacional de Bolivia-Missouri Botanical Garden. 958 pp.

Brako, L. and J.L. Zarucchi. 1993. Catálogo de las angiospermas y gimnospermas del Perú. *Monographs in Systematic Botany from the Missouri Botanical Garden.* Missouri Botanical Garden Press. St. Louis, MO. 1286 pp.

CAR. 1990. Diagnóstico y plan de manejo para la zona de reserva de páramos Documento interno, Corporación Autónoma Reegional de Cundinamarca (CAR), Bogotá.

Castaño, O., E. Hernández, and G. Cárdenas. 2000. Reptiles. En J.O. Rangel-Ch. (Ed.). *Colombia Diversidad Biótica III. La Región Paramuna.* Instituto de Ciencias Naturales, Universidad Nacional de Colombia. Bogotá. pp. 612–616.

Cleef, A.M.1981. The vegetation of the páramos of the Colombian Cordillera Oriental. *Dissertationes Botanicae* 61: 321 pp. J. Cramer, Vaduz. Also published in: El Cuaternario de Colombia 9 (T. Van der Hammen, Ed.). Amsterdam.

Cleef, A.M. and J.O. Rangel-Ch. 1984. La vegetación del Páramo del NW de la Sierra Nevada de Santa Marta. En T. Van der Hammen., A. Pérez-P.y., Pinto-E. (Eds). *La Sierra Nevada de Santa Marta Transecto Buritaca-La Cumbre.* Estudios de ecosistemas tropandinos. Vol. 2. J. Cramer, Vaduz. pp. 203–226.

Cortés-S, S., J.O. Rangel-Ch, and H. Serrano. 2003. Transformación de la cobertura vegetal en la alta montaña de la cordillera Oriental de Colombia. Resúmenes del Segundo Congreso sobre la Conservación de la Biodiversidad y Cuarto Congreso Ecuatoriano de Botánica. Loja, Ecuador.

Cuatrecasas, J. 1934. Observaciones geobotánicas en Colombia. *Trab Mus Nac Cienc Nat Ser Bot* 27: 144 pp. Madrid.

Delgado, A.C. and J.O. Rangel-Ch. 2000. Aves. In J.O. Rangel-Ch. (Ed.). *Colombia Diversidad Biótica III. La región Paramuna.* Instituto de Ciencias Naturales, Universidad Nacional de Colombia. Bogotá. pp. 629–644.

Díaz, S.L. 2002. La vegetación del páramo del Volcán de Doña Ana. Trabajo de grado. Facultad de Ciencias. Universidad del Cauca. Popayán (manuscrito).

Falla, P. and E. Rolón. 2002. Proceso de ocupación y distribución poblacional y calidad de vida de los asentamientos humanos de alta montaña en Colombia. In C. Castaño-U. (Ed.). Páramos y ecosistemas alto andinos de Colombia en condición HotSpot and Global Climatic Tensor. Bogotá. pp. 267–274.

Fernández-A, J.L. 1993. Novedades taxonómicas en *Aragoa* H.B.K. (Scrophulariaceae) y sinopsis del género. *Anales Jard Bot Madrid*, 51(1): 73–96.

Hernández, J. 2002. La vegetación zonal de los páramos de la cordillera Oriental colombiana: síntesis fitosociológica preliminar. Trabajo de grado. Departamento de Biología. Instituto de Ciencias Naturales. Universidad Nacional. Manuscrito.

Jørgensen, P.M. and S. León-Yáñez. (Eds). 1999. Catálogo de las plantas vasculares del Ecuador. *Monographs in Systematic Botany from the Missouri Botanical Garden*. Vol. 75. Missouri Botanical Garden Press. St. Louis, MO. pp. 43–106.

Luteyn, J.L. 1999. Páramos, a checklist of plant diversity, geographical distribution, and botanical literature. *Memoirs of the New York Botanical Garden* 84: 278 pp. New York.

Muñoz, Y., A. Cadena, and J.O. Rangel-Ch. 2000. Mamíferos. In J.O. Rangel-Ch. (Ed.). *Colombia Diversidad Biótica III. La región Paramuna.* Instituto de Ciencias Naturales, Universidad Nacional de Colombia. Bogotá. pp. 599–611.

Parra, L.N. 2002. Análisis facial de alta resolución de sedimentos del Holoceno tardio en el páramo de Frontino, Antioquia. Comunicación interna programa de Doctorado en Biología. Universidad Nacional de Colombia.

Rangel-Ch, J.O. and J. Aguirre-C. 1986. Estudios ecológicos en la Cordillera Oriental Colombiana III. Vegetación de la Cuenca del Lago de Tota (Boyacá). *Caldasia* 15(71–75): 263–312.

Rangel-Ch, J.O., P. Lowy-C, M. Aguilar-P, and A. Garzón-C 1997. Tipos de vegetación en Colombia. In J.O. Rangel-Ch. (Ed) *Colombia Diversidad Biótica*. Vol. 2. Instituto de Ciencias Naturales-IDEAM. Bogotá. pp. 89–367.

Rangel-Ch, J.O. 2000. Elementos para una biogeografía de los ambientes de alta montaña de América Latina. In J. Llorente-B. and J.J. Morrone (Eds). *Introducción a la biogeografía en América Latina*. Publicaciones de la UNAM. México.

Rangel-Ch, J.O. 2000a. Visión integradora sobre la región del páramo. In J.O. Rangel-Ch. (Ed.). *Colombia Diversidad Biotica III. La región Paramuna*. Instituto de Ciencias Naturales, Universidad Nacional de Colombia. Bogotá. pp. 814–836.

Rangel-Ch, J.O. 2000b. Tipos de vegetación. In. J. O. Rangel-Ch. (Ed.). *Colombia Diversidad Biótica III. La región Paramuna*. Instituto de Ciencias Naturales, Universidad Nacional de Colombia. Bogotá. pp. 685–719.

Rangel-Ch, J.O. 2000c. Clima. In J.O. Rangel-Ch. (Ed.). *Colombia Diversidad Biótica III. La región Paramuna*. Instituto de Ciencias Naturales, Universidad Nacional de Colombia. Bogotá. pp. 85–125.

Rangel-Ch, J.O. 2000d. Flora y vegetación amenazada del páramo. In J.O. Rangel-Ch. (Ed.). *Colombia Diversidad Biótica III. La región Paramuna*. Instituto de Ciencias Naturales, Universidad Nacional de Colombia. Bogotá. pp. 785–813.

Rangel-Ch, J.O. 2000. (Ed.). *Colombia Diversidad Biótica III. La región de vida paramuna*. Instituto de Ciencias Naturales — Instituto Alexander von Humboldt. Bogotá. 902 pp.

Salamanca, S., A.M. Cleef, and J.O. Rangel-Ch. 1992. La vegetación del páramo in: S. Salamanca: La vegetación del páramo y su dinámica en el Macizo volcánico Ruiz-Tolima (Cordillera Central Colombiana). Análisis geográficos. Publicación del Instituto Geográfico Agustín Codazzi. (IGAC) 21: 38–63. Bogotá.

Sánchez-M.R. and J.O. Rangel-Ch., 1990. Estudios ecológicos en la cordillera Oriental de Colombia V. La vegetación de los depósitos turbosos de los páramos de los alrededores de Bogotá. *Caldasia*, 16(77): 155–193.

Sturm, H. and J.O. Rangel-Ch. 1985. Ecología de los páramos Andinos: una visión preliminar integrada. *Biblioteca J.J.Triana* No. 9. Instituto de Ciencias Naturales. Bogotá. 292 pp.

Sturm, H. 1998. The ecology of the páramo region in tropical high mountains. Verlag Franzbecker: Hildesheim. Berlín. 286 pp.

Uribe, A. 2003. El páramo no es de las papas. UN Periódico. No. 45.

Van der Hammen, T., J.D. Pabón-C., H. Gutiérrez, and J.C. Alarcón. 2002. El cambio global y los ecosistemas de alta montaña de Colombia. In C. Castaño-U. (Ed.). *Páramos y Ecosistemas Alto Andinos de Colombia en Condición HotSpot and Global Climatic Tensor*. Bogotá. pp. 163–210.

Verweij, P.A. 1995. Spatial and Temporal Modeling of Vegetation Patterns. Ph.D. thesis. ITC-University of Amsterdam: Enschede, The Netherlands. 233 pp.

Witte, H.J.L. 1994. Present and Past Vegetation and Climate in the Northern Andes (Cordillera Central, Colombia): A Quantitative Approach. Ph.D. thesis. University of Amsterdam: Amsterdam. 269 pp.

9 Grazing Impact on Vegetation Structure and Plant Species Richness in an Old-Field Succession of the Venezuelan Páramos

Lina Sarmiento

INTRODUCTION

Páramos occupy the alpine belt of northern South America, between 3000 and 4800 masl. Giant rosettes of the genera *Espeletia*, together with sclerophilous shrubs and bunch grasses, dominate the vegetation. In pre-Columbian times, the páramo was almost exclusively used for hunting and gathering (Wagner 1978), and only after the arrival of the Spanish, and mainly during the 18th century, did it begin to be extensively grazed by introduced domestic animals, mainly cattle, horses, and mules. Consequently, the páramo evolved until recent times without domestic herbivory. Many plant species, mostly the endemic ones, probably did not develop specific adaptations to this kind of disturbance and are potentially sensitive.

The carrying capacity of the Venezuelan páramos is low. The main offering of forage is concentrated in small marshes and fens situated in the valley bottoms or in areas with poor drainage and dominated by palatable grasses and sedges (Molinillo and Monasterio 1997). The more widespread páramo vegetation, in which dwarf shrubs, rosette plants, and tussock grasses predominate, presents a lower availability of forage (Molinillo and Monasterio 1997). In the wetter páramos of Colombia, where the cover of tussock grasses is higher and more continuous than in Venezuela, the palatability of the vegetation is commonly improved by burning (Hofstede et al. 1995), but in the drier páramos of Venezuela, where grasses are less abundant, burning is not practiced, and the strategy of the farmers is to develop a closer relationship between agricultural activities and cattle management, complementing the natural sources of forage with crop residues, fodder, and grazing on fallow plots (Molinillo and Monasterio 2002).

To analyze the human impact on páramo vegetation, it is essential to differentiate the Andean and high-Andean ecological belts (Monasterio 1980). In the Andean belt (3000 to 4000 m), night frosts are concentrated during the dry season, allowing crops to develop during the rainy season. In Venezuela, a rapid process of agricultural expansion is taking place in this belt, with potatoes as the main cash crop and livestock husbandry as a complementary activity. The high-Andean belt (above 4000 m), where frosts occur throughout the year, is not suitable for cropping and is only used for extensive grazing. Nevertheless, these two belts are not managed independently, and continuous animal displacements occur between them. Animals used as draft power in agriculture and milking cows are maintained temporally in the Andean belt, where crop residues and fodder are used to complement their diet (Molinillo and Monasterio 1997; Pérez 2000).

Long-fallow agriculture is still practiced in some areas of the Andean belt. This agricultural system generates a landscape mosaic of areas under cultivation, under natural vegetation and at different stages of the fallow period, which can last from 5 to more than 10 years. Fallow areas are important sources of forage for domestic animals maintained in the agricultural belt (Pérez 2000). Fallow agriculture provides a unique opportunity to analyze the rate and mechanisms of páramo regeneration after agricultural disturbance, an essential knowledge to evaluate the reversibility of human impacts and to design future strategies for páramo restoration and management.

Our general objective is to assess if páramo regeneration after agricultural disturbance is affected by grazing and to evaluate this activity as to whether or not it can be compatible with páramo restoration plans. From the literature, it is well known that herbivory causes a pronounced impact on cover, structure, and diversity of plant communities, affecting the functioning of the ecosystems and the environmental services that they provide (Milchunas et al. 1988; Huntly 1991; Pacala and Crawley 1992; Gough and Grace 1998). Herbivory also affects the rates of succession and can produce divergence in successional pathways (Davidson 1993; Van Oene et al. 1999). Nevertheless, the specific consequences of grazing depend on herbivore density and on the characteristics of each particular system, such as the level of soil fertility, the importance of competition for light as a driving successional force, and the sensitivity and adaptive mechanisms of the dominant and subordinate species. As the vegetation response to grazing depends on so many different factors, it is necessary to perform specific studies in each ecosystem to design particular management strategies to preserve ecosystem biodiversity and functioning.

In the páramos, some studies were carried out on the effect of grazing on vegetation, but most of them were based on comparing vegetation relevés between sites with different grazing intensities. Few data come from experimental exclusions, except the unreplicated 1-year experiment of Molinillo and Monasterio (1997). Moreover, in most of the studies, it is not possible to differentiate the impact of grazing from that of burning. We

did not find specific studies on the effect of grazing on páramo regeneration after agricultural disturbance.

The objective of this study is to assess the impact of grazing on páramo secondary succession, including the effect on (1) general ecosystem attributes such as plant biomass, height, and percentage of bare soil, (2) the life-form spectrum of vegetation, (3) plant species richness, (4) individual plant species, including identification of the more susceptible and tolerant ones in different stages of the succession, and (5) the probability of invasion by introduced species more adapted to this kind of disturbance. With these aims, an exclosure experiment was conducted over a period of 4 years in plots at two different stages of páramo succession.

METHODOLOGY

Study Area

The study was carried out in the Páramo de Gavidia, located in the northern Andes in Venezuela, at 8°40 latitude N and 70°55 longitude W. The area lies within the Sierra Nevada de Mérida National Park, at 3400 masl and is a narrow glacial valley, with well-drained inceptisols (*Ustic Humitropept*) of a sandy-loam texture, low pH (4.25 to 5.5), and high content of organic matter (up to 20%) (Abadin et al. 2002). Agriculture is practiced on steep slopes and also on small colluvial and alluvial deposits in the valley bottom. The precipitation regime is unimodal, with the dry season between December and March. The mean temperature ranges between 9 and 5°C, depending on the altitude, and the mean annual precipitation is 1300 mm.

A present population of 400 inhabitants established the settlement at the end of the 19th century, giving the valley a relatively short land use history (Smith 1995). The land-use system is long-fallow agriculture. Potatoes are grown during an agricultural phase lasting from 1 to 3 years. The agricultural practices include the incorporation of the successional vegetation as a green manure and mineral fertilization with an average dose of 300 kg N ha^{-1} a^{-1}. After cultivation, the fields are abandoned, and the fallow period begins. The current average fallow length is 4.6 years, but there is a large variability, with times ranging from 2 to more than 15 years (Sarmiento et al. 2002). During the fallow period, fields are used for extensive grazing, mainly by cattle and horses.

VEGETATION DYNAMICS DURING OLD-FIELD SUCCESSION: MAIN TRENDS

A previous study on plant succession, carried out by Sarmiento et al. (2003), indicated that, as in other extreme environments, succession in the páramo proceeds as an *autosuccession*; the characteristic species of the mature ecosystem colonize very early and succession takes place more by changes in the abundance of these species than by a true replacement. Only a few herbaceous, mostly introduced species (e.g. *Rumex acetosella*) act as strict pioneers and strongly dominate the early stages. Then they undergo a progressive decline, whereas native forbs (e.g. *Lupinus meridanus*) and grasses (e.g. *Trisetum irazuense*) have their peaks of abundance at intermediate stages (4 to 5 years). The characteristic páramo life-forms, sclerophilous shrubs (e.g. *Baccharis prunifolia, Hypericum laricifolium*) and giant rosettes (e.g. *Espeletia schultzii*), appear very early and gradually increase in abundance, becoming dominant after only 7 to 8 years. Vegetation regeneration takes place relatively fast, but despite a rapid reestablishment of the general physiognomy of the ecosystem, the high diversity of the natural páramo is not reached in the current successional times (Sarmiento et al. 2003).

EXPERIMENTAL DESIGN

Eight areas were selected in different parts of the valley: four had just been abandoned after potato cultivation (early plots), and four had already passed through 5 years of grazed succession (intermediate plots). In each area, an enclosure of 200 m^2 was established and divided into two parts, each of 100 m^2 (20 m × 5 m). One of these parts was excluded from grazing, and the other was grazed for 1 h every 3 weeks, equivalent to a stocking rate of 0.4 cows ha^1, considering that a cow grazes 12 h per day. The experiment lasted 4 years, from February 1998 to November 2001, and, in total,

60 different events of grazing were carried out. Controlled grazing was preferred instead of free grazing, to have an identical stocking rate in all the repetitions.

VEGETATION SAMPLING

Twice a year, during the dry and rainy seasons (in March and October), the vegetation was sampled in the grazed and excluded part of each plot, for a total of eight sequential samplings during the 4 years of the experiment (8 sampling dates × 8 plots × 2 treatments = 128 vegetation relevés). The first sampling was carried out just before the start of the experiment. The point intercept method was used (Greig-Smith 1983). Five parallel lines of 20 m length were located at 1-m intervals. Along these lines a pin (diameter, 2.5 mm) was placed vertically every meter, and the contacts of each species in height intervals of 10 cm were recorded.

Using the data obtained from the point intercept method, the biovolume per species, the percentage of bare soil, and the weighted height of the vegetation were calculated. The biovolume was computed as the sum of all the contacts of the species in the 100 points. The average weighted height of the vegetation was calculated by weighing the number of contacts in each 10 cm by the height of the stratum. The percentage of bare soil was calculated from the points that no species touched.

Slope, stoniness, soil texture, and soil total C and N were also measured to characterize the different plots.

ANALYSIS OF THE DATA

Biovolume data can be transformed into biomass using coefficients for each species. The relative abundance of the species is different when data are expressed in one or the other of these units, as the coefficients to transform biovolume to biomass are different for each species, depending on architecture, wood density, specific leaf area, vertical distribution, etc. These coefficients were established for each species from simultaneous measurements of biovolume and biomass in several plots of 2500 cm^2 (20 plots by species in average), selected to include a large variation in species abundance. The relationship between biovolume and biomass was linear for all the species, and the regression coefficient was always significant. The best correlation was obtained for *Acaena elongata* ($r^2 = 0.90$, $p < 0.0001$) and the worst for *Poa annua* ($r^2 = 0.50$, $p = 0.049$). The coefficients were obtained forcing the linear regression to the origin. Values oscillate in the range from 39 to 1774 g m^{-2}, which means that a biovolume of 1 (100 touches in 100 points) corresponds to a biomass of 1774 g m^{-2} for the species with the largest coefficient (*Espeletia schultzii*). As biovolume can be higher than 1, this coefficient does not represent a top limit to biomass. For some less-abundant species, coefficients were not available, and we used those of the more similar species in terms of architecture.

A repeated-measures statistical analysis (GLM) was carried out to test the overall significance of the differences and to identify the effect of the different factors. The age of the plot (young and intermediate) was considered as the between-subject factor; treatment (grazed and excluded) and time (eight sampling occasions) were the within-subject factors. Additionally, paired t-tests were used to compare the mean values between the grazed and excluded treatments over the 4 years. For these paired tests, each pair consisted of the mean values of the grazed and ungrazed treatments of the same plot for a given variable. Statistical analyses were carried out using SPSS (version 7.5). Biomass data were log +1 transformed for the statistical tests.

An index of damage by grazing (ID) was calculated for the different species from their relative abundance in the grazed (G) and ungrazed treatments (NG):

(NG-G)/G ≤ 0.5
    ID = 2, very positively affected
0.5 < (NG-G)/G < 0.1
    ID = 1, positively affected
0.1 ≤ (NG-G)/G ≤ 0.1
    ID = 0, not affected
0.1 < (NG-G)/G < 0.5
    ID = +1, negatively affected
(NG-G)/G ≥ 0.5
    ID = +2, severely affected

RESULTS

PLANT BIOMASS, VEGETATION HEIGHT, AND PERCENTAGE OF BARE SOIL

The effect of grazing on aboveground biomass, vegetation height, and cover is presented in Figure 9.1, and the results of the repeated-measures analysis is shown in Table 9.1. It can be observed that: (1) The total aboveground biomass was significantly lower in the young, compared to the intermediate plots (age effect). (2) Grazing significantly reduced plant biomass (grazing effect). (3) The effect of grazing was similar in the two successional ages (grazing–age interaction). (4) Biomass changed significantly over time (time effect). (5) The effect of time was different in the two successional ages (time–age interaction). (6) In the grazing treatment, biomass increased at a faster rate than in the excluded one (grazing–time interaction). (7) The effect of grazing over time was similar in both successional ages (grazing–time–age interaction). On the average in the 4 years of the experiment, aboveground biomass was 338 g m^{-2} and 585 g m^{-2} in the grazed and ungrazed young plots, and 606 g m^{-2} and 878 g m^{-2} in the grazed and ungrazed intermediate plots, respectively (Table 9.2). The final biomass in the grazed young plots was higher than the initial biomass in the intermediate plots, indicating that the stocking rate in our experiment was probably lower than that existing before the enclosures were made.

Another clear consequence of grazing was the significant reduction in the weighed height of the vegetation. For this variable, no significant differences were detected between young and intermediate plots (Table 9.1). In the grazed young plots, the vegetation remained very low during the 4 years of the experiment (weighed average = 8.8 cm) compared to the ungrazed plots in which the weighed height increased from 7 to 20 cm. In the intermediate plots, the height increased in both treatments but more in the ungrazed one, passing from 8.5 to 15 cm under grazing and to 19 cm under ungrazed conditions.

The percentage of bare soil was also very sensitive to grazing. At the beginning of the experiment, 57% of the surface was uncovered

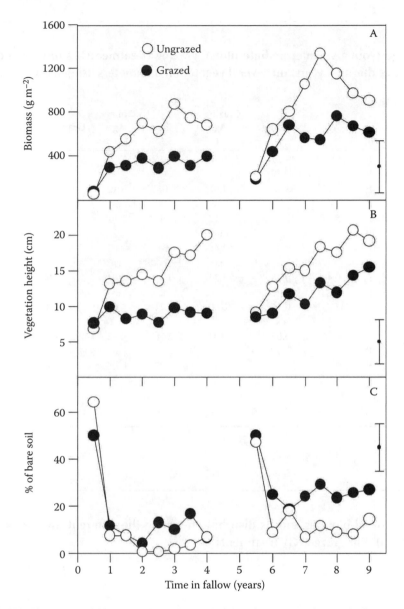

FIGURE 9.1 (A) Aboveground biomass, (B) weighted height of the vegetation, and (C) percentage of bare soil in the excluded and grazed treatments. The bars of error represent the average standard deviation.

in the young plots and 49% in the intermediate plots (Figure 9.1). After 6 months, the percentages of bare soil decreased in all cases but remained higher in the grazed treatment. On average, the percentages of bare soil were 4 and 10 in the ungrazed and grazed young plots, and 11 and 25 in the intermediate ungrazed and grazed plots, respectively. Differences between grazed and ungrazed treatments were significant but not between young and intermediate plots. However, a very significant interaction was found between grazing and time, indicating that the reduction in the percentage of bare soil occurred faster in the ungrazed treatment for both ages. The high percentage of bare soil at the beginning of the experiment in the young plots is due to their recent abandonment after harvest. In the case of the intermediate plots, the high percentage of bare soil at the first sampling date indicates, again, a possible higher grazing pressure before the installation of the experiment.

TABLE 9.1
Effects of age (young vs. intermediate plots), grazing (treatments), and time (consecutive sampling dates during 4 years) on several vegetation parameters using a repeated-measures analysis

Source Parameter	df	Age 1	Grazing 1	Grazing × Age 1	Time 7	Time × Age 7	Grazing × Time 7	Grazing × Time × Age 7
Biomass	F	5.95*	49.29**	0.47	14.46**	2.18*	4.93**	1.47
	P			ns				ns
Height	F	2.91	33.96**	1.61	9.03**	1.12	6.05**	1.94
	P	ns		ns		ns		ns
Percentage of bare soil	F	5.17	32.65**	0.17	13.42**	2.30*	6.20**	1.44
	P	ns		ns				ns
Percentage forbs	F	28.75*[8]	0.05	3.45	6.35**	2.03	2.26*	1.69
	P		ns	ns		ns		ns
Percentage grasses	F	0.40	0.98	3.35	1.53	1.84	5.24**	1.34
	P	ns	ns	ns	ns	ns		ns
Percentage shrubs	F	11.91*	0.36	0.00	17.73**	14.09**	1.05	0.38
	P		ns	ns			ns	ns
Percentage rosettes	F	24.23*[8]	0.05	0.03	2.61*	0.22	1.59	1.97
	P		ns	ns		ns	ns	ns
Species richness	F	5.12*	21.03**	3.43	14.97**	4.55**	2.88*	0.51
	P			ns				ns

*Significant at $p < .05$.
**Significant at $p < .001$.

TABLE 9.2
Total aboveground biomass and its distribution among the different life-forms in the ungrazed (NG) and grazed (G) treatments

	1–4 years		5–8 years	
	NG	G	NG	G
	g m^{-2} (%)	g m^{-2} (%)	g m^{-2} (%)	g m^{-2} (%)
Total aboveground	585[a] (100)	338[b] (100)	878[c] (100)	606[a] (100)
Forbs	370[a] (63[a])	206[b] (61[a])	87[c] (10[b])	62[d] (10[b])
Grasses	123[a] (21[a])	70[b] (21[a])	119[a] (14[a])	50[b] (8[b])
Shrubs	82[a] (14[a])	48[b] (14[a])	358[c] (41[b])	204[d] (34[b])
Giant rosettes	9[a] (2[a])	15[b] (4[a])	314[b] (36[b])	291[b] (48[c])

Note: Values are the average over the 4 years of the experiment, excluding the first sampling.

[a–d] Different letters indicate significant differences between treatments ($p < 0.05$; t-test for dependent samples).

LIFE-FORM SPECTRUM OF THE VEGETATION

The relative contribution of the main life-forms (forbs, grasses, giant rosettes, and shrubs) to the total aboveground biomass is shown in Figure 9.2. In Table 9.1, the results of the repeated-measures analysis are shown and in Table 9.2, the mean values over the study period. The relative contribution of forbs to the total aboveground biomass experienced a clear and significant decrease over time, whereas shrubs and rosettes increased. No significant temporal trends were detected using the repeated-measures analysis in the percentage of grasses (age and time effects not significant).

Despite the reduction in total biomass by grazing, the repeated-measures analysis shows that the effect of grazing on the life-form spectrum was not significant, indicating a proportional reduction in the biomass of the four life-forms. Nevertheless, for forbs and grasses, there is an interaction between grazing and time (Table 9.1). The comparison of the mean values over time (Table 9.2), using a t-test for depen-

dent samples, shows that grazing did not change the perceptual contribution of the different life-forms in the young plots. However, in the intermediate plots, grazing caused a significant reduction in the percentage of grasses (from 14 to 8% of total aboveground biomass) and an increase in giant rosettes (from 36 to 48%).

It is rather surprising that grasses and forbs, the main targets of herbivory, do not experience a more important proportional decrease in biomass. An explanation will arise from the analysis of the response of the individual species.

PLANT SPECIES RICHNESS

The method used to quantify plant species richness (100 points in 100 m^{-2}) underestimates the total number of species in the plot, as curves of numbers of species do not saturate after 100 points (results not shown). Consequently, values have to be interpreted only comparatively. The maximum number of species recorded in a particular plot was 23, which is low compared to the almost 200 species reported for the whole valley.

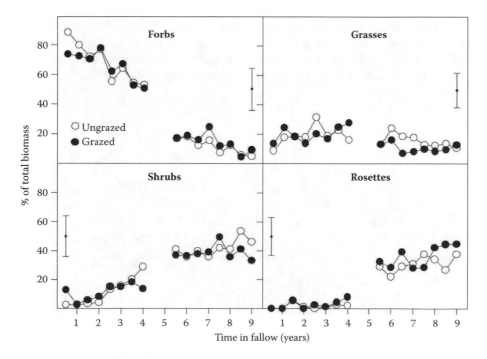

FIGURE 9.2 Percentage of the total aboveground biomass represented by the different life-forms in the excluded and grazed treatments. The bars of error represent the average standard deviation.

There is a significant effect of age and time on plant species richness (Figure 9.3 and Table 9.1), indicating that diversity increases during succession. The rate of increase was significantly higher in the young compared to the intermediate plots. In the intermediate plots, the most important change in the number of species was between the first and the second samplings, when the number of species increased from an average of 8 to an average of 16 as a consequence of fencing out the plots.

Grazing produced a statistically significant but slight reduction in plant species richness (Table 9.1), but it is remarkable that richness did not differ at the last sampling date, suggesting that the effect of grazing at this stocking rate could be only temporal (Figure 9.3). The initial richness of the intermediate plots, at the first sampling date, was lower than at the end point of the young plots, again suggesting a higher grazing pressure before the start of the experiment. Consequently, a bigger effect of grazing on plant richness could be expected at higher grazing pressures.

To analyze the factors that influence plant diversity in this old-field succession, a multiple regression (forward stepwise) was carried out using plant richness as dependent variable, and successional time, percentage of bare soil, total aboveground biomass, weighed height of the vegetation, stoniness, slope, soil texture, total soil nitrogen, and soil total carbon as independent variables. The forward stepwise procedure selected four variables that explained 69% of the variability in plant richness. The included variables were: biomass (B, in g m^{-2}), which explains 47% of the variability, slope of the plot (S, in degrees), which explains an additional 11% of the variability, bare soil (BS, in %), which explains 7%, and successional age (SA, in years), which explains 3.6% more. The inclusion of further variables did not significantly increase the total amount of variance explained. The equation for the multiple regression is:

Richness =
5.02 + 0.06 B + 0.11 S − 0.07 BS + 0.41 SA

A logarithmic function of plant biomass explains more variability (74%) than the multiple lineal regression (Figure 9.4).

RESPONSE OF INDIVIDUAL SPECIES

Over the whole experiment, 61 species were recorded: 17 grasses, 33 forbs, 10 shrubs, and 1 giant rosette. Among these, 28 had a very low abundance and will not be considered further. The successional behavior of the 33 other species and their response to grazing is presented in Table 9.3, including the consumption preference by cattle of the different plant species, according to Molinillo and Monasterio (1997), complemented with personal observations.

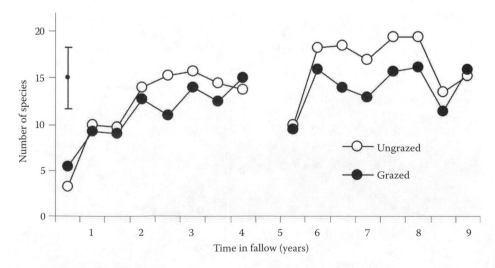

FIGURE 9.3 Species richness in the excluded and grazed treatments. The bar of error represents the average standard deviation.

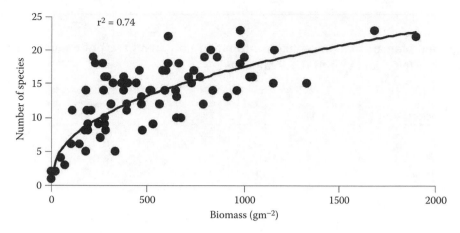

FIGURE 9.4 Relationships between plant biomass and richness using the data of all vegetation samplings. A logarithmic function was adjusted to the points ($p < 0.001$).

There are contrasting successional patterns (Table 9.3 and Figure 9.5). A group of species was more abundant in early succession compared to intermediate succession: *Rumex acetosella*, *Erodium cicutarium*, *Gnaphalium elegans*, *Penisetum clandestinum*, *Bromus carinatus*, *Poa annua*, *Lachemilla moritziana*, and *Lupinus meridanus*; all of them, except the last two, introduced species. Another group of species was more abundant during the intermediate succession: *Espeletia schultzii*, *Acaena elongata*, *Aciachne pulvinata*, *Hypericum laricifolium*, *Oenothera epilobifolia*, *Orthosanthus chimboracensis*, *Brachypodium mexicanum*, and *Nassella linerifolia*, all of them native species. The rest of the species did not present significant differences between young and intermediate plots.

In Figure 9.5, the successional behavior of some representative species can be observed. For example, *Rumex acetosella* decreased regularly with time, with a very significant effect of age and time (Table 9.4). *Lachemilla moritziana* and *Trisetum irazuens* have their peaks of abundance after 2 and 4 years of succession, respectively, with very regular curves of increase and posterior decrease in abundance. Other species, such as *Espeletia schultzii* and *Hypericum laricoides*, showed a progressive and significant increase in abundance with time.

Analyzing the effect of grazing on aboveground biomass (absolute values in Table 9.3), it can be observed that only four species sig-

nificantly increased their biomass and can be considered as promoted by grazing: *Aciachne pulvinata*, *Erodium cicutarium*, *Penisetum clandestinum*, and *Poa annua*. Three of these species are introduced. Twelve species decreased their biomass and can be considered as damaged by grazing: *Acaena elongata*, *Baccharis prunifolia*, *Brachypodium mexicanum*, *Gamochaeta americana*, *Geranium chamaense*, *Hypericum laricifolium*, *Lachemilla moritziana*, *Nassella linerifolia*, *Noticastrum marginatum*, *Rumex acetosella*, *Sisyrinchium tinctorum*, and *Trisetum irazuense*. The remaining 17 species listed in Table 9.3 did not show a significant change in biomass and can be considered as not affected by grazing. In this unaffected group, there are several grasses, such as *Agrostis jahnii*, *Agrostis trichodes*, and *Vulpia myurus*, that are consumed by animals but with an intermediate preference; the only giant rosette recorded, *Espeletia schultzii*, which is not consumed by cattle; a legume, *Lupinus meridanus*, rejected due to its toxic composition; and several forbs that are not consumed by animals, such as *Gnaphalium elegans* and *Gnaphalium meridanum*.

Apart from the absolute changes in biomass, grazing also affected the relative proportion between species (values in parentheses in Table 9.3). These relative changes give additional information concerning the structural transformation of the vegetation. Several

TABLE 9.3

Aboveground biomass and perceptual contribution (in parentheses) of the main species in the ungrazed (NG) and grazed (G) treatments

Species	P	LF	Biomass 1–4 years g m^{-2} (%)		Biomass 5–8 years g m^{-2} (%)		ID
			NG	G	NG	G	
Acaena elongata	3	S	7.8^a (1.2^a)	3.9^b (1.2^a)	68.8^c (7.1^b)	40.2^d (6.6^b)	0
Aciachne pulvinata	5	G	0.1^a (0.0^a)	0.0^a (0.0^a)	7.8^b (0.8^b)	12.9^c (2.1^c)	2
Agrostis jahnii	2	G	22.4^a (3.4^a)	13.9^a (4.1^a)	5.3^{ab} (0.6^{ab})	3.3^b (0.5^b)	0
Agrostis trichodes	2	G	11.9^a (1.8^{ac})	3.2^b (0.9^b)	35.1^a (3.6^{ab})	13.5^a (2.2^c)	+1
Baccharis prunifolia	5	S	67.7^{ab} (10.4^a)	44.0^a (13.0^b)	91.8^b (9.4^{ab})	43.6^a (7.2^{ab})	0
Brachypodium mexicanum	1	G	0.0^a (0.0^a)	0.0^a (0.0^a)	2.9^b (0.3^b)	0.0^a (0.0^a)	+2
Bromus carinatus	1	G	22.1^a (3.4^a)	18.2^a (5.4^a)	4.5^b (0.5^b)	1.1^c (0.2^c)	0
Erodium cicutarium	—	F	0.3^a (0.0^a)	2.6^b (0.8^b)	0.0^a (0.0^a)	0.0^a (0.0^a)	2
Espeletia schultzii	5	R	10.8^a (1.6^a)	15.2^a (4.5^b)	349.1^b (35.7^b)	291.4^b (48.0^c)	2
Gamochaeta americana	4	F	8.1^a (1.2^a)	2.6^b (0.8^{ab})	6.0^a (0.6^b)	2.6^a (0.4^b)	+1
Geranium chamaense	3	F	16.0^a (2.5^a)	7.1^{bc} (2.1^{ab})	8.4^b (0.9^{bc})	3.5^c (0.6^c)	+1
Gnaphalium elegans	4	F	3.5^{ab} (0.5^{ab})	0.3^a (0.1^a)	0.0^b (0.0^b)	0.0^b (0.0^b)	+2
Gnaphalium meridanum	4	F	1.4^a (0.2^{ab})	0.5^a (0.2^{ab})	0.9^a (0.1^b)	1.1^a (0.2^a)	1
Hypericum laricifolium	5	S	16.0^a (2.5^a)	2.7^b (0.8^b)	210.2^c(21.5^c)	108.0^d (17.8^c)	+1
Lachemilla moritziana	3	F	42.5^a (6.5^a)	33.0^b (9.8^b)	13.3^c (1.4^c)	8.7^d (1.4^c)	1
Laennecia mima	5	F	2.5^a (0.4^a)	1.1^a(0.3^a)	0.2^a (0.0^a)	0.1^a (0.0^a)	+1
Lupinus meridanus	5	F	14.1^a (2.1^a)	7.4^{ab} (2.2^a)	2.2^b (0.2^b)	1.5^b (0.3^b)	0
Nassella linerifolia	1	G	1.2^a (0.2^a)	0.0^b (0.0^b)	30.4^c (3.1^c)	1.0^a (0.2^a)	+2
Nassella mexicana	2	G	4.6^a (0.7^a)	0.0^a (0.0^a)	1.4^a (0.1^a)	0.2^a (0.0^a)	+2
Nassella mucronata	1	G	1.1^a (0.2^a)	0.1^a (0.0^a)	0.0^a (0.0^a)	0.0^a (0.0^a)	+1
Noticastrum marginatum	—	F	0.9^a (0.1^{ab})	0.3^a (0.1^a)	2.8^b (0.3^b)	1.7^a (0.3^{ab})	+1
Oenothera epilobifolia	—	F	0.0^a (0.0^a)	0.8^a (0.2^{ab})	3.3^b (0.3^b)	3.5^b (0.6^c)	2
Orthosanthus chimboracensis	5	F	0.5^a (0.1^a)	0.4^a (0.1^a)	5.7^b (0.6^b)	4.5^b (0.7^b)	1
Oxylobus glanduliferus	5	F	0.4^b (0.1^{ab})	0.1^a (0.1^b)	2.1^{ab} (0.1^a)	1.8^b (0.3^a)	0
Paspalum pygmaeum	1	G	0.2^a (0.0^a)	0.3^a (0.1^a)	0.5^a (0.1^a)	1.0^a (0.2^a)	2
Penisetum clandestinum	1	G	0.1^a (0.0^a)	10.1^b (3.0^b)	0.0^a (0.0^a)	0.0^a (0.0^a)	2
Poa annua	2	G	2.8^a (0.4^a)	5.2^b (1.6^b)	0.0^c (0.0^c)	0.0^c (0.0^c)	2
Rumex acetosella	3	F	320.4^a(49.7^a)	148.2^b (43.5^a)	44.9^c (4.6^b)	31.4^d (5.1^b)	0
Sisyrinchium tinctorum	5	F	5.7^a (0.9^a)	1.8^{bc} (0.5^a)	8.6^a (0.9^a)	3.7^c (0.6^a)	+1
Stevia elatior	5	F	1.1^a (0.2^a)	0.6^a (0.2^a)	1.2^a (0.1^a)	0.5^a (0.1^a)	0
Stevia lucida	5	S	0.0^a(0.0^a)	0.0^a (0.0^a)	15.4^b (1.6^b)	9.5^b (1.6^b)	0
Trisetum irazuense	1	G	42.8^a (6.6^a)	3.5^b (1.0^b)	28.2^a (2.9^a)	2.5^b (0.4^b)	+2
Vulpia myurus	3	G	17.0^a (2.6^a)	8.2^a (2.4^a)	12.2^a (1.2^a)	9.4^a (1.5^a)	0

Note: The values are averages for the 4 years of the experiment. The values of the palatability index were taken from Molinillo and Monasterio (1997), and completed or modified using personal observations.

P is an index of palatability in a relative scale 1 = preferred, 2 = good, 3 = regular, 4 = insufficient, 5 = rejected. Life-form (LF) abbreviations are F = forb, G = grass, S = shrub, R = giant rosette. The index of damage (ID) is: +2 = very positively affected, +1 = positively affected, 0 = not affected, 1 = negatively affected, and 2 = very negatively affected.

[a–d]Different letters indicate significant differences between treatments ($p < .05$, t-test for dependent samples).

trends are possible: (1) a reduction in biomass not accompanied by a reduction in the relative contribution of the species, (2) a reduction in biomass and in the relative contribution of the species, (3) a reduction in biomass but an increase in the relative contribution of the spe-

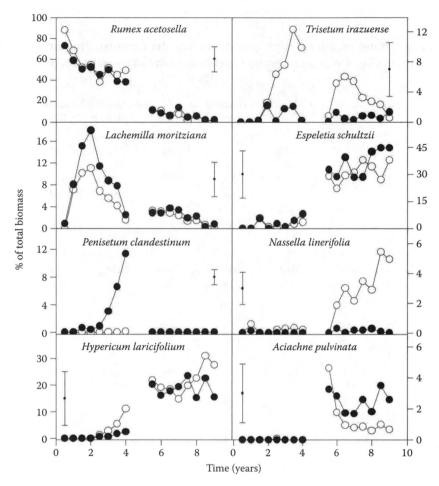

FIGURE 9.5 Percentage of the total biomass of some representative species along the 4 years of the study in the excluded and grazed treatments. The bars of error represent the average standard deviation.

cies, (4) an increase in the absolute and proportional biomass, and (5) no change in biomass but an increase in proportion. Situation 1 indicates that the species is consumed or damaged as a function of its biomass, without a preferential positive or negative selection. This is the case of *Rumex acetosella* (see also Figure 9.5), *Acaena elongata*, *Baccharis prunifolia*, and *Hypericum laricoides* (Figure 9.5), among others. Situation 2 indicates a preferential consumption or damage, as is the case with only three species: *Trisetum irazuens* (Figure 9.5), *Brachypodium mexicanum*, and *Nassella linerifolia* (Figure 9.5), all tall grasses with a very high palatability and accessibility to animals. Situation 3 indicates a little negative selection and is only found in the case of *Noticastrum marginatum*.

Situation 4 indicates that the effect of grazing is positive, as in the case of *Poa annua*, *Aciachne pulvinata* (Figure 9.5), *Erodium cicutarium*, and *Penisetum clandestinum* (Figure 9.5). Situation 5 indicates that the species is not consumed or damaged by animals but indirectly favored as its proportion in the total biomass increased. This is the case of *Espeletia schultzii* (Figure 9.5), whose biomass remained constant but increased its relative contribution from 36 to 48% in the intermediate plots. Another species with the same behavior is *Oenothera epilobifolia*, a prostrate forb.

The last column of Table 9.3 presents the index of damage by grazing. The more fragile species are grasses with a high palatability, such as *Trisetum irazuense*, *Brachypodium mexi-*

TABLE 9.4
Effects of age (young vs. intermediate plots), grazing (treatments), and time (consecutive sampling dates during 4 years) on several vegetation parameters using a repeated-measures analysis

Parameter	df	Age 1	Grazing 1	Grazing[a] Age 1	Time 7	Time[a] Age 7	Grazing[a] Time 7	Grazing[a] Time[a] Age 7
Rumex	F	85.5**	0.17	4.31	6.49**	2.32*	2.41*	1.67
acetosella	P		ns	ns				ns
Lachemilla	F	10.48*	7.47*	2.13	11.28**	1.57	1.14	0.34
moritziana	P		ns			ns	ns	ns
Trisetum	F	0.04	11.84*	0.28	1.51	3.87*	1.12	3.84*
irazuense	P	ns		ns	ns		ns	
Penisetum	F	3.38	3.61	3.61	4.35*	4.35*	4.27*	4.27*
clandestinum	P	ns	ns	ns				
Espeletia	F	45.58**	0.03	0.07	2.49*	0.31	1.60	2.25*
schultzii	P		ns	ns		ns	ns	
Hypericum	F	11.41*	3.50	0.35	4.21*	1.72	2.26*	0.51
laricifolium	P		ns	ns		ns		ns
Nassella	F	2.86	5.10*	2.91	0.29	0.54	1.26	1.36
linerifolia	P	ns	ns	ns	ns	ns	ns	ns
Aciachne	F	1.16	1.33	1.51	1.29	1.15	0.84	1.27
pulvinata	P	ns	ns	ns	ns	ns	ns	ns
Poa annua	F	15.19*	3.67	3.44	9.90**	9.89**	3.17*	2.9*
	P		ns	ns				

*Significant at $p < .05$.
**Significant at $p < .001$.

canum, and *Nassella linerifolia*, which are preferentially consumed by cattle but apparently do not have efficient mechanisms to resist this kind of disturbance. The shrub *Hypericum laricoides* also appears as a fragile species, probably due to trampling.

Data in Table 9.3 also suggest that the introduced species *Bromus carinatus* and *Penisetum clandestinum*, and in a lesser measure, the native species *Agrostis jahnii*, are good sources of forage. These species have a high palatability but do not suffer significant damage when grazed.

Some species that are not negatively affected by grazing are *Espeletia schultzii*, *Rumex acetosella*, *Paspalum pygmaeum*, *Stevia lucida*, and *Vulpia myurus*. *Espeletia schultzii* is rejected by cattle and is not sensitive to trampling. *Paspalum pygmaeum* evades grazing by its creeping habit. *Rumex acetosella*, a European weed, is not preferentially selected but, as

it is the most abundant species at the beginning of the succession, it represents an important percentage of animal diets. Nevertheless the growth form of this species (a rhizomatous herb) allows it to tolerate grazing.

Table 9.3 shows that the lack of response to grazing of the grasses, as a life-form, is due to a very contrasting response of the individual species. Caespitose grasses (such as *Penisetum clandestinum* and *Aciachne pulvinata*) and creeping grasses (such as *Poa annua* and *Paspalum pygmaeum*) are favored by grazing, probably because they are tolerant, as does *Penisetum clandestinum*, or because they have mechanisms of evasion, as does *Aciachne pulvinata*, a thorny prostrate species, or *Paspalum pygmaeum*, a very small species that concentrates its biomass in 1 or 2 cm above the soil surface. On the other hand, tall grasses (such as *Trisetum irazuense* and *Bromus carinatus*) are preferentially consumed.

DISCUSSION

The aboveground biomass after 8 to 9 years of succession (606 and 878 g m^{-2} on average for the grazed and ungrazed treatment, respectively) lies in the low part of the range reported by Hofstede (1995) for several Colombian páramos (735 to 3486 g m^{-2}). This difference can be explained considering that our plots are still in a relatively early successional phase and that Venezuelan páramos are drier and probably less productive than the Colombian ones. Our estimations are closer to the values reported by Ramsay and Oxley (2001) for a grassland páramo in Ecuador (800 g m^{-2}), but this kind of páramo does not have giant rosettes and shrubs, which account for an important part of the total aboveground biomass in late succession in Venezuela. In the same area of our study, Montilla et al. (2002) measured, using harvest techniques, a total aboveground biomass of 952 g m^{-2} in a 12-year successional plot, in the same order of magnitude as the figures obtained using the biovolume–biomass coefficients in the 9-year plots, a result that partially validates our method.

One of the most noticeable effects of grazing in this secondary succession was the reduction of plant aboveground biomass under extensive grazing. Other studies in the páramo confirm this trend; for example, Hofstede (1995) reported a total aboveground biomass of 3486 g m^{-2} in an undisturbed Colombian páramo, compared to 2567 g m^{-2} in a similar but extensively grazed area. This 26% reduction in biomass can be compared to the 30 to 40% reduction found in this study. Furthermore, Molinillo and Monasterio (1997) also reported an increase in biovolume of 52% after 1 year of grazing exclusion for a Venezuelan rosette–shrub páramo community.

A decrease in biomass is not an obvious or generalized response of vegetation to grazing. For example, in natural alpine grasslands in the Alps, Körner (1999) found that very extensive grazing had a positive effect on biomass due to the stimulation of nutrient cycling. An increase in biomass, production, richness, or other ecosystem properties under moderate disturbance is reported in many ecosystems and is explained by the intermediate disturbance hypothesis.

However, at a high intensity of disturbance, the normal response is a reduction in biomass due to the diminution of the LAI and of the photosynthetic capacity of the vegetation. In the case of the páramo, we have no evidences of a possible augmentation of biomass under very extensive grazing, but our data suggest that the deleterious effect occurs at relatively low grazing pressure.

The stocking rate of this experiment, estimated at 0.4 cows ha^{-1}, can be considered as high compared to the carrying capacity of 0.1 cows ha^{-1} reported by Molinillo and Monasterio (1997) for a drier rosette–shrub páramo and to the mean animal stocking rate of the valley estimated in 0.13 cows ha^{-1} (Pérez 2000). Nevertheless, we consider that the effective stocking rate was not as high as it seems. According to Schmidt and Verweij (1992), the daily dry matter intake by adult cows grazing in páramo ecosystems is around 11.8 kg. Assuming 12 h of grazing per day, the consumption in 1 h can be estimated in 980 g, or 9.8 g m^{-2} for a 100-m^2 plot. This figure corresponds to a consumption of 2 to 4% of the total aboveground bio-mass per month, which is not a very high proportion of the total biomass. In addition, the rotative grazing method (only 1 h each, 3 weeks) probably allows vegetation to recover between two consecutive events of grazing. Furthermore, our results suggest that the stocking rate before the installation of the fences was higher than under the grazing treatment (lower biomass and richness and higher percentage of bare soil in the initial intermediate plots compared with the final point of the grazed young plots). Further research is needed to arrive at more conclusive results concerning the response of biomass and other attributes to different intensities of grazing in this ecosystem.

The decrease in plant biomass observed under this stocking rate can have different consequences for the functioning and stability of the ecosystem, but an accurate prediction is difficult to make because of the existence of compensatory mechanisms and nonlinear processes in ecosystems. As a first approximation, we can predict that a reduction in biomass, mainly in the photosynthetic one, may produce a reduction in net primary production due to less light interception. Consequently, a reduced amount of necromass would be incorporated into the soil, leading to a progressive depletion in soil organic matter (SOM), which in turn can reduce the soil's capacity to retain water and nutrients. However, trends in SOM are only detectable in mean and long-term studies, due to the large amount initially present in these mountain soils. A decrease in SOM, together with the increase in the proportion of bare soil under grazing, could favor erosive processes and could negatively affect the sustainability of the system. This is the case documented by Podwojewski et al. (2002) in an intensive sheep-grazed páramo in Ecuador, where a dramatic decrease of SOM took place as a consequence of the decrease in plant biomass, the exportation of sheep manure, and by soil surface erosion favored by the increased percentage of bare soil and the mechanical effect of animals. Nevertheless, in our case, contrabalancing processes can also be taking place; for example, if the species with higher nitrogen content were consumed preferentially, litter quality and also decomposition rates would decrease, compensating for the reduction in the amount of litter. No definitive conclusions concerning the effect of grazing on ecosystem properties can be derived from this experiment, and further experimental work is necessary to test the consequences of changes in quality and quantity of plant biomass. Also, a simulation approach could give further insights.

Only small changes were detected in the life-form spectrum of the vegetation. Possible trends toward a reduction in the proportion of grasses and an increase in the proportion of giant rosettes were observed at the end of the experiment in the intermediate plots. In a Colombian páramo, Hofstede (1995) observed an increase in the density and biomass of giant rosettes (*Espeletia hartwegiana*) as a consequence of grazing and in absence of burning. However, Verweij and Kok (1992) reported different results for the same species. Vargas et al. (2002) reported a deleterious effect of grazing on *Espeletia killipii*. These inconsistencies can be related to different grazing intensities or to the interaction between grazing and burning.

The classification of the species in four life-forms is too general to analyze the structural

changes produced by grazing, as species belonging to the same life-form can present different responses. A more detailed classification is necessary to assess the effect of grazing on functional types of plants, which should take into consideration more specific traits influencing the way plants are affected by grazing (palatability, mechanical fragility, grazing defenses, etc.). This approach could be interesting when comparing different sites in terms of ecological equivalents. Using a more detailed classification of grasses, the diminution of tall and short grasses, and the increase in creeping and rhizomatous grasses can be identified as grazing effects. Other authors also reported the replacement of tall and tussock grasses by a short carpet grass vegetation in páramo ecosystems (Verweij and Budde 1992; Hofstede 1995; Podwojewski et al. 2002).

Independently of grazing, the abundance of giant rosettes and shrubs increases in the succession and, consequently, the offer of forage decreases. Early plots seem to be more suitable for grazing than intermediate and late ones, and decisions by farmers concerning the duration of the fallow period probably take this aspect into consideration.

In this experiment, grazing produced a slight but consistent reduction in plant diversity. In literature, different responses of plant richness have been reported, depending on the intensity of grazing and on the characteristics of the species that conform the community. Körner (1999) reported, for an alpine grassland in the Alps, a positive effect of extensive grazing, whereas Podwojewski et al. (2002) reported a diminution of the number of species under intensive grazing in the páramo. In our experiment, the abundance of some of the species was dramatically reduced, but they did not disappear completely, explaining the small changes observed in plant richness. A stronger effect would be expected with higher grazing pressures.

The observed reduction in biodiversity is in accordance with the hypothesis that in poor environments such as the páramo, grazing reduces plant richness, whereas in rich environments, plant richness can be increased (Milchunas and Lauenroth 1993). Grazing can promote plant biodiversity by balancing competitive interactions between species, reducing competition for light, promoting dispersion, and creating more recruitment opportunities for subordinate species (Berendse 1985; Milchunas and Lauenroth 1993; Bakker 2003). For example, Bakker (2003) found, in a grassland in the Netherlands, a negative correlation between the height of the vegetation and plant richness and a positive correlation between richness and the percentage of bare soil, indicating that in this environment, competition for light and the existence of opportunities for establishment are the main factors controlling plant richness. However, in the páramo, the situation seems to be the opposite. Plant richness is positively correlated to plant biomass and negatively correlated to the amount of bare soil, indicating that other factors are controlling plant biodiversity. Light competition does not appear to be an important factor controlling diversity, probably because even without grazing, there is a significant proportion of bare soil. In general, the effect of herbivory on plant species diversity is mainly determined by the response of the subordinate species. In the páramo, grazing seems to enhance the abundance of the dominant species that are less palatable and to promote the extinction of some of the subordinate species that are more palatable. On the other hand, regeneration niches do not seem to be limiting, and herbivores are not expected to increase the opportunities of colonization. In this páramo, succession grazing seems to promote extinction without favoring colonization.

At the level of individual species, the effect of grazing in this old-field succession is clear; it promotes some species, damages others, and does not affect a third group. This differential effect can be related to the palatability of the species, their adaptations to tolerate or avoid herbivory, and their mechanical fragility. According to Körner (1999), the impact of grazing on alpine vegetation is much more severe by trampling than by direct consumption due to the extreme sensitivity of shrub communities. The sensitivity of *Hypericum laricifolium* observed in this study and by Molinillo and Monasterio (1997), and explained by the fragility of its branches, supports this affirmation.

The positive effect of grazing on some páramo species, mainly creeping and prostrate life-forms such as *Aciachne pulvinata* and *Lachemilla orbiculata*, is widely recognized in the literature (Verweij and Budde 1992; Hoftede 1995). Also, the positive impact on some introduced species, such as *Poa annua*, *Taraxacum officinaris*, and *Rumex acetosella*, is reported in other studies (Velázquez 1992; Verweij and Budde 1992; Pels and Verweij 1992; Podwojewski et al. 2002).

The positive response to grazing of the introduced species and their strong dominance in early succession, documented in a previous study by Sarmiento et al. (2003), indicate that current management favors the invasion of the páramo by ruderal species. These probably have more adaptations to herbivory, because they evolved in other environments in which this kind of disturbance is common.

As a conclusion, it can be said that at the moderate stocking rate used in this experiment, some negative effects of grazing were detected, but they are less important than expected, considering the short history of grazing of this ecosystem. A certain level of grazing is probably compatible with the restoration of the páramo ecosystem without severely threatening plant biodiversity. Nevertheless, additional studies are necessary to evaluate more accurately the impact of grazing on SOM and on the long-term stability of the system, and to determine the appropriate stocking rate.

SUMMARY

An exclosure experiment was carried out to analyze the effect of grazing on plant biomass and vegetation composition during secondary succession in a Venezuelan páramo. Four young plots (never in fallow) and four intermediate plots (5 years in fallow) of 200 m² each were fenced and divided into two parts; one was excluded and the other was grazed during 4 years using a stocking rate equivalent to 0.4 cows ha⁻¹. The vegetation was sampled twice a year using the point intercept method. At this stocking rate, grazing produced a reduction of 30 to 40% of total aboveground plant biomass. Vegetation height was reduced in the same order, and the percentage of bare soil increased significantly. Despite the reduction in aboveground biomass, the life-form spectrum of the vegetation was only slightly affected, indicating that grazing impact was almost homogeneous for the different life-forms. A possible augmentation in the percentage of giant rosettes in the intermediate plots was detected at the end of the experiment in the grazed treatment. This is probably due to the total rejection of rosettes by cattle and to the low impact of trampling on this life-form. Grazing also reduced plant species richness slightly, but significantly, and a more severe effect could be expected from a higher grazing pressure. The response of individual species was very clear. An index of damage allowed classifying them into the following: damaged by grazing (e.g. *Trisetum irazuense*, *Nassella linerifolia*, *Brachypodium mexicanum*, etc.), unaffected (e.g. *Acaena elongata*, *Baccharis prunifolia*, *Lupinus meridanus*, etc.), and positively affected (e.g. *Aciachne pulvinata*, *Espeletia schultzii*, *Penisetum clandestinum*, etc.). Some explanations of the individual responses are advanced based on the preferences by cattle, plant architecture, and sensitivity to trampling.

ACKNOWLEDGMENTS

This research was supported by the International Foundation for Science (Grant C/2668) and by the EU project TROPANDES (ICI8-CT98-0263). The author received a fellowship from the Wageningen Institute for Environment and Climate Research for the redaction of the paper. Thanks to A. Escalona, N. Marquez, A. Olivo, C. Molina, A. Berg, and B. Briceño for their participation in the fieldwork and in botanical identifications. J.K. Smith and L.D. Llambí also helped in the selection and installation of the plots. Specials thanks to Cristobal, Alfonso, Gregorio, Sra. Rosa and Sra. Hilbina of the páramo de Gavidia for handling the animals that grazed the plots.

References

Abadín, J., González-Prieto, S.J., Sarmiento, L., Villar, M.C., and Carballas, T. (2002), Successional dynamics of soil characteristics in a long fallow agricultural system of the high tropical Andes, *Soil Biology and Biochemistry*, 34(11): 1739–1748.

Bakker, E. (2003), Herbivores as Mediators of Their Environment: the Impact of Large and Small Species on Vegetation Dynamics, Ph.D. thesis, Wageningen University, Wageningen.

Berendse, F. (1985), The effect of grazing on the outcome of competition between plant species with different nutrients requirement, *Oikos*, 44: 35–39.

Davidson, D.W. (1993), The effects of herbivory and granivory on terrestrial plant succession, *Oikos*, 68: 23–35.

Gough, L. and Grace, J.B. (1998), Herbivore effects on plant species diversity at varying productivity levels, *Ecology*, 79: 1586–1594.

Greig-Smith, P. (1983), *Quantitative Plant Ecology*, University of California Press, Berkeley.

Hofstede, R.G.M. (1995), Effects of Burning and Grazing on a Colombian páramo Ecosystem. Ph.D. thesis, University of Amsterdam, Amsterdam.

Hofstede, R.G.M., Modragon, M.X., and Rocha, C.M. (1995), Biomass of grazed, burned and undisturbed páramo grasslands, Colombia, Aboveground vegetation, *Artic and Alpine Research*, 27: 1–12.

Huntly, N.J. (1991), Herbivores and the dynamics of communities and ecosystems, *Annual Review of Ecology and Systematics*, 22: 477–503.

Körner, C. (1999), *Alpine Plant Life: Functional Plant Ecology of High Mountain Ecosystems*, Springer-Verlag, Berlin.

Milchunas, D.G. and Lauenroth, W.K. (1993). Quantitative effects of grazing on vegetation and soils over a global range of environments. *Ecological Monograph*, 63: 327–366.

Milchunas, D.G., Sala, O.E., and Lauenroth, W.K. (1988), A generalized model of the effects of grazing by large herbivores on grassland community structure. *American Naturalist*, 132: 87–106.

Molinillo, M. and Monasterio, M. (1997), Pastoralism in páramo environments: practices, forage, and impact on vegetation in the Cordillera de Mérida, Venezuela. *Mountain Research and Development*, 17(3): 197–211.

Molinillo, M. and Monasterio, M. (2002), Patrones de vegetación y pastoreo en ambientes de páramo, *Ecotropicos*, 15(1): 19–34.

Monasterio, M. (1980). Las formaciones vegetales de los Páramos venezolanos, in Monasterio, M. (Ed.), *Estudios Ecológicos en los Páramos Andinos*, Ediciones de la Universidad de los Andes, Mérida, pp. 93–159.

Montilla, M., Monasterio, M., and Sarmiento, L. (2002), Dinámica sucesional de la fitomasa y los nutrientes en parcelas en sucesión-regeneración en un agroecosistema de páramo, *Ecotropicos*, 15(1): 75–84.

Pacala, S.W. and Crawley, M.J. (1992), Herbivores and plant diversity, *American Naturalist*, 140: 243–260.

Pérez, R. (2000), Interpretación ecológica de la ganadería extensiva y sus interrelaciones con la agricultura en el piso agrícola del Páramo de Gavidia, Andes venezolanos, M.Sc. thesis, Postgrado de Ecología Tropical, Facultad de Ciencias, Universidad de los Andes, Mérida, Venezuela.

Podwojewski, P., Poulenard, J., Zambrano, T., Hofstede, R. (2002), Overgrazing effects on vegetation cover and properties of volcanic ash soil in the páramo of Llanguahua and La Esperanza (Tungurahua, Ecuador), *Soil Use and Management*, 18: 45–55.

Ramsay, P. and Oxley, R.B. (2001), An assessment of aboveground net primary productivity in Andean grasslands of central Ecuador, *Mountain Research and Development*, 21: 161–167.

Sarmiento, L., Smith, J., and Monasterio, M. (2002), Balancing conservation of biodiversity and economical profit in the agriculture of the high Venezuelan Andes: are fallow systems an alternative? In Körner, C. and Spehn, E. (Eds.), *Mountain Biodiversity — A Global Assessment*, Parthenon, pp. 285–295.

Sarmiento, L., Llambí, L.D., Escalona, A., and Marquez, J. (2003), Vegetation patterns, regeneration rates and divergence in an old-field succession of the high tropical Andes, *Plant Ecology*, 166: 63–74.

Schmidt, A.M. and Verweij, P.A. (1992), Forage intake and secondary production in extensive livestock systems in páramo, in Balslev, H. and Luteyn, J.L. (Eds.), *páramo: An Andean Ecosystem under Human Influence*. Academic Press, London, pp. 197–210.

Smith, J.K. (1995), Die Auswirkungen der Intensivierung des Ackerbaus im Páramo de Gavidia — Landnutzungswandel an der oberen Anbaugrenze in den venezolanischen Anden, Diplomarbeit, University of Bonn, Germany.

Van Oene, H., van Deursen, M., and Berendse, F. (1999), Plant-herbivore interaction and its consequences for succession in wetland ecosystems: a modeling approach, *Ecosystems*, 2: 122–138.

Vargas, O., Premauer, J., and Cardenas, C. (2002), Efecto del pastoreo sobre la estructura de la vegetación en un páramo humedo de Colombia, *Ecotropicos*, 15: 35–50.

Velázquez, A. (1992), Grazing and burning in grassland communities of high volcanoes in Mexico, in Balslev, H. and Luteyn, J.L. (Eds.), *páramo: An Andean Ecosystem under Human Influence*, Academic Press, London, pp. 231–241.

Verweij, P.A. and Budde, P.E. (1992), Burning and grazing gradients in páramo vegetation: initial ordination analyses, in Balslev, H. and Luteyn, J.L. (Eds.), *páramo: An Andean Ecosystem under Human Influence*, Academic Press, London, pp. 177–195.

Verweij, P.A. and Kok, K. (1992), Effects of fire and grazing on *Espeletia hartwegiana* populations, in Balslev, H. and Luteyn, J.L. (Eds.), *páramo: An Andean Ecosystem under Human Influence*, Academic Press, London, pp. 215–229.

Wagner, E. (1978), Los Andes Venezolanos, arqueología y ecología cultural, *Ibero-Amerikanisches Archiv*, Neue Folge, Berlin, 4(1): 81–91.

10 Vegetation and Grazing Patterns in Andean Environments: A Comparison of Pastoral Systems in Punas and Páramos

Marcelo Molinillo and Maximina Monasterio

INTRODUCTION

The Spanish conquest generated profound transformations in Andean cultures and environments. The introduction of new plant and animal species and the implementation of new technologies had a variety of impacts on the environment and on traditional land use practices (Thomas 1979; Little 1984). One of the consequences was the change in traditional grazing patterns, which had developed during the millennia of interactions of human populations with mountain environments. The success of these strategies was based on a thorough knowledge of these environments and of the genetic makeup of domesticated Andean animals, resulting from thousands of years of selection (Brush 1984).

In the punas of the Central Andes, the most important change was the replacement of native llamas and alpacas by animals introduced from Europe (mainly ovines and bovines). In some cases, the new animals were incorporated into the management systems within the framework of ancient practices and ideologies (Merlino and Rabey 1983; Rabey and Merlino 1988). However, in extreme highland areas, the lack of adaptation of the introduced livestock to the demanding climatic conditions constituted a barrier for the replacement of domesticated camelids (Wheeler 1988). Even so, the Spanish conquest was the starting point of an expansion process of ovines and bovines in the Central Andes that has been linked to profound impacts on the original vegetation and soils, and on the community organizations and grazing strategies (Flores Ochoa 1988a, b).

The Spanish conquest also introduced cattle into the páramo of the northern Andes, which was unfamiliar with large herds grazing, as in the puna region. This process was linked to the introduction of new crops and cereal technologies (Monasterio 1980). However, in a few hundred years, new grazing patterns were developed, strongly linked to the agricultural calendar with a typical agropastoral strategy (Molinillo 1992; Molinillo and Monasterio 1997).

The use of natural resources with "traditional" animals and technologies, adapted to local environmental conditions, has generally been associated with stable production systems of high productivity (Little 1984), whereas the use of exotic animals in the Central Andes has been linked to overgrazing. Even though this has been considered an ancient problem (Browman 1974), overgrazing produced large-scale soil erosion problems and drastic changes in vegetation structure, mainly during the postconquest era (Eliminar las Citaas de Koford 1957; Mann 1968 cited by Thomas and Winter-halder, 1976).

The exclusive use of sheep in the altiplano region has produced changes in the vegetation, related to the local extinction of palatable pasture species with high nutritional value (Millones 1982). This has been explained in terms of the inability of introduced animals to adapt to the characteristics of Andean ecosystems (Flores Ochoa 1979; Millones 1982). Similarly, toward increasing dominance of grasses and shrubs of a low forage value and a low vegetation cover in the punas of northwestern Argentina have been attributed to overgrazing by exotic animals (Solbrig 1985).

The magnitude of the environmental impact of new grazing regimes on the páramos and the punas depends, to a large extent, on the spatial patterns of livestock movement, which have been, in turn, controlled by the spatial and temporal distribution of natural forage and the animals' ability to utilize pastoral space. The efficient use of pastoral space depends on the grazing strategy used for exploiting the forage resources of a given region. This pastoral strategy includes the daily and seasonal patterns of animal movements on pastoral space and the animals' requirements. Efficiency in the use of pastoral space has direct environmental consequences. A low space-use efficiency leads to the concentration of livestock in the few places that meet animal requirements and/or intentional alteration of vegetation to increase the surface area of optimal sites.

In this chapter, pastoral strategies, space-use efficiency, and their relationship with dominant vegetation patterns are compared in three Andean systems: alpacas and llamas in the punas of the Ulla-Ulla altiplano (Bolivia), ovines in high-altitude grasslands of the Cumbres Calchaquies (northern Argentina), and cattle in the páramos of the Cordillera de Mérida (Venezuela).

METHODS

To analyze the use of pastoral space, an attribute known as the *capacity* of a given territory for the development of a specific activity is calculated (Cendrero 1982; Gómez Orea 1992). To evaluate a territory's capacity for grazing, geographic information systems (GISs) and multi-criteria evaluation (MCE) techniques were used. MCE techniques are based on the use of matrices of alternative criteria and the assignment of weights to each criterion, developing a hierarchy. The criteria were expressed as thematic maps whose units represent the different alternatives offered by a given territory (Eastman et al. 1995; Barredo 1996). Then, the grazing capacity was calculated for each region through a linear weighted sum. In all cases, the criteria to assess the maximum grazing capacity were: (1) areas with the highest forage quality, (2) areas with the shallowest slopes, (3) areas closest to water sources, and (4) areas with the highest accessibility for grazing. Based on the following criteria, thematic maps were drawn: forage supply, slopes, distance to water sources, and distance to human settlements. For mapping, different value scales and weight hierarchies were used, depending on the conditions of each region and the characteristics of the animals. In each case, pastoral space was expressed as the percentage of the total area where the highest values of grazing capacity are found.

Forage supply maps were obtained from vegetation maps using the following equation:

$$OF = \cfrac{\cfrac{[(WP_1 \times FC_1) + (WP_2 \times FC_2)] \times [TFC]}{SF}}{100}$$

where, WP_1 is the weight for palatability in scale 1, WP_2 is the weight for palatability in scale 2, FC_1 is the forage cover of palatability 1, FC_2 is the forage cover of palatability 2, SF is an index of the seasonality of forage, and TFC is the total forage cover for each vegetation unit.

Information on palatability was obtained from different sources: preference sampling of animals in the Venezuelan páramos (Molinillo 1992; Molinillo and Monasterio 1997); specific secondary sources for Ulla-Ulla in Bolivia (Bárcena, 1988; La Fuente et al., 1988), and Tacanas in Argentina (Molinillo 1988, 1990, 1993) and general secondary sources for the puna region (Tapia and Flores Ochoa, 1984; Ruiz and Tapia, 1987).

Vegetation maps were obtained by processing a Landsat 6 image from 1996 (Venezuela) and a Spot image from 1986 (Argentina) and by simplifying the vegetation map for the Ulla-Ulla region (Seibert 1994). In every case, field controls were taken through random sampling of each vegetation unit using 0.25-m^{-2} quadrats for swards and 1-m^{-2} quadrats for shrubland and grassland areas. In each quadrat, species cover, forage supply, and other variables such as dry matter, bare soil, soil humidity, and stoniness were measured. To analyze the relationship between animal management and the vegetation structure, multivariate analysis (principal components analysis, PCA) was used. The analysis combined the data from field quadrats with grazing distribution data. Slope maps for the three regions were obtained from digital elevation models derived from the official cartography at a 1:100,000 scale (Argentina and Venezuela) and from the 1:50,000 topographic map of Ulla-Ulla (Bolivia) by Seibert (1994). The water sources and human settlements distance maps were drawn by calculating buffer zones between hydrographic elements (rivers and lagoons) and human settlements.

To assign weights for factors and alternatives, matrices were built using the analytic hierarchies method (Saaty 1980; Barredo 1996). Here, weights are assigned based on pair comparisons, so that a relative weight is assigned to each factor in relation to all other factors. This method provides more realism than a simple procedure without weights. By assigning weights to each factor in the different regions, the influence of these factors on the particular grazing systems in that region could be evaluated.

RESULTS AND DISCUSSION

GRAZING AREAS AND ANIMAL MANAGEMENT

The Ulla-Ulla altiplano region (above 4000 masl) in the Bolivian puna is characterized by a cold and dry climate with strong winds (dry and very cold). Rainfall is concentrated in the summer months (less than 500 mm per year), showing marked interannual variability. The vegetation shows marked differences between *bofedales* (irrigated areas) and dry plains. In the inundated bofedales, species with high forage

value dominate (*Distichia* spp., *Lachemilla* spp., *Plantago* sp., etc.). In contrast, the dry plains are covered by relatively isolated bunch-grasses (*Stipa* sp., *Festuca* sp., *Calamagrostis* sp., etc.) and cushions (*Arenaria* sp., *Aciachne* sp., etc.) growing on soils of limited pedogenetic development. Hence, these areas are only used during the wet season. The transition zones between wet and dry areas create a complex vegetation mosaic controlled by seasonal water availability. The Aymara communities utilize the bofedale and its surrounding areas for herding llamas and alpacas. Each family has an average of 200 animals, with twice the number of alpacas than llamas. They also herd cattle and sheep, but in fewer numbers. Alpaca herds graze in the bofedales most of the year, separated in groups of males and females. Animal loads in these sites can reach more than two alpaca units per hectare. The llamas graze the plains and neighboring sierras (highlands), particularly during the wet season (summer) when they benefit from the growth of seasonal pastures. During the dry season (winter), grazing of female llamas and alpacas is mainly on the irrigated bofedales. Those communities without access to irrigated bofedales must take their animals to natural pastures in high-mountain areas for green forage vegetation. In a few cases, forage (mostly oats) is cultivated in small plots near the farmers' houses and then stored.

In the Cumbres Calchaquíes region of northwestern Argentina, the high belts (between 2500 and 4000 masl) are characterized by a climate with marked seasonality: temperate and humid summers (December to March) and cold, dry winters (June to September). Vegetation types include high-altitude grasslands (dominated by *Festuca* spp.), swards in the upper limits of *Alnus acuminata* forest, and both puna and prepuna shrublands. The traditional grazing system is based on seasonally migratory ovine and bovine cattle. Inter-Andean valleys above 3500 masl that are covered with grasslands are used during the wet season, whereas slopes with pastures and swards in the forest limits (between 2500 and 3000 masl) are used during the dry season. Within each vegetation belt, a temporary settlement is established, allowing better control of animals during grazing. With the arrival of summer, animals are concentrated in the pastures and swards of the high inter-Andean valleys, areas where animal loads can reach two sheep units per hectare in 4 to 5 months. Low temperatures and dry conditions have made herders look for better forages in the low wooded slopes, where they spend the rest of the year.

In the Sierra de la Culata region in Cordillera de Mérida (western Venezuela), an agro-pastoral system (potato/cereal agriculture and bovine grazing) has been developed in the páramos in a cold and humid climate (above 3500 masl). This environment shows a less discontinuous rainfall distribution pattern than the puna, though, there is a clearly defined dry season. In the slopes of the Andean belt, between 3700 and 4100 masl, the dominant vegetation is an *Espeletia schultzii* rosette–shrubland community. Higher up, in the high-Andean belt, giant rosettes of *E. timotensis* dominate, interspersed with areas of sparse vegetation cover in the periglacial desert. Both in lower and high-Andean belts, swards of Gramineae and Cyperaceae in the valley bottoms and slope landings are distributed in patches along watercourses. These sites are the main grazing areas. Animals move between sward patches, establishing circuits in which vegetation can regenerate after short periods of intensive grazing. Here, animal loads can reach 0.4 bovine units per hectare. In contrast to the situation in Bolivia and Argentina, there are no human settlements in these grazing areas. Animals are taken above the agricultural belt to the páramo areas several hours away (by horse or on foot) and controlled once or twice each month. In the agricultural zone (3000 to 3700 masl), animals find supplementary forage in fallow areas and in the stubble of potato and wheat crops, or in plots with oats or grass. Animal movement between vegetation belts is strongly linked to the agricultural calendar.

USE OF PASTORAL SPACE

Grazing capacity maps in the three regions indicate that, in general, the areas with the highest forage supply correspond to the high-capacity areas (Figure 10.1). Bofedale sites located near human settlements between watercourses on shallow slopes show the highest-capacity values in the Ulla-Ulla region. This results in a pastoral space of 60% of the total surface area for llamas and only 40% for alpacas. In the Tacanas region (Argentina), the high-capacity class is occupied by swards on slopes and the summits of the Tacanas River sources, where temporary and permanent settlements for grazing are located. In this case, pastoral space corresponds to 25% of the total surface area. In the Sierra la Culata region (Venezuela), the highest-capacity class is occupied by the swards, rosette–swards, and the marshes in shallow slope areas in the valley bottom, crossed by watercourses that are less than 5 km away from human settlements. These places correspond to only 20% of the total surface area.

In the Ulla-Ulla altiplano, a more limited pastoral space for alpacas can be explained by their requirement for tender and green fodder, which can only be found in the bofedales. Even though the alpacas also eat the dry and hard fodder characteristic of the puna (such as tussocks of *Festuca* sp. and *Stipa* sp.), they cannot eat it continuously without the risk of malnutrition and negative effects on the quality of their delicate wool (Palacios Ríos 1988). Alpacas are taken out of the bofedales and swards only during the wet season. In some cases, this is done to remove them from flooded or humid areas to avoid diseases. Consequently, their permanent corrals are located at the border of the bofedales, and the size of the bofedales determines the number of alpacas that a family or a community can maintain (as these are the only means of survival for the animals during the dry season). The situation with llamas is different. Their wide feeding range and their ability to digest the dominant hard fodder in the puna slopes (Genin et al. 1994) allows them to use the pastoral space better. Hence, there are only two strategies available to widen the alpaca's pastoral space: to look for natural bofedales on the steep areas (dependent on glaciers) and to increase the size of the bofedales through irrigation technology.

In the Sierra de la Culata (Cordillera de Mérida), limitations, as to the use of the páramo pastoral space for bovines, have meant that animals are restricted almost exclusively to valley bottom areas with sward patches of high forage

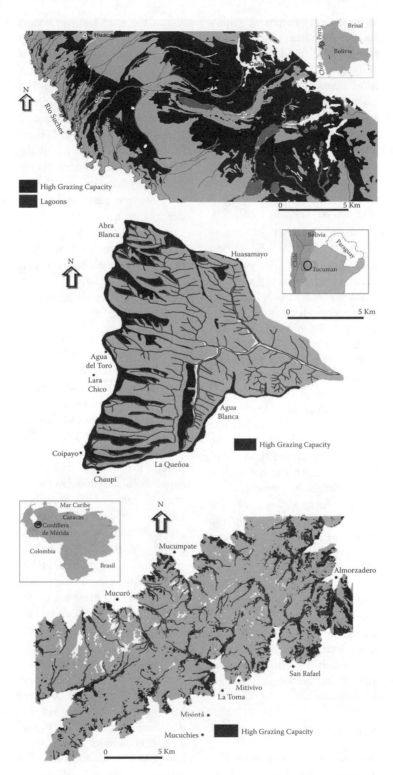

FIGURE 10.1 Areas with high grazing capacity in the study area. Top: Punas of the Ulla-Ulla altiplano (Bolivia). Center: forest and highland pastures of the Tacanas watershed (Argentina). Bottom: páramos of the Sierra de la Culata (Venezuela).

quality. They can graze very extensively only on the shrubland and rosette areas that dominate most slopes between 3700 and 4100 masl. In the páramos, the forage quality limitations for cattle are much more severe than in the northern Argentinean highlands. Here, the mosaic of sward patches and marshes with high grazing capacity is used intensively by cattle. However, supplemental food from the agricultural belt is almost indispensable to sustain cattle grazing in the páramos, especially during the dry season when their rangeland forage capacity decreases abruptly (Molinillo and Monasterio 1997).

In the Tacanas, the small amount of acreage available for bovine and ovine cattle grazing can be explained by their limited ability to consume the dominant forage (Molinillo 1990, 1993). To widen pastoral space, shepherds use several practices. Intentional fire setting, particularly in the pastures of *Festuca* sp. and *Stipa* sp., is a common practice. The objective is to increase the ratio of tender to dry fodder so that the animals can use (at least temporarily) vast pastureland extensions, which would not be accessible in the mature state of vegetation. Another common strategy used is to change pasture areas daily and seasonally. Both of these practices are aimed at maximizing the utilization of forage resources, given the constraints imposed by grazing animals. For example, the shepherds of the Lara River valley (near the Tacanas River watershed) transit during the summer daily by several routes with their animals, which they alternate or temporarily change within their pastoral spaces so as to maximize availability of the different vegetation types available (Molinillo 1988). Seasonal animal movements also allow for the use of forests, highland pastures, puna shrublands, and prepuna vegetation, widening pastoral spaces considerably. Even so, animals still concentrate in the best forage areas, located near watercourses on shallow terrains.

GRAZING AND VEGETATION PATTERNS

In all three cases discussed in the preceding text, the size of the pastoral space has been strongly influenced by vegetation structure and its composition, particularly in those areas that maintain the highest grazing loads. Figure 10.2

shows the four groups in the Ulla-Ulla region classified according to the relation between the floristic composition and the environmental and management variables. Within this organizing pattern, the most meaningful variable is the soil water, which forms a gradient in axis I, and the fodder quality and grazing density, which form a gradient in axis II. The flooded bofedales (group 1) with high grazing density and fodder quality, the scarce bofedales (group 2) with fodder quality from intermediate to poor condition, the transition pastures (group 3) with low fodder quality and little grazing density, and the vegetation of plains (group 4) with high proportion of bare soil and little water supply on the ground are each characterized by a floristic composition, mostly derived from the interaction between environmental conditions and grazing frequency.

The floristic composition in the bofedales is related to their high soil water levels and moderate to high animal densities. Under these conditions, species such as *Distichia filamentosa*, *D. muscoides*, *Lachemilla diplophylla*, and *Plantago tubulosa* dominate. The fact that in bofedales where grazing has been excluded, other species such as *Calamagrostis ovata* and *Juncus arcticus* dominate over *Distichia* spp. supports this conclusion (Seibert 1994). At the other extreme of the humidity gradient, in dry flatland areas with relatively high grazing levels, the dominant species are *Stipa ichu*, *S. brachyphylla*, *Bougueria andicola*, *Cardionema* sp., *Paronychia andina*, and *Selaginella peruviana*. Human disturbance and grazing has been linked with the presence of *Stipa ichu* (Seibert 1994). These are low-carrying-capacity areas or areas under regeneration that were probably subjected to llama grazing. The rest of the sites are transitional forms between dry flatlands and bofedales, both in terms of water availability and grazing levels. When the water supply to the bofedales is neglected (through site abandonment or deficiencies in the maintenance of irrigation channels, for example), their floristic composition changes to plant formations dominated by *Azorella diapensioides*, *Plantago rígida*, *Werneria apiculata*, *Festuca andicola*, *Gentiana sedifolia*, and *Lachemilla pinnata* in the humid areas, and *Aciachne pulvinata*, *Mul-*

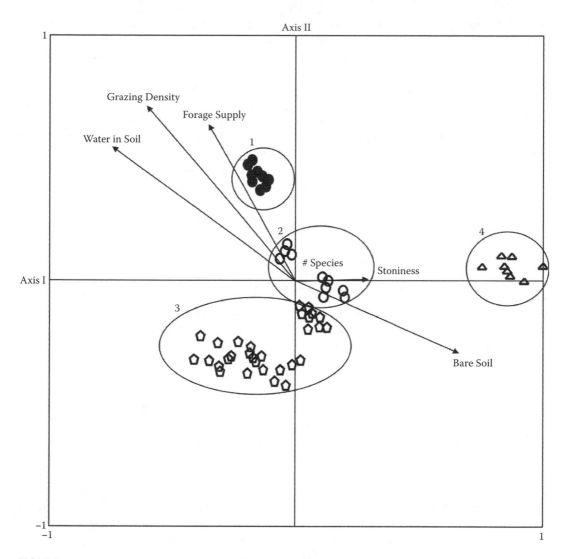

FIGURE 10.2 Ordination diagram (PCA) of 60 sampling quadrats in terms of their floristic composition in the Ulla-Ulla altiplano (Bolivia). On Axis I is a gradient of bare soil and soil water. On Axis II is a gradient of forage quality and grazing density. (()) high and very high for alpacas and llamas. *Group 1:* bofedales inundated of *Distichia filamentosa, Distichia muscoides, Lachemilla diplophylla, Plantago tubulosa, Caltha sagitata. Group 2:* abandoned bofedales dominated by *Calmagrostis* sp., *Azorella diapensioides, Festuca andicola, Plantago rigida, Werneria apiculata, Gentiana sedifolia, Geranium sessiliflorum, Ranunculus cymbalaria. Group 3*: transition zones dominated by *Festuca andicola, Werneria apiculate, Lachemilla pinnata, Eleocharis tucumanesis, Scirpus* sp., *Aciachne pulvinata, Mullenbergia* sp., *Calandrinia acaulis. Group 4*: Plains dominated by *Stipaichu, S. brachphylla, Bourgueria andicola, Cardionema* sp., *Paroychia andina, Selaginella peruviana.*

lenbergia sp., *Scirpus* sp., and *Calamagrostis minima* in the dry areas.

In the extensively grazed areas of northwestern Argentina, grazing and (almost certainly) fire have created mosaics of low pastures and swards in the forest belt, highland pastures, and prepuna shrublands. In Figure 10.3, the organizing pattern classifies seven groups according to their floristic function and altitude gradient (axis I), and to the pasture quality and grazing density (axis II). Taking into account only the influence of grazing upon the vegetation, the features to be highlighted are high-quality fodder and grazing density (group 4), the patches of higher-altitude swards and marsh (group 5), and the low grasses on slopes with floristic composition due to grazing pressure and availability of water on the ground (group 6).

These grazing-induced swards are dominated by the preferred forage species, resistant to moderate to high seasonal animal loads. In the Tacanas River watershed (below 3200 masl), permanently humid soil and a high grazing density of ovines and bovines most of the year, are associated with swards of *Alchemilla pinnata*, *Hypoxis* sp., *Juncus* sp., *Tripogon spicatus*, *Bromus unioloides*, *Vulpia bromoides*, *Hypochoeris meyeniana*, and *Paspalum pygmaeum*. In the high valleys of the Lara River watershed (3200 to 3800 masl) moderate to high ovine loads during the summer are associated with swards of *Alchemilla pinnata*, *Eleocharis* sp., and *Koeleria permollis* along the river banks and the low pastures of *Stipa uspallatensis*, and *Festuca lilloi* on the humid slopes and the summit surfaces.

Finally, in the páramos of the Sierra de la Culata region in Venezuela, the pattern in Figure 10.4. shows five major groups according to the floristic composition, water in the soil, fodder quality (axis I), and grazing density (axis II). The grasses and the swamps (group 1), flooded most of the year, have very good fodder quality but low grazing density because of difficult accessibility. The degraded swards (group 3) with colonizers that, in the past, were subjected to high grazing pressure and showed little water provision for their recovery, and the continuous grasses (group 5), with high fodder quality, that get flooded only part of the year, and have high and seasonal grazing density. The moderate to high seasonal bovine loads that concentrate in shallow valley bottoms near the water sources are associated with swards of *Calamagrostis mulleri*, *Carex albolutescens*, *Lachemilla* spp., *Muehlenbergia ligularis*, and *Agrostis breviculmis* (group 5). The presence and frequency of animals on these swards is strongly linked to the agricultural calendar and the supplementary forage offered by the agricultural belt. The establishment of an equilibrium between water availability (allowing growth after grazing) and the type and duration of animal loads favor the maintenance of good-quality forage species in swards. A drastic decrease in grazing in the marshy inundated areas changes species composition toward the dominance of *Carex albolutescens*, *C. humboldtiana*, and *Juncus* sp. (group 1). Areas that remain inaccessible for several years can favor the establishment of tussock grass species, such as *Festuca tolucensis* and *Calamagrostis ligulata* (group 2). On the contrary, swards with limited water availability and high animal accessibility are dominated by native weeds and exotics such as *Rumex acetosella*, *Aciachne pulvinata*, *Acaulimalva acaule*, and *Geranium* spp. (group 3). Low forage supply and abundant cattle dung (past grazing indicator) determine the low present-grazing loads in these units. In all three cases analyzed, no significant relationship has been found between species richness and the vegetation units under grazing. Table 10.1 shows correlation of the variables with the first three axes of the ordination analyses. In no case does the variable number of species reach significant values and, as can be seen in Figure 10.2 to Figure 10.4, this variable does not correlate with the grazing-density variable. In the Tacanas region (Argentina) and La Culata (Venezuela), the impact of moderate to heavy grazing on low pastures and swards is associated with an increase in the number of native weeds and exotic species. Hence, species richness tends to be maintained at similar levels as that of the other vegetation units.

THE USE OF PASTORAL SPACE AND ITS GRAZING IMPACT

The pastoral space of the livestock farming systems analyzed is intimately related to the impact of grazing on vegetation structure. In the Ulla-Ulla altiplano, the wide pastoral space of llamas has allowed the development of livestock farming activities with minor impacts on the dominant vegetation (at least in recent times), which is in sharp contrast with the Andean systems based on exotic animals. The establishment of bofedales through irrigation has allowed the maintenance of species with high nutritional value and resistance to grazing loads. Here, changes in vegetation composition

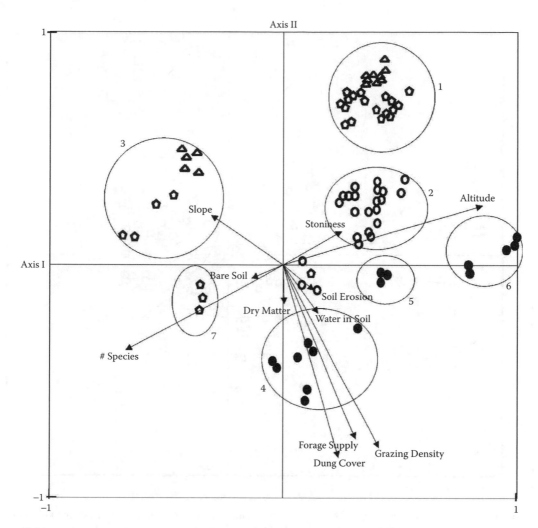

FIGURE 10.3 Ordination diagram (PCA) of 85 sampling quadrats in terms of their floristic composition for vegetation units of *Alnus* forest and highland pastures between 2000 and 3800 m asl at the Cumbres Calchaquies (Tucuman, Argentina). Axis I is a gradient of height, Axis II is a gradient of forage quality, grazing density and dung quantity. *Group 1:* shrublands dominated (>3000 masl). *Parastrephia phylicaeformis, Chiliotrichiopsis keidelli, Adesmia horridiuscula, Baccharis polifolia, Fabiana densa, Senecio argophylloides* y herbs like: *Glandularia incise, Paronichya setigera, Bidens* sp. *Group 2:* grasslands and shrubland-grasslands (>3000 m asl). *Stipa leptostachya, Festuca weberbaueri, F. hieroyimii, Parastrephia phylicaeformis, Chiliotrichiopsis keidelli, Solanum* sp., *Astragalus* sp., *Geranium* sp., *Plantago* sp., *Tarasa* sp., *Perezia* sp. *Group 3:* Forests and shrublands (>3000 m asl). *Siegesbeckia jorulensis, Calamagrostis poligama, Bidens andicola Duchesnea* sp., *Pteris* sp., *Dunalia brachiacantha, Tibouchina paratropica, Ichnanthus minarun, Pennisilum latifolium, Jungia pausiflora, Adiantus* sp., *Ophryosporus* sp., *Digitaria ternata, Setaria geniculata, Hyptis mutabilis. Group 4:* swards (2500 to 3200 m asl). *Alchemilla pinnata, Hypoxis* sp., *Juncus* sp., *Tripogon spicatus, Bromus unioloides, Plantago* sp., *Vulpia bromoides, Hypochoeris meyeniana, Paspalum pygmaeun. Group 5:* swards and marshes (>3200 m asl). *Eleocharis* sp., *Koeleria Permolis, Gamochaete* sp. *Group 6:* low pasturelands (>3200 m asl). *Stipa uspallatenis, Festuca Iilloi, Boulesia* sp., *Geranium* sp. *Group 7:* disturbed vegetation near pens dominated by weeds: *Lepidium bonariensis, Cynodon dactylon, Dichondra repens, Amaranthus quitensis, Calycera* sp., *Tagetes* sp., *Erodium* sp., *Cuphea* sp., *Rumex* sp., *Cirsium* sp.

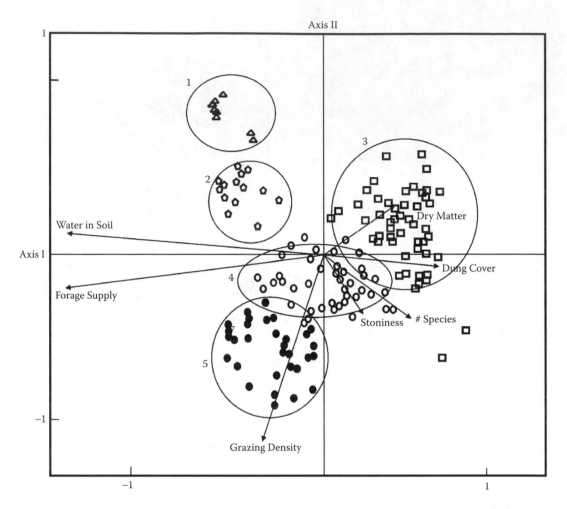

FIGURE 10.4 Ordination diagram (PCA) of 200 sampling quadrats in terms of their floristic composition in páramo swards between 3800 and 4200 m asl at the Cordillera de Mérida (Venezuela). Axis I is a gradient of soil water availability and forage quality, Axis II is a gradient of grazing density. *Group 1*: Inundated swards and marshes dominated by *Carex alboutescens, C. humboldtiana,* and *Juncus* sp. *Group 3*: Degraded swards dominated by *Rumex acetosella, Aciachne pulvinata, Acaulimalva* sp. *and Geranium* spp. *Group 5*: Continuous swards of *Calamagrostis mulleri, Carex albolutescens, Lachemilla* spp., *Muehlenbergia ligularis. Group 2 and 4*: Transition swards between inundated, degraded, and continuous swards.

are associated, to a larger extent, with irrigation technologies rather than with animal loads.

The introduction of bovines in páramos without a history of grazing by large herbivore herds, as was the case in Venezuela, or grazing with new animals (ovines, which replaced llamas) in the mountains of Argentina, has been associated with reduced pastoral spaces and changes in the structure and species composition of the dominant vegetation. In the highland grasslands of the Tacanas watershed and La Culata páramos, limited pastoral spaces translate into higher animal concentration on the few

sites where plant species composition responds almost exclusively to grazing loads and water availability. These areas where grazing has induced the formation of swards are characterized by favored forage species that can resist moderate loads, but require temporary grazing exclusion and water to recover. The grazing strategies developed have favored the permanence of these swards, critical for successful livestock farming. However, these delicate equilibria can be easily disrupted by an increase in animal loads or frequency of grazing events.

TABLE 10.1

Correlation of variables with the first three axes of the ordination analyses with the correspondence analysis for the sites on the Ulla-Ulla Plateau (Bolivia), the Tacanas River Basin (Argentina), and the Sierra de La Culata (Venezuela)

Variables/Axes	Ulla-Ulla (Bolivia) Figure 10.2			Tacanas (Argentina) Figure 10.3			La Culata (Venezuela) Figure 10.4		
	1	2	3	1	2	3	1	2	3
Water in soil	0.7268[a]	0.5316	0.2580	0.1366	0.1937	0.2209	0.8423[a]	0.0518	0.0836
Forage supply	0.3487	0.6151[a]	0.5440	0.3106	0.7101[a]	0.0971	0.8788[a]	0.1073	0.1031
Alpaca density	0.5928	0.6944[a]	0.2573	—	—	—	—	—	—
Llama density	0.5769	0.6172[a]	0.1219	—	—	—	—	—	—
Sheep density	—	—	—	0.4007	0.7429[a]	0.1231	—	—	—
Cattle density	—	—	—	—	—	—	0.2281	0.6418[a]	0.2224
Stoniness	0.2982	0.0034	0.1446	0.2543	0.1364	0.1615	0.1367	0.1676	0.6089[a]
Species number	0.0118	0.0002	0.0276	0.6619[a]	0.3509	0.3239	0.3133	0.1796	0.2217
Bare soil	0.6460[a]	0.2845	0.3858	0.1296	0.0612	0.1694	0.0530	0.0446	0.0211
Dung cover	—	—	—	0.2351	0.7796[a]	0.2313	0.3814	0.0399	0.0163
Dry matter	—	—	—	0.0101	0.1621	0.4145	0.2593	0.1243	0.1026
Slope	—	—	—	0.3077	0.2031	0.4448	—	—	—
Altitude	—	—	—	0.8351[a]	0.2349	0.1889	—	—	—
Soils erosion	—	—	—	0.1310	0.0993	0.2310	—	—	—

[a]Significant correlations.

In other mountains of the world, the establishment and maintenance of swards has been directly linked with animal movements and their preferred grazing areas (Brasher and Perkins 1978; O'Connor 1978; García Gonzáles et al. 1990). These grazing-induced swards can represent a permanent increase in forage supply, species diversity, and stability as long as they are subjected to controlled grazing in which loads are regulated. In other words, a lower efficiency in the use of pastoral space must correlate with strong regulations on grazing, as overgrazing can too easily result on sites with inherent limitations.

Grazing history, the type and response of vegetation, and soil water availability (Milchunas et al. 1988) seem to be critical factors in explaining the causes of impacts on vegetation of these Andean livestock-farming systems. Pronounced changes on vegetation structure have also been reported in other Andean mountains where livestock farming shows limited pastoral spaces. This has been the case in the fire-grazing strategy for the exploitation of forage resources in the highlands of the páramos of Colombia (Hofstede, 1995; Verweij, 1995)

and Ecuador (Hess, 1990; White and Maldonado, 1991; Laegaard, 1992). Under the influence of the fire-grazing cycle, the páramo grasslands were transformed into pasture mosaics of different structure and composition, in which ecological dynamics are linked to the frequency and intensity of the cycle (Molinillo and Monasterio 2003). Therefore, these transformations have produced an increase in diversity on a regional scale, because of the decrease in dominance of large tussock grasses and the increase of sward and herbaceous species. In these mosaics, the swards offer the highest forage supply and diversity levels, but there is an increase in the presence of native and exotic weeds (Verweij 1995). However, although these swards are relatively grazing-resistant if exposed to frequent and heavy loads, they can give way to open herbaceous formations with a low forage quality and a sharp decrease in species diversity.

In contrast with these strategies to increase the use of available pastoral space in the Sierra de la Culata páramos, the relationship between livestock farming and agriculture is responsible for the extra forage supply, resulting in less severe environmental consequences. Sometimes, fodder and cereal cultivation, crop residues, fallow and stubble fields, and natural vegetation areas can all reduce cattle loads on the páramos at critical times. Similarly, in the Cumbres Calchaquíes, a strategy based on animal management (seasonal migrations and daily changes in grazing paths), and not environmental transformations, has allowed moderation of the impact of grazing on the sites with the best natural forage supply. However, in some of these places, the lack of grazing control has led to erosive processes and the predisposition of slopes to mass-movement processes that are related to wet cycles in the local climate (Molinillo 1993).

Finally, in the Ulla-Ulla altiplano, an efficient water management for irrigation and good organization of community work have favored the expansion and maintenance of large bofedale extensions that sustain high animal loads, especially during the dry season (when, due to lack of forage in the plains, alpacas join the llamas in these wetlands). However, the sustainability of the traditional system is endangered by damage to natural pastures in the plains (Alzérreca et al. 1981; Seibert 1994), the increasing concentration of grazing in bofedale areas, and a decrease in water management efficiency. In this context, farmers perceive the conservation of their vicuña populations (Ulla-Ulla is one of the largest vicuña reserves) as an extra source of pressure on the fragile system. Hence, a series of measures is necessary before the cultural and natural systems deteriorate even further. This should include the restoration of the natural pastures in the plains; better organization of water management systems; and a direct involvement of farmers in managing the vicuña, promoting a more efficient management of their populations and the generation of resources to allow a decrease in grazing pressure.

CONCLUSIONS

Grazing systems with introduced cattle and a short grazing history (Argentina and Venezuela) show a reduced pastoral space compared to the traditional camelid grazing in the puna (Bolivia), particularly of llamas. The lowest pastoral space values found for alpacas correspond to the grazing of animals with a more specialized diet that are adapted to less demanding environmental conditions. Grazing-impact magnitudes of introduced animals depended largely on the spatial patterns developed by the animals, which are controlled by the spatial and temporal distribution of natural forage and the animals' ability to exploit the dominant vegetation. The difficulties encountered by animals to efficiently obtain and consume the dominant vegetation were a major barrier in the introduction of livestock farming; this limited the available grazing areas in these environments. Animal concentration in optimal sites has been associated with the formation of grazing-induced swards. These swards, similar to puna bofedales, combine good water availability and the dominance of good forage species resistant to grazing. Above moderate loads, successional trends show the colonization of native and exotic weeds, and a decrease in forage supply. This negative impact can only be decreased through a strong control of grazing, using animal dispersion practices (Cumbres Calchaquíes in Argentina), a decrease in animal loads in critical periods and extra forage supplies (Sierra de la Culata in Venezuela), or the expansion and maintenance of optimal sites through irrigation technology (the Ulla-Ulla altiplano in Bolivia).

SUMMARY

Pastoral strategies, space-use efficiency, and their relationship with dominant vegetation patterns are compared in three Andean systems: alpacas and llamas in the punas of the Ulla-Ulla altiplano, ovines in the high-altitude grasslands of the Cumbres Calchaquies (northern Argentina), and cattle in the páramos of the Cordillera de Mérida (Venezuela). GIS and multicriteria assessment methods are used to map the vegetation and its capacity for grazing. Multivariate analysis techniques are used for analyzing the vegetation structure and the effects of management variables. The results show that pastoral systems with more recently introduced animals (Argentinean grasslands and the Venezuelan páramos) have limited available grazing area, related to the concentration of the animals in a few optimal sites and the limited capacity of the animals to utilize effectively the dominant vegetation. These sites are characterized by high ground water availability and the dominance of species with high palatability and grazing resistance. The same characteristics are also important for the use of the bofedales in the Bolivian altiplano for alpacas, whereas the diverse diet of llamas allows them to use the widest pastoral space of the three systems analyzed. The management strategies for animal dispersion, increased forage supply, or vegetation structure alterations that could contribute to the decrease in the environmental impact of grazing in the three regions are discussed.

ACKNOWLEDGMENTS

This research is part of the Interdisciplinary Project for Sustainable Agricultural Managment in the High Andes of Mérida, Venezuela, financed by the AGENDA PAPA-CDCHT (CV1 PIC-C0201) of the Universidad de los Andes. Páramo communities in the project area (La Toma, Misintá, and El Banco) offered essential support during sampling and mapping. The organization Program Andes Tropicales (PAT) also offered logistic support during sampling and mapping. We also wish to thank Dr. Luis D. Llambí for his help with the translation of the manuscript.

References

Alzérreca, H., Cordero, R., Lara, R., and Rivero, V. (1981). Ensayo de recuperación de praderas naturales para camélidos en Ulla-Ulla, Bolivia. Informe de investigaciones agropecuarias, Instituto Nacional de Fomento Lanero, serie de estudios especializados, La Paz, Bolivia.

Bárcena, E. (1988). Composición botánica de la dieta al pastoreo de las alpacas (*Lama pacos*) en dos épocas diferentes del año. Informe de investigaciones agropecuarias, Instituto Nacional de Fomento Lanero, serie de estudios especializados, La Paz, Bolivia.

Barredo, J.I. (1996). *Sistemas de información geográfica y evaluación multicriterio en la ordenación del territorio*. RAMA Editorial, Madrid.

Brasher, S. and Perkins, D.F. (1978). The grazing intensity and productivity of sheep in the grassland ecosystem. In Heal, O.W. and Perkins, D.F. (Eds.). *Production Ecology of British Moors and Montane Grasslands*. Springer-Verlag, Berlin, pp. 354–374.

Browman, D.L. (1974). Pastoral nomadism in the Andes. *Current Anthropology*, 15(2): 188–196.

Brush, S.B. (1984). El ambiente natural y humano de los Andes Centrales. In Baker, P.T. (Ed.), *Informe sobre los conocimientos actuales de los ecosistemas andinos*. Vol. 2, Montevideo, UNESCO.

Cendrero, A. (1982). *Técnicas e instrumentos de análisis para la evaluación, planificación y gestión del medio ambiente*. CIFCA, Serie Opiniones No. 6.

Eastman, R.J., Jin, W., Kyem, P., and Toledano, J. (1995). Raster procedures for multi-criteria/multi-objective decisions. *Photogrammetric Engineering and Remote Sensing*, 61(5): 537–539.

Flores Ochoa, J.O. (1979). Desarrollo de las culturas humanas en las altas montañas tropicales (estrategias adaptativas). In Salgado-Labouriau, M.L. (Ed.). *El medio ambiente Paramo*. Ediciones centro de Estudios Avanzados, Caracas, Venezuela, pp. 225–234.

Flores Ochoa, J.O. (1988a). Cambios en la Puna. In Flores Ochoa, J.O. (Ed.), *Llamichos y paqocheros. Pastores de llamas y alpacas*. Centro de estudios andinos, Cuzco, Peru, pp. 273–293.

Flores Ochoa, J.O. (1988b). Distorsiones en el uso del ecosistema de la Puna y los programas de cooperación técnica. In Flores Ochoa, J.O. (Ed.), *Llamichos y paqocheros. Pastores de llamas y alpacas*. Centro de estudios Andinos, Cuzco, Peru, pp. 255–271.

García-González, R., Hidalgo, R., and Montserrat, C. (1990). Patterns of livestock use in time and space in the summer ranges of the western Pyrenees: a case study in the Aragon Valley. *Mountain Research and Development*, 10: 241–255.

Genin, D., Villca, Z., and Abasto, P. (1994). Diet selection and utilization by llama and sheep in a high altitude-arid rangeland of Bolivia. *Journal of Range Management*, 47(3): 245–248.

Gómez Orea, D. (1992). Evaluación de impacto ambiental. Editorial agrícola española, Madrid.

Hess, C.G. (1990). Moving up – moving down: agropastoral land use patterns in Ecuadorian Páramos. *Mountain Research and Development*, 10(4): 333–342.

Hofstede, R. (1995). Effects of Burning and Grazing on a Colombian Paramo Ecosystem. Wotro, ICG, The Netherlands.

Laegaard, S. (1992). Influence of fire in the grass páramo vegetation of Ecuador. In Balslev, H. and Luteyn, J.L. (Eds.), *Paramo. An Andean Ecosystem under Human Influence*. Academic Press, pp. 151–170.

La Fuente, A., Velasco, A., and Alzérreca, H. (1988). Evaluación de la productividad de campos nativos de pastoreo en Ulla-Ulla. I Reunión Nacional en Praderas Nativas de Bolivia. PAC — Programa de Autodesarrollo Campesino, La Paz, pp. 56–66.

Little, M.A. (1984). Poblaciones humanas de los Andes. Las Ciencias humanas como base para la planificación de las investigaciones. In Baker, P.T. (Ed.), *Informe sobre los conocimientos actuales de los ecosistemas Andinos*. Vol 1. ROSTLAC, Montevideo, Uruguay, pp. 83–129.

Merlino, R.J. and Rabey, M.A. (1983). Pastores del altiplano andino meridional: religiosidad, territorio y equilibrio ecológico. *Allpanchis* 21: 149–171.

Milchunas, P.G., Sala, O.E., and Lauenroth, W.K. (1988). A generalized model of the effects of grazing by large herbivores on grassland community structure. *American Naturalist*, 132: 87–106.

Millones, J.O. (1982). Patterns of land use and associated environmental problems of the Central Andes: an integrated summary. *Mountain Research and Development*, 2(1) : 49–61.

Molinillo, M. (1988). Aportes a la ecología antropológica de las Cumbres Calchaquíes: Usos de los recursos naturales en el Valle de Lara. Master thesis, Universidad Nacional de Tucumán, Tucumán, Argentina.

Molinillo, M. (1990). Uso de los recursos naturales y su impacto en ecosistemas de montaña de la cuenca del río Tacanas, Tucumán, Argentina. Informe final al consejo nacional de investigaciones científicas y técnicas (CONICET), Tucumán, Argentina.

Molinillo, M. (1992). Pastoreo en ecosistemas de páramo: estrategias culturales e impacto sobre la vegetación en la cordillera de Mérida, Venezuela. Master thesis, Universidad de Los Andes, Mérida, Venezuela.

Molinillo, M. (1993). Is traditional pastoralism the cause of erosive processes in moutain environments? The case of Cumbres Calchaquies in Argentina. *Mountain Research and Development*, 13(2): 189–202.

Molinillo, M. and Monasterio, M. (1997). Pastoralism in paramo environment: practices, forage and vegetation impact in the Cordillera of Merida, Venezuela. *Mountain Research and Development*, 17(3): 197–211.

Molinillo, M. and Monasterio, M. (2003). Patrones de vegetación y pastoreo en ambientes de páramo. *Ecotropicos*, 15(1): 17–32.

Monasterio, M. (1980). Poblamiento humano y uso de la tierra en los altos Andes de Venezuela. In Monasterio, M. (Ed.), *Estudios ecológicos en los Páramos Andinos*. Universidad de Los Andes, Mérida, Venezuela, pp. 170–198.

O'Connor, K.F. (1978). The rational use of high mountain resources in pastoral systems. The use of high mountains of the world. IUCN, New Zealand, pp. 169–183.

Palacios Rios, F. (1988). Tecnología del pastoreo. In Flores Ochoa, J. (Ed.), *Llamichos y paqocheros. Pastores de llamas y alpacas*. Centro de estudios Andinos, Cuzco, Peru, pp. 87–100.

Rabey, M.A. and Merlino, R.J. (1988). Un sistema de control ritual — rebaño entre los pastores del sur de los Andes Centrales. In Flores Ochoa, J. (Ed.), *Llamichos y paqocheros. Pastores de llamas y alpacas*. Centro de estudios andinos, Cuzco, Peru, pp. 113–120.

Ruiz, C. and Tapia, M. (1987). Producción y manejo de forrajes en los Andes del Perú. PISA (INIPA-CIID-ACDI), Lima, Peru.

Seibert, P. (1994). The vegetation of the settlement area of the Callawaya pand the Ulla-Ulla highlands in the Bolivian Andes. *Mountain Research and Development*, 14(3): 189–211.

Solbrig, O.T. (1985). Los Andes meridionales y las Sierras Pampeanas. In Baker, P.T. (Ed.), *Informe sobre los conocimientos actuales de los ecosistemas Andinos*. Vol. 4. ROSTLAC, Montevideo, Uruguay.

Tapia, M. and Flores-Ochoa, J. (1984). *Pastoreo y pastizales de los Andes del sur del Perú*. Programa colaborativo de apoyo a la investigación de rumiantes menores, Perú.

Thomas, B.R. and Winterhalder, B.P. (1976). Physical and biotic environment of southern highland Perú. In Baker, P.T. (Ed.), *Man in the Andes: A Multidisciplinary Study of High-Altitude Quechua*. Dowden, Hutchinson and Ross, Pennsylvania, pp. 21–59.

Thomas, B.R. (1979). Effects of change on high mountains human adaptive patterns. In Webber, P.J. (Ed.), *High Altitude Geoecology*. AAAS Symposium 12, Westview Press, Colorado, USA, pp. 139–181.

Verweij, P.A. (1995). Spatial and Temporal Modelling of Vegetation Patterns. Burning and Grazing in the Paramo of Los Nevados National Park, Colombia. ITC Publication 30, ITC, Enschede, The Netherlands.

Wheeler, J. (1988). Llamas and alpacas of South America. In Thonsen, G. and Weide, K. (Eds.), *17th Seminar for Veterinary Technicians*, Nevada, USA, pp. 301–310.

White, S. and Maldonado, F. (1991). The use and conservation of natural resources in the Andes of Southern Ecuador. *Mountain Research and Development*, 11(1): 37–55.

11 Grazing Intensity, Plant Diversity, and Rangeland Conditions in the Southeastern Andes of Peru (Palccoyo, Cusco)

Jorge Alberto Bustamante Becerra

INTRODUCTION

In the high-elevation (3900 to 4800 m) grasslands of the Andes, known as the *puna*, extensive grazing land areas have been utilized by rural farmers (campesinos) for over 10,000 years (Burger, 1992; Burns, 1994). Troll (1968) classified the *puna* into three provinces: the moist puna, the dry puna, and the desert puna. Precipitation in the puna is concentrated in a single wet season (between October and April), is of variable length, and ranges from 150 in the desert puna to 1200 mm.a^{-1}, in the moist puna belt (Molina and Little, 1981). Evaluation of puna grassland characteristics requires information on both soil and vegetation. These grasslands are characterized by large variations in time and space. Classification of grasslands into range sites, habitat types, or some other unit of landscape is an attempt to deal with spatial variation (Pamo et al., 1991).

The *puna* has a distinct vegetation type that is found predominantly in Andean Peru, but also extends into adjacent areas such as Bolivia, north of Chile, and northwestern Argentina. Weberbauer (1936) distinguished four major vegetation formations in the moist puna: puna mat, bunchgrass formation, moor grasslands, and the vegetation of rocks and stone fields. Floristically, the moist and dry puna are closely related. Evergreen shrubs are more common in the dry puna (Weberbauer, 1936; Wilcox et al., 1987). In the desert puna, vegetation cover is lower and is dominated by shrubs. Examples of vegetation changes because of human impact are the elimination of *Polylepis* forests (Simpson, 1979) in much of the puna and proliferation of *Opuntia floccosa* Salm-Dyck (Molina and Little, 1981).

Regarding the use of the puna, indigenous culture developed highly productive and sustainable agriculture based on efficient soil and water management and the integration of crops and livestock (Tapia Nunex and Flores Ochoa, 1984). However, the growing human population has increased the demand for land and food. Traditional production systems have broken down or been forgotten, and puna resources are being degraded by grazing herds of domestic llamas, alpacas, goats, and sheep, as well as by people gathering wood for fuel. Introduced and invasive species, as well as uncontrolled fires, also cause environmental problems (Tapia Nunex and Flores Ochoa, 1984).

Grazing has traditionally been viewed as having a negative impact on the subsequent rate of energy capture and primary production within grazing systems through a series of direct and indirect effects on plant growth (Heitschmidt and Stuth, 1991). Direct effects of grazing are those associated with alterations in plant physiology and morphology resulting from defoliation and trampling (Caldwell, 1984). Grazing also indirectly influences plant performance by altering microclimate, soil properties, and plant competitive interactions (Woodmansee and Adamsen, 1983).

The value of grasslands to agricultural interests commonly depends on the quality and quantity of forage produced. This is reflected indirectly in the capacity of the range to produce livestock (carrying capacity). Forage production can be expressed in terms of range conditions; in general, the more the forage produced on a given site, the better the range conditions (Humphrey, 1962). On the other hand, there is often a general relationship between range conditions and stages in secondary plant succession. Thus, in general, the better the conditions, the more advanced the successional stage. To assist in determining the range condition class for a range site, plant species are grouped as decreasers, increasers, or invaders, based primarily on the response to grazing intensity (Humphrey, 1962; Lacey and Taylor,

2003). *Decreasers* are highly productive, palatable plants that grow under low grazing intensity. These plants decrease in relative abundance under continued intensive grazing. *Increasers* are less productive and less palatable plants that also grow in the original climax community. They tend to "increase" and take the place of the decreasers that weaken or die due to heavy grazing, drought, or other range disturbances. *Invaders* are native or introduced plants that are rare in the climax plant community. They invade a site as the decreasers and increasers are reduced by grazing or other disturbances. A relationship between the grazing intensity, range conditions, and the relative proportion of decreasers, increasers, and invaders for a hypothetical grassland site is shown in Figure 11.1. Botanical composition and species diversity have been reported to change with the degree of utilization in degraded grasslands

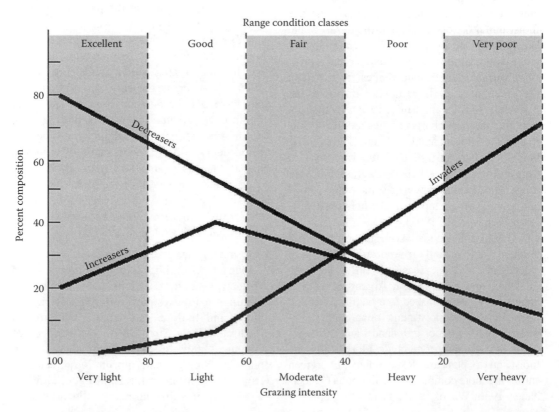

FIGURE 11.1 Relationship between intensity of grazing, range condition, and percentage of decreasers, increasers, and invaders. (Modified from Stoddart et al, 1975.)

(Flórez et al., 1985). For example, high-quality grasses that are preferred by grazing animals tend to disappear, whereas the growth of annuals that have thorns and that contain tannins tends to increase in the course of degradation (Belsky, 1992).

Based on the information given in the preceding text, my hypothesis was that the grazing system of Andean pastoralists in the puna (3950 to 5000 m asl) is characterized by moderate grazing intensity and intermediate frequency of disturbance that favor high plant diversity. The main objective of this study was to relate grassland species diversity to different rangeland conditions and the main environmental and socioeconomic factors.

STUDY AREA AND METHODS

STUDY AREA

The study area of approximately 9786 ha was the peasant community of Palccoyo, District of Checacupe, Province of Canchis in the Department of Cusco, Peru (14°03 S, 71°21 W). Palccoyo is located approximately 128 km from the city of Cusco. Elevation ranges from 3950 to 5000 m, and the main village is at 4100 m. Topography consists of both gentle and rugged mountainous terrain. Palccoyo is in the upper land of the Vilcanota valley, located on the southeastern cordillera of the Andes in the dry puna belt, as classified by Troll (1968). According to Holdridge's classification, the life zones present in Palccoyo are:

1. *Subtropical mountain–humid forest*: Elevation ranges from 3950 to 4050 m, precipitation ranges from 500 to 1000 mm per year, and the average monthly temperature ranges from 13 to 15°C. Vegetation is composed of perennial grasses, forbs, some shrubs, and tree remains of *Escallonia resinosa* and *Escallonia myrtilloides*. Agriculture (cultivation and pastoralism) is the main activity.
2. *Very humid paramo–subtropical sub-Andean*: Elevation ranges from 4050 to 4550 m, precipitation ranges from 500 to 1000 mm per year, and the average monthly temperature ranges from 6 to 12°C. Vegetation is composed of bunchgrass formation, and pastoralism is the main activity.
3. *Pluvial tundra–subtropical Andean*: Elevation ranges from 4550 to 4900 m, precipitation is above 500 mm per year, and the average monthly temperature ranges from 1.5 to 3°C. Vegetation is composed of bunchgrass formation; tufted grasses are also important components. Pastoralism is the main activity.
4. *Subtropical nival*: Elevation is above 4900 m, precipitation is above 500 mm per year, and the average monthly temperature is below 1.5°C. Vegetation is almost absent, with the exception of several lichens and mosses. Alpaca herders do not use this zone for grazing in the Palccoyo area.

LIVESTOCK HOLDING

The population of the Palccoyo peasant community is 834 inhabitants (INEI, 1993), distributed in 161 families, with 5.2 persons per family, of which 3.3 are children. Livestock breeding is the main activity, but people also grow potatoes (more than 15 native varieties), native varieties of tubers (oca, olluco, and añu), and edible roots to feed themselves, to exchange, and to sell any surplus (Bustamante Becerra, 1993). Family-owned flocks consist of alpacas, sheep, llamas, horses, and some cattle. Families of the Palccoyo community (45 in total) were classified into three socioeconomic levels (high, medium, and low), according to the number of livestock owned. Livestock possession varied considerably within the community (Table 11.1). Families of a high or medium level have on an average four species of livestock: alpacas, sheep, llamas, and horses; very few at the high level also own cattle. Families of a low level usually have three species: alpacas, sheep, and horses. Livestock possession in the Palccoyo community showed a clear differentiation between the three levels, mainly depending on the tenure of bofedales, which are

TABLE 11.1
Number of grazing animals per socioeconomic level in Palccoyo

Socioeconomic Level	Number of Families	Percentage	Number of Grazing Animals (OU) per Household	Total (OU)
High	25	15.53	245.94	6,148.50
Medium	71	44.10	137.86	9,788.06
Low	65	40.37	36.76	2,389.40
Total	161	100.00	420.56	18,325.96

Note: OU is ovine unit

essential for the feeding of alpacas and sheep during the dry season.

Grazing system, and the spatial and chronological arrangement, were determined by surveying 15 families of high, medium, and low socioeconomic levels (45 families in total), and subsequent *in situ* visual checking (Bustamante Becerra, 1993).

Spatial and Chronological Arrangement of the Natural Grasslands

The two main spatial arrangements in Palccoyo are the grasslands of the low part (altitudinal range from 4000 to 4250 m) and high part (from 4400 to 4800 m) of the community. Both parts are mainly natural dry grasslands and bofedales. The grasslands of these parts are better defined in four classes (Table 11.2), as follows:

1. *Natural grasslands of low parts* are located close to the small settlements and main village of the community, and are characterized by the small size of crop plots combined with a rotational pattern of crops and natural grasslands. Undesirable species, such as *Astragalus garbancillo, Astragalus unioloides*, and *Oenothera multicaulis*, are indicators of overgrazing and are common in several of these grasslands.
2. *Bofedales of low parts* are located in the middle of the low parts and also close to the small settlements and main village. Good conditions and plant cover characterize these sites.

3. *Natural grasslands of high parts* are located far from the village, on the steep slopes of the mountains, and are placed on the high parts of the community. Here, shelters and corrals can be found, with herders (pastoralists) also remaining during pasturing, close to their grazing animals.
4. *Bofedales of high parts* are located at the foot of mountains of the high parts. Corrals and shelters are close to bofedales and, generally, on the gentle slopes of the mountains, whose peaks are often covered by snow.

Grazing Systems

The grazing system is continuous, with seasonal rotation of the grasslands of Palccoyo. The local people's knowledge of the puna environment allows for spatial and chronological arrangements throughout the year. The first period of pasturing starts in December and lasts until the end of May. During this season, livestock grazing occurs in range sites of the low part, where the grasslands are in good condition. Plant cover is as good a parameter of rangeland conditions as plant vigor and forage species composition (Bustamante Becerra, 1993). The second period of pasturing starts in June, when livestock are transferred from the grasslands of the lower to the higher parts. The livestock stays there for 6 months (until November). This is when the bofedales are of significant importance as they sustain grazing during the critical dry season.

TABLE 11.2
Range conditions, range site extension, and estimated carrying capacity (CC) according to the range condition and actual stocking rate (SR) of the Palccoyo community range sites

Site	Altitude (m)	Range Condition	CC OU/ha/yr	Range Site Extension ha	Range Site Extension Percentage	Range Site CC
Juque[eb]	4,600	53.72 Fair	1.5	5,185.5	77.6	7,778.2
Occojuque[e,a]	4,500	70.55 Good	3.0	317.0	4.74	951.0
Jawacholloca[d,b]	4,400	53.89 Fair	1.5	119.0	1.78	178.5
Uracholloca[d,b]	4,350	56.50 Good	3.0	315.6	4.72	946.9
Chullunquiani[d,b]	4,250	52.23 Fair	1.5	525.0	7.86	787.5
Antakarana[c,b]	4,200	37.86 Poor	0.5	148.4	2.22	74.2
Huayllapampa[c,a]	4,000	64.94 Good	3.0	72.0	1.08	216.0
Total natural grassland				6,682.50	100	
CC (OU/ha/yr) and total CC			1.64			10,932
SR (OU/ha/yr) and total livestock number			2.74			18,326

Note: Range site CC is estimated as follows = (CC × range sites), where CC is expressed as OU/ha/yr, and range sites as ha. Total CC is estimated as the grand total of the seven range site CCs. OU, ovine unit.

[a] Grassland with high humidity or wetland named *bofedales*.
[b] Grassland with little or absent moisture, named *semiarid grasslands*.
[c] Range site of the community lowland.
[d] Range site of the community midland.
[e] Range site of the community upland.

METHODS

In the study area, seven range sites were identified and measured from visual interpretation (texture and tonality) of an aerial photograph (scale 1:25,000) and a map (Sicuani, sheet, scale 1:100,000) both of 1975 (Oficina de Catastro Rural, 1976) and subsequent fieldwork, to produce the range site map. The *range site* is defined as a large area of natural grasslands with similar environmental characteristics and used as rangeland (Flórez et al., 1992; Young, 1997; Pamo et al., 1991).

To understand the grasslands of the Palccoyo community better, three altitudinal classes and two soil humidity classes were identified. Altitudinal classes of grasslands were upland (above 4500 ma sl), midland (4250 to 4500 m asl), and lowland (below 4250 m asl). Soil humidity classes were humid (i.e., bofedales) and dry grasslands. According to these criteria (altitude, soil humidity), seven range sites were identified, as shown in Table 11.3.

The following abiotic parameters were sampled during the survey at each range site: soil texture, depth, humidity, altitude, and slope. Soil texture was recorded by the "fell" method, using the soil texture triangle and soil depth following the procedures proposed in the Soil Survey Manual (Soil Survey Division Staff, 1993). Species composition was measured using the nearest-point sampling method (Owensby, 1973). Point samples were recorded along a 100-m transect at 1-m intervals (100 point/transect). Plant species names and features such as bare ground and the presence of rock, litter, and moss were recorded at each point. Plant cover for each species was calculated as the percentage of direct hits per transect. Therefore, each transect will always have 100 registers (points). Seven sites in the study area were sampled, each with three transects. These samplings were repeated at three different dates: November 1992, January 1993, and May 1993.

To determine the range condition (or vegetation condition) at each range site, four rating criteria were used in the site-potential approach,

TABLE 11.3

Identification of the seven range sites according to altitudinal and classes humidity

	Lowland[c] (below 4250 m)	Midland[d] (4250–4500 m)	Upland[e] (above 4500 m)
Dryland[b]	Antakarana[cb]	Jawacholloca[db] Uracholloca[db] Chullunquiani[db]	Juque[eb]
Humid land[a]	Wayllapampa[ca]		Occojuque[ea]

[a] Grassland with high humidity or wetland named *bofedales*.
[b] Grassland with little or absent moisture, named *semiarid grasslands*.
[c] Range site of the community lowland.
[d] Range site of the community midland.
[e] Range site of the community upland.

based on Humphrey (1962) and Flórez et al. (1992): (1) *Composition of desirable species*, (2) *Forage species*, (3) *Plant vigor*, and (4) *Erosion* (Table 11.4). Range condition was calculated as 0.5 (1) + 0.2 (2) + 0.1 (3) + 0.2 (4).

1. *Composition of desirable species* is the most important of the various criteria employed. The total plant cover, within reach of livestock, was subdivided by forage value based on desirable (decreasers), less desirable (increasers), and undesirable (invaders) species. These classes were determined from specialized literature on grassland species palatability for alpacas and sheep in the Andean region (Contreras, 1967; Antezana, 1972; Peña, 1970; Montufar, 1983; La Torre, 1963; Sanches, 1966; Farfan, 1981; Reiner and Bryant, 1986; Bryant and Farfan, 1984; Reiner, 1985). Composition of desirable species was determined by registering the percentage of desirable species.
2. *Forage species* is usually identified as the percentage of ground surface covered by the current year's growth of desirable and less desirable species.
3. *Plant vigor* of two key forage species is a useful indicator of range conditions. Vigor is determined by comparing the heights of ten plants from

the area being rated with ten of the same species identified as vigorous and flourishing, located in ungrazed areas.

4. *Erosion* is an indirect measure of vegetal cover and was determined by registering bare soil, rock, and pavement on the transect on each range site sampled.

The checklist of species composition, palatability of grassland species, and results of the four criteria for range conditions for the study area is given in Bustamante Becerra (1993).

The three assessments of range conditions correspond to the beginning of the wet season (November), the peak of the wet season (January), and the beginning of the dry season (May).

The land use factor of grasslands is defined as the relationship between the *stocking rate* (SR) and *carrying capacity* (CC) of the grassland. *Stocking rate* is the number of specific kinds and classes of animals grazing on a unit of land for a specified period (Society for Range Management, 1989). Both SR and CC are expressed as ovine units per hectare per year (OU/ha/a). One OU is defined as a sheep of 35 kg in the Andean region (Leon Velorie and Izquierdo Cadena, 1993; Flórez et al., 1992). CC is the maximum stocking rate possible that does not damage range conditions and maintains or improves vegetation or related resources. This may vary from year to year in the same area because of fluctuating forage production. In the Andean region, according to Flórez et al. (1992), range sites with excellent conditions have a carrying capacity of 4 OU/ha/a, good conditions have 3 OU/ha/a, fair conditions have 1.5 OU/ha/a, poor conditions have 0.5 OU/ha/a, and very poor conditions have 0.25 OU/ha/a.

Nomenclature of plant species follows McBride (1936) and Tovar (1960, 1965, 1972). Species identification was confirmed at the Herbarium of the San Antonio Abad University in Cusco, Peru, based on collected samples (Bustamante Becerra, 1993). Species-relative abundance and Shannon species diversity index [H' = sum of $pi \cdot \ln pi$] (Magurran, 1988; Whittaker, 1972) were determined by calculating the frequency of each plant species (pi = proportion of points along each transect at which species i was recorded). H' measures how many different species are in an ecological system and how many of each species are present. Plant species richness (S = number of species sampled per transect) and evenness of species abundance (Pielou's J index = H'/ln S where ln in S=H' max, or the maximum possible diversity when all species are represented by the same number of individuals) were also calculated for each transect.

Spatial distribution of plant species was analyzed by correspondence analysis (CA) (Hill and Gauch, 1980; Pielou, 1984) to determine clustering (assemblage) of species and samples along ordination gradients, represented as ordination axes. Variables amounting to 21 (seven range sites assessed at three different dates) and 62 cases (species) were analyzed. One measure of the importance of the ordination axis is the eigenvalue (λ) of CA, which is equal to the (maximized) dispersion of species scores in the ordination axis (ter Braak, 1995). Values above 0.5 often denote a good separation of the species along the axis. The first five ordination axes were correlated (using Spearman rank order correlations at p-level < 0.05) with environmental variables (altitude, slope, soil depth, and texture). Afterwards, a multiple linear regression analysis between axis and environmental variables that presented significant correlation (p-level < 0.05) was carried out to determine how environmental variables explain the spatial distribution of plant species along a

defined ordination axis. The relationship between indicators of plant diversity (Shannon diversity, species richness, and evenness) and range condition (of seven range sites) was analyzed by Spearman rank order correlations at p-level < 0.05.

RESULTS AND DISCUSSION

PLANT COMPOSITION

The most important families in the Palccoyo area were Poaceae (24.19% of the total species), Asteraceae (17.74%), Gentianaceae (9.68%), and Cyperaceae (8.06%). The remaining families represented 40.33% of the total (Bustamante Becerra, 1993). The number of species and the percentage of herbaceous species, graminoids, and Gramineae species are listed in Table 11.5. The bofedales range sites showed a greater number of graminoid species than semiarid range sites, whereas semiarid range sites showed a greater number of Gramineae species than bofedales range sites.

PLANT COVER

The highest value of vegetation cover corresponded to a range site with greater moisture — bofedales (Occojuque, 100%, Table 11.6), and the lowest values represented a range site located in a semiarid area (Antakarana, 73%). The study area, as a whole, had a high vegetation cover (92%) during the wet season.

RANGE CONDITION

Soil conditions and plant cover of seven range sites are shown in Table 11.7. The best range sites are the bofedale sites (Occojuque and Huayllapampa) because of their good edaphic characteristics for the development of natural grasslands (loamy soil texture, immense depth, and slight inclination). Bofedales located in the highest parts are humid throughout the year because of seepage of groundwater, precipitation in the wet season, and melting snow in the dry season.

The sites with lower range values, such as Antakarana and Juque, lack water sources that would allow for better range conditions. Another important factor determining range

TABLE 11.4

Classification of vegetation conditions utilized to classify Andean natural grasslands, using four criteria

1. Composition of Desirable Species

Percentage	Score (0.5) = [(Percentage of Desirable Species)]
70 to 100	35.0–50.0
40 to 69	20.0–34.5
25 to 39	12.5–19.5
10 to 24	5.0–12.0
0 to 9	0.0–4.5

2. Forage Species

Percentage	Score (0.2) = [(Percentage of Forage Species)]
90 to 100	18.0–20.0
70 to 89	14.0–17.8
50 to 69	10.0–13.8
40 to 49	8.0–9.8
less than 40	0.0–7.8

3. Plant Vigor

Percentage	Score (0.1) = [(Percentage of Plant Vigor)]
80 to 100	8.0–10.0
60 to 79	6.0–7.9
40 to 59	4.0–5.9
20 to 39	2.0–3.9
less than 20	0.0–1.9

4. Erosion

Percentage	Score (0.2) = [(100-%]
10 to 0	18.0–20.0
30 to 11	14.0–17.8
50 to 31	10.0–13.8
60 to 51	8.0–9.8
more than 60	0.0–7.8

5. Range Condition

Total Score	Quality
79 to 100	Excellent
54 to 78	Good
37 to 53	Fair
23 to 36	Poor
0 to 22	Very poor

Total score = 0.5 (1) + 0.2 (2) + 0.1 (3) + 0.2 (4)

Source: From Flórez et al. (1992).

TABLE 11.5
Some characteristics of plant composition found in the Palccoyo community, using seven range sites

Site	Species Number	Herbaceous (%)	Graminoids (%)	Gramineaes (%)
Juque[e,b]	27	66.77	7.41	25.82
Occojuque[e,a]	23	73.91	13.04	13.05
Jawacholloca[d,b]	21	57.14	9.52	33.33
Uracholloca[d,b]	23	73.91	4.35	21.74
Chullunquiani[d,b]	23	73.91	4.53	21.56
Antakarana[c,b]	13	53.85	0	46.15
Huayllapampa[l]	17	64.71	17.65	17.64
Total	62	64.52	11.29	24.19

TABLE 11.6
Plant cover found in the Palccoyo community

Site	Altitude	Beginning of Wet Season	Peak of Wet Season	End of Wet Season	Average
Juque[e,b]	4600	92.59	87.92	84.58	88.36
Occojuque[e,a]	4500	99.63	99.59	99.94	99.72
Jawacholloca[d,b]	4400	97.63	94.94	90.93	94.50
Uracholloca[d,b]	4350	97.62	90.27	88.94	92.28
Chullunquiani[d,b]	4250	94.61	95.63	87.63	92.62
Antakarana[c,b]	4200	78.64	85.96	73.29	79.30
Huayllapampa[c,a]	4000	99.97	98.63	98.61	99.07
Average		94.38	93.28	89.13	92.26

Note: Using seven range sites measured at the beginning of the wet season (November 1992), in the middle of the wet season (January 1993), and at the beginning of the dry season (May 1993).

[a] Grassland with high humidity or wetland named *bofedales*.

[b] Grassland with little or absent moisture, named *semiarid grasslands*.

[c] Range site of the community lowland.

[d] Range site of the community midland.

[e] Range site of the community upland.

condition of a site is the proximity to small settlements and the main village, as animals frequently consume the grass at these places, contributing to the process of vegetation degradation at these range sites.

Range conditions at the beginning and peak wet season were good (63 and 65 points, respectively) but range conditions at the beginning of the dry season were significantly lower ($p < 0.01$) and declined to almost fair conditions (56 points). This decrease of the range condition in the dry season is well known in the Andean region (Molinillo and Monasterio, 1997; Bryant and Farfan, 1984; Tapia Nunez and Flores Ochoa, 1984). Therefore, bofedales

become important in the dry season when the range condition of semiarid grasslands degrades.

CARRYING CAPACITY AND ACTUAL LAND USE

According to Table 11.2, the livestock number for Palccoyo was 18,326 OU in total. With a total grassland area dedicated to animal food production of 6,682.5 ha, the resulting stocking rate was 2.74 OU/ha/a. Dry or semiarid range sites represented 94.15% of the total grasslands, whereas bofedales represented 5.82%.

TABLE 11.7
Soil conditions and plant cover found in the Palccoyo community at the seven study range sites

Site	Humidity	Altitude (masl)	Soil Texture	Soil Depth (cm)	Slope (cm)	Plant Cover (November)	Plant Cover (January)	Plant Cover (May)	Plant Cover (Average)
Juque[e,b]	Dry	4600	Clay loam	30	37	58.65	60.01	53.72	57.46
Occojuque[e,a]	Humid	4500	Silt loam	150	15	76.16	75.54	70.55	74.08
Jawacholloca[d,b]	Dry	4400	Loam	29	25	60.26	62.43	53.89	58.86
Uracholloca[d,b]	Dry	4350	Loam	49	17	60.03	64.78	56.50	60.44
Chullunquiani[d,b]	Dry	4250	Loam	57	30	56.80	60.94	52.23	56.66
Antakarana[c,b]	Dry	4200	Loam	55	5	56.33	60.82	37.86	51.67
Huayllapampa[c,a]	Humid	4000	Silt loam	80	15	72.15	69.72	64.94	68.94
Average						62.91	64.89	55.67	61.16

Note: Plant cover was measured at three different dates: beginning of the wet season (November 1992), peak of the wet season (January 1993), and beginning of the dry season (May 1993).

[a] Grassland with high humidity or wetland named *bofedales.*
[b] Grassland with little or absent moisture, named *semiarid grasslands.*
[c] Range site of the community lowland.
[d] Range site of the community midland.
[e] Range site of the community upland.

The total carrying capacity of the Palccoyo community, on the other hand, was only 10,932.3 OU/ha/a or 1.64 OU/ha/a. The land use factor (expressed as the relationship of stocking rate and carrying capacity) of Palccoyo was therefore not appropriate, because the stocking rate (2.74 OU/ha/a) is much greater than the carrying capacity (1.64 OU/ha/a). This relationship implies a future degradation of grassland ecosystems by overgrazing because of overstocking.

The situation of overstocking was even worse due to the seasonally uneven distribution of the livestock. The fact that livestock remains on lowland grasslands of the community during the favorable wet season while the uplands remain without livestock resulted in undergrazing of natural high-elevation grasslands. This situation (undergrazing) is reverted in the dry season with an additional reduction of range conditions of the grasslands by the movement of livestock from lowland to upland grasslands of the community, thus causing an even stronger decrease in carrying capacity than with a constant stocking rate throughout the year.

RELATIONSHIP BETWEEN SPATIAL DISTRIBUTION OF PLANT SPECIES AND MICROENVIRONMENTAL VARIABLES

The results of spatial distribution of species by correspondence analysis (CA) showed that the first two CA ordination axes, 1 and 2, denoted good separation (λ) of the species along their axes, $\lambda_1 = 0.72$ and $\lambda_2 = 0.58$, respectively. On the other hand, correlation analysis between ordination axes and environmental variables showed that the axes 1, 2, and 5 are correlated (p-level < 0.05) as follows: axis 1 showed significant correlation with soil texture (0.8882) and depth (0.5072), axis 2 with slope (0.7399) and soil depth (0.4993), and axis 5 with altitude (0.6487).

Therefore, spatial distributions of grassland species in the puna are significantly related to environmental gradients. Similar results have been found by Cingolani et al. (2003) in a mountain in central Argentina, where topographic and edaphic parameters were related to species distributions; Adler and Morales (1999) demonstrated in a site at northwestern Argentina that environmental variables explained

22% of the variation in species composition between assessed sites; Bustamante Becerra (2002) related, in an Andean region of southeastern Peru, the distribution of grassland species with soil depth and soil moisture.

RELATIONSHIP BETWEEN PLANT DIVERSITY AND RANGE CONDITIONS

A comparative analysis between range conditions and species diversity indices showed a significant and negative correlation ($r = 0.896$) between range conditions and species evenness. Cingolani et al. (2003) obtained similar results in Argentine granite grasslands, where grazing intensity also increased species evenness. In general, range sites with poor range conditions showed low species richness ($S = 10$) and Shannon diversity ($H' = 2.10$), whereas those with fair and good range conditions showed high species richness ($S = 20.67$) and Shannon diversity ($H' = 2.79$). Bustamante Becerra (2002) and Wilcox et al. (1987) found similar results in the south and Central Andes of Peru. The pattern observed here closely follows Huston's (1979) dynamic equilibrium model for species diversity, which states that diversity is controlled by the rate of competitive displacement among species and forces that prevent equilibrium (any disturbance that reduces population size). This model predicts that for communities with low or intermediate growth rates (such as those of the puna), diversity will be reduced at high frequencies of disturbance (heavy grazing or poor condition) by the reduction or extinction of populations unable to recover from the disturbances. Also, at low frequencies of disturbance (moderate grazing or good conditions), diversity will be lower because of competitive displacement. Diversity is highest at intermediate frequency of disturbance (intermediate grazing or fair condition). Thus, if one assumes that frequency and severity of grazing disturbance are low on the range sites with good conditions, intermediate on the range sites with fair conditions, and high on the range sites with poor conditions, then the dynamic equilibrium model for species diversity explains the pattern of observed diversity.

CONCLUSIONS

The main factors determining the condition of grassland vegetation in the puna are soil humidity and grazing pressure. Grazing varies not only in numbers but also in temporal and spatial distribution of livestock (alpacas, sheep, and llamas). The grazing system in Palccoyo is determined by the dry and wet (bofedales) range sites of upland and lowland areas with 6 months (dry season) of continuous pasturing in the upland areas and the other 6 months (wet season) of the year in the lowland areas. In Palccoyo, the natural grassland is overgrazed. The calculated land use factor showed that the stocking rate (2.7 OU/ha/a) is almost twice as much as the carrying capacity (1.5 OU/ha/a). As a consequence, livestock feeding is affected by overgrazing, especially in the dry season, in which range conditions of semiarid sites vary from fair to poor. In addition, overgrazing is recognizable by the presence of nonpalatable species such as *Aciachne pulvinata* and *Astragalus garbancillo*, which are dominant in most of the evaluated vegetal communities. Species diversity patterns were explained best by Huston's (1979) model for species diversity. Species diversity was highest on the range site that experienced intermediate disturbance (fair range conditions).

SUMMARY

This study was carried out in Palccoyo, in the high-elevation (3950 to 5000 m) grasslands of the Andes, puna, Southeastern Peru, looking at the impact of grazing intensity on range conditions and plant diversity in the upper-Andean grasslands. The relationships between the stocking rate, carrying capacity of the grasslands, indicators of plant diversity, and microenvironmental variables were analyzed. Vegetation surveys were undertaken using the point transect method; interviews on the grazing system and household surveys on the socioeconomic background were conducted. Alpacas and sheep (140 OU per family) were the principal grazing animals. The actual overall stocking rate was 2.71 OU/ha/a, with a carrying capacity of only 1.5 OU/ha/a, resulting in overgrazing. Overgrazing was also evident by the presence of some indicator plants (nonpalatable species such as

Aciachne pulvinata and *Astragalus garbancillo)* that were dominant in most of the evaluated plant communities. The grazing system was continuous and seasonal, with some rotation. During the dry season (6 months), grazing animals grazed in the upland areas, mainly on wetlands (bofedales), whereas during the wet season, livestock was mainly concentrated in the lowland areas. This rotation, in addition, led to temporal undergrazing of the highland sites, decreasing the range condition of the upland wetlands. In general, sites with poor range conditions had a lower species richness (10 species) compared to sites with fair and good range conditions (20 species on average). Plant diversity of the grasslands was highest on the range site that experienced intermediate disturbance (intermediate grazing or fair range condition).

References

Adler, P.B. and Morales, J.M. (1999). Influence of environmental factors and sheep grazing on an Andean grassland. *J Range Manage* 52(5): 471–481.

Antezana, C. (1972). Estado y Tendencia de las Pasturas Alpaqueras en el Sur-Oriente Peruano. Agricultural engineer's thesis. University of San Antonio Abad of Cusco, Peru.

Belsky, J. (1992). Effects of grazing, competition, disturbance and fire on species composition and diversity in grassland communities. *J Veg Sci*, 3: 187–200.

Bryant, F.C. and Farfan, R.D. (1984). Dry season forage selection by alpaca *(Lamapacos)* in southern Peru. *J Range Manage*, 37: 330–333.

Burger, R.L. (1992). *Chavin and the Origins of Andean Civilization.* Thames and Hudson, New York.

Burns, K.O. (1994). *Ancient South America.* Cambridge University Press, London.

Bustamante Becerra, J.A. (1993). Intensidad de Pastoreo en Comunidades Altoandinas — Caso de Palccoyo — Canchis — Cusco, Peru. Biologist Thesis. University of San Antonio Abad of Cusco, Peru.

Bustamante Becerra, J.A. (2002). Community spatial structure of the Andean natural pastures in the Manu National Park. 45th IAVS Symposium of the International Association for Vegetation Science. Porto Alegre, Brazil.

Caldwell, M.M. (1984). Plant requirements for prudent grazing. In *Developing Strategies for Rangeland Management.* Westview Press, Boulder, CO, pp. 117–152.

Cingolani, A.M., Cabido, M.R., Renison, D., and Solis, V.N. (2003). Combined effects of environment and grazing on vegetation structure in Argentine granite grasslands. *J Veg Sci* 14(2): 223–232.

Contreras, E. (1967). Estudios de las Principales Forrajeras Naturales en Puno. Base para la Alimentación de los Auquenidos. Agricultural engineer's thesis. University of San Antonio Abad of Cusco, Peru.

Farfan, F. (1981). Soportabilidad Pecuaria de los Pastos Naturales de cuatro Comunidades Campesinas de Pisaq. Zootechnical engineer's thesis. University of San Antonio Abad of Cusco, Peru.

Flórez, A., Malpartida, E., Bryant, F.C., and Wiggem, E.P. (1985). Nutrient content and phenology of cool-season grasses of Peru. *Grass Forage Sci*, 40: 365–369.

Flórez, A., Malpartida, E., and San Martin, F. (Eds.). (1992). *Manual de Forrajes para Zonas Aridas y Semiáridas Andinas.* Red de Rumiantes Menores, Lima, Perú.

Heitschmidt, R.K. and Stuth, J.S. (1991). *Grazing Management: An Ecological Perspective.* Portland, Timber Press, Portland, OR.

Hill, M.O. and Gauch, H.G. Jr. (1980). Detrended correspondence analysis: an improved ordination technique. *Vegetatio* 42: 47–58.

Humphrey, R. (1962). *Range Ecology.* University of Arizona, The Ronald Press, NY.

Huston, M. (1979). A general theory of species diversity. *Am Naturalist*, 113: 89–101.

INEI, Instituto Nacional de Esladistica e Informatica, 1993. IX Censo Nacinal de Poblción y IV de vivenda. INEI, Lima, Peru.

Lacey, J. and Taylor, J.E. (2003). *Montana Guide to Range Site, Condition and Initial Stocking Rates.* MSU Extension Service, Montana State University.

La Torre, W. (1963). Valor Nutritivo de Algunas Plantas Forrajeras. Agricultural Engineer's thesis. University of San Antonio Abad of Cusco, Peru.

León-Velarde, C. and Izquierdo-Cadena, F. (1993). Producción y utilización de los pastizales de la zona Andina: Compendio. Red de Pastizales Andinos (REPAAN), Quito, Ecuador.

Magurran, A.E. (1988). *Ecological Diversity and Its Measurement.* Princeton University Press, Princeton, New Jersey.

McBride, J.F. (1936). *Flora of Peru.* Field Mus. Nat. Hist. Bot. Serv., Pub. 351, Vol. 13, Chicago.

Molina, E.G. and Little, A.V. (1981). Geoecology of the Andes; the natural science basis for research planning. *Mount Res Dev*, 1: 115–144.

Molinillo, M. and Monasterio, M. (1997). Pastoralism in paramo environments: practices, forage, and impact on vegetation in the Cordillera of Merida, Venezuela. *Mount Res Dev*, 17(3): 197–211.

Montufar, E. (1983). Análisis Químico de Suelos y Análisis Bromatológico de las Especies Nativas de la Comunidad de Accocunca (dist. Ocongate, Prov. Quispicanchis y Dpto. Cusco). Zootechnical Engineer's thesis. University of San Antonio Abad of Cusco, Peru.

Oficina de Catastro Rural, (1976). Checacupe (Sicuani), Sheet 26u-III-SE, map 1:25 000. Ministerio de Agricultura. Lima, Peru.

Owensby, C.E. (1973). Modified step-point system for botanical composition and basal cover estimates. *J Range Manage*, 26: 302–303.

Pamo, E.T., Pieper, R.D., and Beck, R.F. (1991). Range condition analysis: Comparison of 2 methods in southern New Mexico desert grasslands. *J Range Manage*, 44(4): 374–378.

Peña, E. (1970). Estudio y evaluación de pastos naturales en la zona de llacturqui (prov. Grau. Dpto. Apurimac). Agricultural Engineer's thesis. University of San Antonio Abad of Cusco, Peru.

Pielou, E.C. (1984). *The Interpretation of Ecological Data. A Primer on Classification and Ordination.* John Wiley & Sons, New York.

Reiner, R.J. (1985). Nutrition of Alpacas Grazing High Altitude Rangeland in Southern Peru. Ph.D. thesis. Texas Tech University, Lubbock, TX.

Reiner, R.J. and Bryant, F.C. (1986). Botanical Composition and Nutritional Quality Diets in Two Andean Rangeland Communities of Alpaca. *J Range Manage*, 39(5): 424–427.

Sanches, L. (1966). Gramíneas del Valle de Paucartambo. Agricultural Engineer's thesis. University of San Antonio Abad of Cusco, Peru.

Simpson, B. (1979). A revision of the genus Polylepis (Rosaceae: Sanguisorbeae). *Smithsonian Contr Bot* 43: 1–62.

Society for Range Management (1989). *A Glossary of Terms Used in Range Management.* 3rd ed., Denver, CO, p. 14.

Soil Survey Division Staff (1993). *Soil Survey Manual.* USDA Agric. Handb. 18, U.S. Gov. Print. Office, Washington, D.C., U.S.A.

Stoddart, L.A., Smith, A.D., and Box, T.W. (1975). *Range Management.* McGraw-Hill, New York.

Tapia Nunez, M.E. and Flores Ochoa, J.A. (1984). *Pastoreo de los Andes del Sur del Peru.* Small Ruminants Collaborative Research Program. U.S.A.I.D. Lima, Peru.

ter Braak, C.J.F. (1995). Ordination. In Jongman, R.H.G., ter Braak, C.J.F., and Van Tongeren, O.F.R. (Eds.), *Data Analysis in Community and Landscape Ecology.* Cambridge University Press, Cambridge, pp. 91–169.

Tovar, O. (1960). Revisión de las especies peruanas del género *Calamagrostis. Mem. Mus. His.*"Javier Prado," N. 11:1–91, Lima.

Tovar, O. (1965). Revisión de las especies peruanas del género *Poa. Mem. Mus. His. "Javier Prado,"* 15:1–67. Lima.

Tovar, O. (1972). Revisión de las especies peruanas del género *Festuca. Mem. Mus. His. "Javier Prado,"* 16:1–95. Lima.

Troll, C. (1968). The cordilleras of the tropical Americas; aspects of climate, phytogeographical and agrarian ecology. In Troll, C. (Ed.), *Geoecology of the Mountainous Regions of the Tropical Americas.* Ferd. Dummlers Verlag, Bonn.

Weberbauer, A. (1936). Phytogeography of the Peruvian Andes. In McBride, J.F. (Ed.), *Flora of Peru.* Field Mus. Nat. His. Bot. Ser., Pub. 351, Vol. 13, Chicago.

Whittaker, R.H. (1972). Evolution and measurement of species diversity. *Taxon* 21: 213–251.

Wilcox, B.P., Bryant, F.C., and Belaun, V.F. (1987). An evaluation of range condition on one range site in the Andes of Central Peru. *J Range Manage*, 40(1): 41–45.

Woodmansee, R.G. and Adamsen, F.J. (1983). Biogeochemical cycles and ecological hierarchies. In Lowrance, R.R., Todd, R.L., Asmussen, L.E., and Leonard, R.A. (Eds.), *Nutrient Cycling in Agricultural Ecosystems.* Georgia Agr. Exp. Sta., Athens, USA.

Young, K.R. (1997). Wildlife conservation in the cultural landscapes of the Central Andes. *Landscape and Urban Planning* 38: 137–147.

12 Importance of Carrying Capacity in Sustainable Management of Key High-Andean Puna Rangelands (*Bofedales*) in Ulla Ulla, Bolivia

Humberto Alzérreca, Jorge Laura, Freddy Loza, Demetrio Luna, and Jonny Ortega

INTRODUCTION

Native pastoral landscapes (*canapas*) of the *bofedal* type, also called *vegas, oconales, turberas*, and other names, are natural or artificial rangelands (Erikson, 2000) that may be permanently or seasonally humid; vegetation cover is principally pulviniform, which is adapted to the high groundwater level and differing water quality and distribution and is strongly influenced by climatic conditions and management history. The bofedales represent a very important resource for the pastoral economy of the altiplano and the high-Andean regions of Bolivia. In general, they are ecosystems of great biological and hydrological value. They are the habitat for numerous species of plants and animals, some of which are endemic, and they function as regulators for water flow by retaining water during the wet season and releasing it during the dry season.

A study of the classification and distribution of bofedales in 9,294,519 ha (100%) in the system of Lake Titicaca and Lake Poopó, Río Desaguadero, and Salar de Coipasa, reported 1586 units with a total area of 102,341 ha (1.1%). A classification system with ten categories was proposed based on a combination of the following criteria: height above sea level, hydrological regime, and soil salinity. By area, the hydromorph acidic upper-Andean bofedales stand out with 21,618 ha (21.1%), the Altiplano hydromorph alkaline bofedales with 29,474 ha (28.8%), and the altiplano hydromorph acidic bofedales with 20,101 ha (19.6%). By hydrological regime, the permanently humid bofedales (hydromorph or udic) cover an area of 80,218 ha (78%), and the seasonally humid bofedales (mesic or ustic) cover 22,123 ha (22%); sites may vary in size within a wide range of between 0.4 ha for both bofedales types to 2552 ha for the hydromorph and 3401 ha for the mesic type (Alzérreca et al., 2001a). Other authors suggest the classification of bofedales into: hydric, with *Deyeuxia chrysantha*; hydromorph, with *Distichia muscoides* and *Oxychloe andina*; mesic, with *Carex incurva* and *Werneria pygmaea*; and saline mesic, with *Deyeuxia* spp. (Troncoso, 1982a, b; De Carolis, 1982).

These ecosystems are extremely fragile, and drastic changes to the water regime (e.g. diversion of water for other uses) and agricultural use result in rapid and irreversible destruction of the habitat. Furthermore, minor changes in climate, water quantity, and management may result in drastic changes in species composition and plant diversity, a more severe microclimate, and failure of the traditional pasturing system, among other consequences (Liberman, 1987; Seibert, 1993; Messerli et al.,

1997). At present, the increase in demand for water to satisfy the needs of economic and demographic development constitutes a more serious and immediate danger for the development of sustainable use of these ecosystems than does inappropriate grazing. Despite the importance of the bofedales for Andean cattle ranching — in particular, of alpacas and llamas — there exists scant information about the response of the vegetation to grazing in the high puna (Bradford et al., 1987; De Carolis, 1982; Alzérreca et al., 2001; Farfán et al., 2000).

The bofedales are critical components of Andean pastoral production because they provide forage throughout the year. In zones with unimodal rainfall, with a wet season and a very distinct dry season, forage of sufficient quality from other sources is only available during the wet season, which makes the bofedales the only source of fodder of appropriate quality for animal nutrition during the dry season (Buttolph and Coppock, 2001; Scoones, 1991). Some of the more than 2,398,000 domesticated camelids, including all alpacas (around 400,000 animals) and the introduced vicuñas and ruminants (ovine, bovine, and equine), obtain part of their fodder from the bofedales. There are around 53,000 families of camelid breeders, some of whom are totally or partially involved in the use and management of bofedales.

It is also recognized, although not sufficiently documented, that the ecological degradation of some bofedales is a consequence of grazing mismanagement; for example, overstocking of animals, continuous grazing, and mixed herds (including sheep, which are considered harmful to the bofedales, and in some cases including pigs, which can have catastrophic effects). However, problems with land ownership (Caro, 1992; Buttolph, 1998; Coppock et al., 2002) and the decrease in the principal water source for the bofedales, the glaciers of the cordillera (Vuille et al., 2001), have also been mentioned. In this context, Seibert (1993) indicated that the present vegetation cover in Ulla Ulla is the result of former anthropogenic activities, in particular, grazing and burning. Other authors have shown that the degradation of the pastoral ecosystems of the Andes took place a long time ago and has created the present, more stable, state that has a high grazing tolerance (Ellenberg, 1979; Buttolph, 1998; Browman, 1974).

The notable tolerance of the bofedales and adjacent rangelands to grazing and the climate is related to its 1000-year-old pastoral history (Kent, 1988; Wheeler, 1991). The vegetation has adapted to grazing and to the cold by developing physiological and morphological characteristics that make it more tolerant to these factors, such as prostrate and rosette life-forms, small, often pubescent leaves, and a notable capacity to resprout. Specifically for the bofedales, changes in ecological character by grazing should not be underestimated simply because they affect small areas, as they are a continuous source of forage production and, for that reason, are inevitably subject to intensive use (Dodd, 1994). Other authors are not convinced of the negative effects of grazing in altiplano pastures (Buttolph, 1998; Genin, 1997; Alzérreca, 1982; Coppock, 2001).

This study is a contribution to increasing our limited understanding of the effects of pastoral management in the bofedales, the main objective being to determine the forage balance and the influence of other management factors on the present condition of the bofedales in the upper-Andean zone of Ulla Ulla, Bolivia. The hypothesis formulated to fulfill this objective is that differences in grazing intensity do not affect the pasture vegetation of the hydromorph bofedales and, therefore, changes to management practices do not affect bofedal vegetation.

METHODS

Two administrative units (ranches) in the pampa (prairie) and two in the mountain range (cordillera) in the Ulla Ulla zone were chosen for their similarity in potential production and socioeconomic differences in management (Figure 12.1). The Ulla Ulla zone is situated in the highmontane puna in the upper-Andean ecological belt. The climate is subhumid and very cold. The annual mean precipitation is 550 mm, mean temperature is 4.4°C, relative humidity is 51%, and temperatures are below freezing for around 230 d/a^{-1}. Politically, the zone is situated in a protected area with permitted traditional use (Area Natural de Manejo Integrado Nacional Apolobamba) in the municipality of Pelechuco,

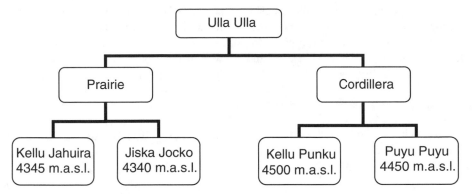

FIGURE 12.1 Organigram showing the different levels of approximation used in the study. (1) General level, which covers the Ulla Ulla zone, an extensive area where camelid cattle are bred; (2) physiographic level, which includes two units: cordillera and prairie (pampa); and (3) ranch level, the basic unit of study; two ranches (administrative units) were chosen in the cordillera and two in the prairie.

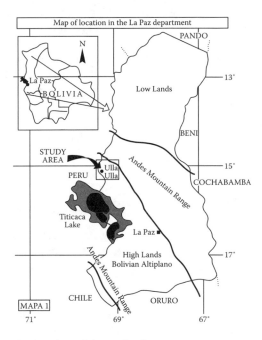

FIGURE 12.2 Map showing the location of the study site.

in the provinces of Bautista Saavedra and Franz Tamayo in the Department of La Paz (Figure 12.2). Each administrative unit (UADM) consists of a production body with a defined territory in which one or more families determine the management of their natural resources. The UADMs, locally called *ranches*, were characterized by the features described in the following subsections.

VEGETATION

Various techniques from preliminary definitions of units of vegetation, based on satellite images to intensive field sampling of the vegetation (from January 18 to January 28, 2001), using the point intercept method (Bonham, 1989) were employed. At least 6 transects of 100 sampling points were established in each floristic association. The floristic associations were subsequently grouped by rangeland type into hydric bofedales and hydromorphic, mesic, and arid rangelands. Species that were not identified in the field were sampled and pressed for later identification in the lab as well as in the Herbario Nacional de Bolivia.

DIVERSITY INDICES

The diversity indices were estimated from the overall means for each type of bofedal. The Shannon–Wiener index was used to quantify species diversity:

$$H' = -\sum_{i=1}^{S}(p_i)\,(\ln p_i)$$

where S = number of species, p_i = relative abundance of the ith species expressed as the proportion of total cover, and $\ln$ = natural logarithm.

The Berger–Parker index was used to determine dominance (d):

$$d = \frac{\text{Total number of species}}{\text{Total number of the most abundant species}}$$

The index proposed by McIntosh was used to calculate distribution:

$$E_m = \frac{N - U}{N - \left(\dfrac{N}{\sqrt{S}}\right)}$$

where N = number of individuals, S = number of species, and $U = \sqrt{S_{n_i}^2}$.

Floristic species richness was considered as the total number of species N in the community.

CARRYING CAPACITY

Pastoral value, a global index of canapas quality, was estimated using floristic composition (vegetation cover) as an indicator of quantity and an index of the forage quality of the component species of the rangeland (raw protein, digestibility, energetic content, cell walls, acceptability, and availability) (Daget and Pois-

sonet, 1971; Troncoso, 1982a). Carrying capacity (CC) is given in alpaca units (UAL), which correspond to an adult alpaca of 47 kg live weight that consumes 2.5% of its own weight in fodder per day.

STOCKING RATE

Stocking rate was determined from a census of the cattle in the UADM and by conversion of the data into UAL. The stocking rate of previous years was reconstructed from interviews with the farmers. The annual grazing cycle was determined by following the grazing herd and by interviewing farmers.

RESPONSE OF THE BOFEDALES TO MANAGEMENT

An important part of this study was to make a critical revision of previous works relating to the recovery of the bofedales and adjacent meadows, and to evaluate the grazing trials set up by Loza (2001) in 1999. The parameters determined were floristic cover, composition, and yield.

THE ECOLOGICAL CONDITION OF RANGELANDS

The ecological condition of the rangelands was estimated from the presence of palatable (desirable) species, poorly palatable (little desirable) species, and unpalatable (undesirable) species, which were classified as such by their ecological response to grazing. The state of the soil was also used to determine the condition of the rangelands. The condition reflects the present state of health of the rangeland with respect to animal production. The concepts of carrying capacity and plant succession were used as indicators of the potential production of the bofedales, under the assumption that the dynamics of these ecosystems correspond to that of a system in equilibrium, which responds to anthropogenic disturbance caused by management practices (Clements, 1916; Dyksterhuis, 1958). The value of this index is very limited in ecosystems in disequilibrium, in which the ecosystem dynamics depend more on the prevalent climate than on management practices (Bartels et al., 1993). In this respect, alternative models of ecosystem dynamics for systems in disequilibrium have been elaborated, which allow the generation of new ideas on different management practices and the evaluation of their sustainability (Westoby et al., 1989; Laycock, 1991; Ellis and Swift, 1988; Ellis, 1960; Dodd, 1994; Behnke and Scoones, 1993). In the subhumid, semiarid, and arid altiplano, it is possible that models of rangelands in dynamic equilibrium and in disequilibrium coexist, depending on the predominance of one or the other type of soil humidity and its periodicity; the mesic, hydric, and hydromorph canapas types tend to react as systems in equilibrium, whereas the arid types react as systems in disequilibrium.

RESULTS AND DISCUSSION

The ranches in the cordillera have hydromorph bofedale (udic), arid rangelands (*totorillares*), and small areas of hydric bofedales. The administrative unit in Kellu Punku has a total area of 270 ha and 131 ha in Puyu Puyu. In the prairie, there is also a unit of the mesic bofedal type (*ustic*); the total area of Kellu Jahuira was 666 ha, which was similar to the 591 ha in Jiska Joko. The relative proportion of hydromorph bofedal to the total area of the unit is variable: 54% in Puyu Puyu, 4% in Kellu Punku and Jiska Joko, and 18% in Kellu Jahuira. These relative proportions become more similar when other types of bofedales are included, resulting in 50% bofedales area in each unit except in

Kellu Punku where there was only a slight increase to 4.4% (Table 12.1).

The rangeland soils in all units were acidic (pH 4.8 to 6.1), and the texture was coarse with variants of limestone and clay. The pH of the water was also acidic in the units of the cordillera (4.8 to 6.5) and in some of the units of the pampa (6.2 to 8.0). The availability of water was greater in the bofedales of the cordillera (116 to 578 l/s^{-1}.) than in the pampa (52 to 72 l/s^{-1}.).

The types of rangeland listed in Table 12.1 are a typical example of the variety of available forage sources in the upper-Andean pastoral production units in Ulla Ulla. This demonstrates the variation in plant species composition along a humidity gradient: from hydrophilic plants (such as *Myriophyllum* spp., *Potamogeton* spp., *Lilaeopsis* spp., and *Lachemilla diplophylla* in permanently wet sites, to *Deyeuxia chrysantha* and *D. eminens* in the hydrophile–bofedal ecotone; *Distichia muscoides*, *Oxychloe andina*, and *Plantago tubulosa* in permanently humid areas that do not become inundated; *Werneria pygmaea*, *Lachemilla aphanoides*. and *Deyeuxia* sp. in hydromorph soils saturated with surface drainage; *Plantago tubulosa* and *Gentiana prostrata* in unsubmerged hydromorph soils with superficial groundwater level; *Festuca rigescens*, *Festuca dolichophylla*, and *Deyeuxia curvula* in temporarily humid sites with deep soils and shallow groundwater level; and *Pycnophyllum* sp., *Scirpus rigidus*, and *Aciachne pulvinata* in arid canapas.

TABLE 12.1
Types and areas of *Canapas* given in administrative units (UADM) in cordillera and prairie

Physiography	Cordillera		Prairie	
Types of Rangeland per UADM	Kellu Punku	Puyu Puyu	Kellu Jahuira	Jiska Joko
Hydromorph bofedal (wetland), *udic*	10.8	70.7	114.0	23.8
Mesic bofedal, *ustic*	0.0	0.0	148.7	292.6
Hydrophile bofedal	1.2	7.7	20.7	2.9
Arid rangeland, *totorillaces*	258.0	52.6	382.0	271.8
Total area of the administrative unit	270.0	131.0	665.4	591.1

Note: The unit used is hectare.

VEGETATION COVER

There were no statistically significant differences ($p > .05$) in vegetation cover, cover of palatable species, or area covered by water in the *mesic* bofedales, even though the area covered by water was more than twice as high in Kellu Punku than in Puyu Puyu. The opposite was true for the values of unpalatable species. Graminaceae and the like (Juncaceae and Ciperaceae) dominated in Puyu Puyu, whereas forbs dominated in Kellu Punku. In general, the bofedale in Kellu Punku appeared to be in better condition than the bofedale in Puyu Puyu, which seems to be due to the presence of greater quantities of water, even though the difference is not significant ($t = 0.199$). However, better management in this Kellu Punku is possible, as the water content of the organic layer is greater than that of the other unit, and the less intensive use permits recovery (Table 12.2).

Distichia muscoides and *Oxychloe andina* contribute the majority of the cover by grasses (52.3%) in Puyu Puyu; the former species is of fair forage quality and the latter of low quality; these species are also present in low densities in Kellu Punku, which positively influences the condition, forage value, and CC of the bofedale in Kellu Punku.

There were no significant differences in total ground cover between the hydromorph (udic) bofedales in Kellu Jahuira and Jiska Joko. However, the cover of palatable species is significantly greater ($t = 0.001$) in Kellu Jahuira than in Jiska Joko; the opposite is true for the vegetation cover of little-desirable species and of grasses (Table 12.2).

ECOLOGICAL CONDITION AND CARRYING CAPACITY (CC) OF THE UNITS

In the cordillera, the hydromorph bofedales in Kellu Punku have greater indicator values than those in Puyu Puyu. Nevertheless, these differences, except for a difference in the score for condition, disappear when all the rangelands are included in the calculation of these values. This is because of the incorporation of arid rangelands, which are much more extensive in Kellu Punku (258 ha) than in Puyu Puyu (52.6 ha). In addition, they have a slightly inferior

pastoral value and CC (1.56 UAL/ha in Kellu Punku and 1.57 UAL/ha in Puyu Puyu), which is sufficient to reduce the total CC to an overall slightly lower value in the UADM in Puyu Puyu. These data suggest better management of the arid rangeland in Puyu Puyu, but mismanagement of the key rangelands, the bofedales.

The indices of condition, pastoral value, and CC in the prairie units are higher in Jiska Joko than in Kellu Jahuira. These data suggest a better state of health of the bofedal in Jiska Joko, considering that the availability of water is very similar between bofedales. At the level of the UADM also, the indices are greater in Jiska Joko, which implies that the incorporation of other rangeland types into the UADM, to calculate the adjustment (per area) of the mean, also results in higher values. This may indicate that the rangelands of this UADM are better managed. In both physiographic zones, and at the bofedal and UADM level, the prairie unit in Jiska Joko shows greater values than the rest (Table 12.3).

The CC was estimated for the entire area of land covered by vegetation and accessible for grazing by cattle; a characteristic of small units of production with intensive use is that, in general, the entire area is grazed, provided nothing limits the access of cattle to the pastures. However, considering that 4 to 8% of unused area is recommended in these cases, the CC is expected to decrease by this percentage, and the discrepancy will increase with the stocking rate, as discussed in the following text.

DIVERSITY INDICES

In the cordillera sites, the indices for diversity and floristic species richness were similar but those for dominance and distribution were different (Table 12.4). The higher value for dominance in Puyu Puyu was attributed to the presence of the species *Distichia muscoides* (23.2%), *Oxychloe andina* (9.3%), and *Aciachne pulvinata* (8.1%) in the plant community, in which species of prostrate growth and low pastoral value predominate. *Aciachne pulvinata* (8.8%), of poor pastoral value, and the palatable *Werneria pygmaea* (13.0%) are also dominant in the bofedale at Kellu Jahuira, but to a lesser extent. The species distribution

TABLE 12.2
Values of ground cover in hydromorph bofedales in the Cordillera and prairie (pampa)

Units Detail	Cordillera			Prairie		
	Kellu Punku	Puyu Puyu	Statistic	Kellu Jahuira	Jiska Joko	Statistic
Vegetation cover (%)	63.0 (8.13)	59.9 (3.84)	$t = 0.198$, $df = 141$	69.4 (5.22)	75.8 (3.99)	$t = 1.6578$, $df = 118$
Palatable species (%)	38.7 (0.40)	15.3 (0.23)	$t = 0.206$, $df = 18$	29.6 (2.21)	61.3 (4.02)	$t = 0.001$, $df = 88$
Poorly palatable species (%)	23.3 (0.34)	45.0 (1.12)		39.5 (0.85)	14.5 (2.00)	
Unpalatable species (%)	1.0	0.0		0.3	0.0	
Covered by water (%)	62.9 (1.81)	28.3 (4.02)	$t = 0.199$, $df = 89$	22.7 (1.70)	19.3 (4.10)	
Organic layer cover (%)	21.3 (2.75)	10.9 (2.02)		7.1 (0.53)	4.83 (1.46)	
Grasses, etc. (%)	26.1 (0.28)	52.3 (0.97)		48.8 (3.64)	32.8 (0.76)	
Forb cover (%)	36.9 (0.42)	7.6 (0.14)		20.6 (1.53)	42.5 (0.50)	

Note: Values in parentheses are standard errors of means.

TABLE 12.3
Forage importance of (Hydromorphic) bofedales and other rangelands

Units	Cordillera		Prairie	
Detail	Kellu Punku	Puyu Puyu	Kellu Jahuira	Jiska Joko
Score of condition (REP)[a]	57.0 (good)	46.8 (fair)	55.37 (good)	73.04 (good)
Pastoral value (1–100) of bofedales	10.70	9.60	9.30	14.70
Carrying capacity (CC) in UAL/ha	2.40	2.12	2.90	3.24
Score of condition UADM[b]	47.2 (fair)	43.4 (fair)	48.27 (fair)	59.57 (good)
Pastoral value of UADMs	7.3	8.7	7.22	9.25
CC of the UADMs in UAL/ha	1.62	1.91	1.76	2.04

[a]REP = ecological response to grazing.
[b]UADM = administrative unit.

TABLE 12.4
Comparison of diversity indices of the bofedal vegetation

Units	Cordillera			Prairie		
Detail	Kellu Punku	Puyu Puyu	Statistic	Kellu Jahuira	Jiska Joko	Statistic
Diversity (Shannon–Wiener H)	2.68	2.28	$t = 2.999$, $df = 116$	2.10	2.70	$t = 3.523$, $df = 89$
Dominance (Berger–Parker) d	0.21	0.37		0.45	0.13	
Distribution (McIntosh's)	0.84	0.71		0.64	0.94	
Richness	29	25		26	19	

in Puyu Puyu is, therefore, more uniform than in Kellu Punku. There are no significant differences ($p > .05$) between the diversity indices per ranch in the two sites (Table 12.4). The greater dominance index in the bofedales at Kellu Jahuira (0.45) is due to the relatively high contribution of 31.2% to the total vegetation cover by the species *Distichia muscoides*, which also causes a species distribution at this site (0.64) that is less uniform than in the bofedal at Jiska Joko (0.94). At the ranch level, there is a notable difference in species richness between Jiska Joko, with only 19 species, and the other units; the bofedales with the best management, therefore, has the lowest species richness. Apparently, moderate grazing favors plants with higher growth forms, which compete advantageously with the smaller species characteristics of bofedales. In contrast to this, overused bofedales favor the growth of smaller species (Table 12.4)

SOCIOECONOMIC CHARACTERISTICS AND FORAGE BALANCE

Land ownership differs greatly between units. Kellu Punku is managed by a single family, with a herd size of 1004 UAL, the area of hydromorph bofedale in the property of 10.8 ha, and a total unit area of 270 ha. In contrast, in Puyu Puyu, the UADM is managed by 23 families with a per capita herd size of 52 UAL, 3.07 ha of hydromorph bofedale, and only 5.7 ha of total rangeland per family (Table 12.5). This is insufficient land and cattle to provide a living solely from ranching and, consequently, there is little interest and incentive to manage the rangelands better, which is manifested in the overstocking that the rangelands are subjected to in Puyu Puyu (7.2 UAL/ha) and an apparently greater degradation of bofedal resources than in Kellu Punku. This may also have contributed to the greater decline in UAL numbers in Puyu Puyu between 1996 and 2001.

Herd composition, once the data were converted to UALs, was 742 alpacas, 118 llamas, and 144 sheep in Kellu Punku, and 1155 alpacas and 41 sheep in Puyu Puyu. The alpaca–sheep combination is considered to create increased competition for forage, as both species prefer common forage plants. There were no differences in the grazing period in the bofedales between the sites; grazing took place between April and December in both cases. Grazing in the bofedales is therefore not continuous, and there is a recovery period during the rainy season from January to March, when forage of high nutritional value is available in the arid rangelands, and pests, diseases, and accidents in the very humid and contaminated environment of the bofedales can be avoided. The use of species with different grazing habits, high animal density, and the lack of prolonged periods of recovery create conditions favoring the presence of parasites in rangelands (Table 12.5). In summary, the forage resources did not meet the demand in either UADM of the cordillera during the sampling period, and the discrepancy was greater in Puyu Puyu. Data of stocking rate dynamics suggest that this overuse was continuous between May and November. The greater quantity of desirable forage plants and the bet-

ter condition and higher grazing value of the hydromorph bofedales in Kellu Punku stemmed from a combination of better management and greater water availability than in Puyu Puyu.

In the prairie, despite the greater numbers of proprietor families and the smaller per capita rangeland area in Kellu Jahuira than in Jiska Joko, the difference between the CC and the stocking rate was not very important, even though it was less than the CC in both units. The greatest difference was the decrease in animal numbers between 1996 and 2001: 978 UALs in Jiska Joko and 102 in Kellu Jahuira. This important reduction in grazing pressure in Jiska Joko may have positively influenced the improvement of the rangelands. The herd composition and the grazing periods in hydromorph bofedales are similar between units, being continuous in Jiska Joko and with a recovery period in February in Kellu Jahuira. In summary, the forage resource availability during the study period was lower than the demand, with similar values in both units. Animal dynamics data indicate that this disequilibrium is common. Other indicators suggest that the better condition of the rangelands in Jiska Joko was related to better management. Of all the ranches, Jiska Joko decreased stocking rate the most drastically in the period from 1996

TABLE 12.5
Additional data characterizing the production units in the Cordillera and the prairie

Units	Cordillera		Prairie	
Detail	Kellu Punku	Puyu Puyu	Kellu Jahuira	Jiska Joko
Number of families	1.0	23.0	6.0	13.0
Bofedal area (hydromorph type, Table 12.1) per family [ha]	10.8	3.07	20.43	1.83
Rangeland area per family in UADM [ha]	270	5.7	111.1	45.47
Herd size per family	1004	52	228.5	108.77
Stocking rate (CA) per UADM [UAL/ha]	3.72	9.13	2.06	2.39
Carrying capacity (CC) of the UADM [UAL/ha]	1.62	1.91	1.76	2.04
Difference CC CA in UADM [UAL/ha]	2.09	7.22	0.3	0.35
Stocking rate in 1966 per UADM [UAL]	1492	1794	1473	2392
Stocking rate in 2001 [UAL]	1004	1192	1371	1414
Difference in CA between 2001 and 1996 [UAL]	488	602	102	978
Grazing period in (hydromorph) bofedales	April–Dec	April–Dec	March–Feb	March–Jan
Herd composition	alp., she., lla.	alp., she.	alp., she.	alp., she.

Note: alp. = alpaca; she. = sheep; lla. = llama.

to 2001, and it was the ranch with the best indicator values of management. It seems that this regulation of stocking rate according to CC was the measure with the greatest positive impact on the vegetation of the UADM.

STUDY ZONE

In general, a deficit in available forage occurred in all four units, and this deficit was variable, with values ranging from 0.30 to 7.22 UAL/ha, with higher values in the cordillera than in the prairie. The adjusted mean of the excess load over all units is 1.16 UAL/ha (Table 12.6), showing that there was an excess of more than 1918 UAL in the four units in 2001, even though there had been a significant ($p = .0001$) decrease of 2166 UAL between 1996 and 2001. This theoretical calculation of forage balance does not take into account the consumption of forage by other herbivores in the zone. The vicuñas in particular increased in numbers to 8299 individuals by 2001. The calculation also neglects seasonal changes in forage resource availability.

When the observed overload is compared with the grazing intensity and these data are related to the state of the rangelands, a negative relationship is seen at the hydromorph bofedal level ($r = 0.76$, $p = .029$) and at the level of the UADMs ($r = 0.56$, $p = .016$). The average condition was taken as a measure of the present state of the rangelands because of a distinct management history, and grazing intensity was taken as a point measure for the year 2001. The values suggest that the present degradation is the result of high grazing intensity. Nevertheless, according to the data of animal population dynamics, the grazing intensity was even higher in the 5 years preceding 2001 (Figure 12.3).

The decrease in stocking rate was common in all units and can, therefore, be considered a consequence of the interaction between mismanagement of rangelands (overstocking) and a short cycle of low precipitation between 1996 and 1999 (El Niño–ENSO [El Niño–Southern Oscillation] year in 1997 to 1998, with precipitation of less than half the historical average). If the rangelands had been in better condition, they could have tolerated less drastic adjustments to the stocking rate in periods of crisis, but as they were not, and because other sources of forage were lacking, the situation escalated and the mortality increased. The owners found themselves forced to decrease the stocking rate dramatically, albeit to levels that, following our theoretical calculation, were still insufficient to create an equilibrium between the stocking rate and CC.

Grazing intensity in the bofedales and adjacent rangelands may be higher than usually reported when seasonal variation in the availability of forage under a more or less constant stocking rate is considered. Consequently, overstocking occurs in the dry season even if this is not the case during the wet season. The stocking rate and CC are usually estimated for the rainy season, but neither index is adjusted for the seasonal variation in forage availability. In a conservation management scenario, the stocking rate and CC estimated from the season with the lowest forage availability should be used, but this would not meet the economic needs of the cattle ranchers. High grazing intensity of forage plants in bofedales that do not have a period of dormancy during the dry season decreases their physiological activity with the low winter temperatures and, therefore, affects normal development. Indirect evidence for this

TABLE 12.6
Carrying capacity deficit in UAL at different levels

Zone	UAL/ha	Physiography	UAL/ha	Administrative Unit	UAL/ha	ha
		Prairie	0.32	Kellu Jahuira	0.30	665.4
Ulla Ulla	1.16			Jiska Joko	0.35	591.1
		Cordillera	3.77	Kellu Punku	2.10	270.0
				Puyu Puyu	7.22	131.0

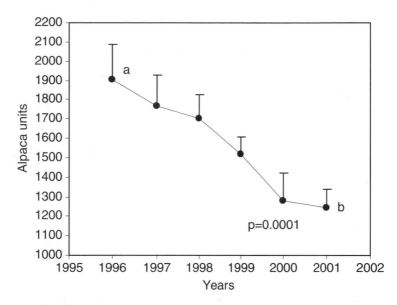

FIGURE 12.3 Annual changes to the mean animal population in the four units studied. Significant changes ($p < .001$) were detected between 1996 and 2001, which were attributed to the strong decrease in carrying capacity (shortage of fodder) because of several dry years (especially 1997 to 1998, an El Niño–ENSO year) in overused rangelands. Therefore, there was an increase in mortality, parasitosis (sarna), accidents (animals drowned while trying to graze hydrophile forage), and the removal of animals from the herd; some animals migrated to areas with greater forage availability (Perú). In 1996, the Cattle Ranchers Association distributed alfalfa hay to mitigate the forage shortage in the rangelands. In 2001, precipitation increased, which is reflected in a lesser decrease in the animal population. The rangelands subjected to high grazing pressure during the crisis period between 1996 and 2000 were bofedales.

can be obtained from studying the dynamics of the annual weight increment of the livestock.

There are significant differences in the forage yield per month in the bofedales of the Cordillera Oriental (Alzérreca et al., 2001a; ABTEMA/ORSTOM, 1998; Moreau et al., 1997). October is the month with the lowest forage availability and a CC of 1.9 UAL/ha, followed by the period from May to September and November (mean CC of 2.82 UAL/ha); there is frequently a surplus of forage during the rainy season between December and April with a CC of 3.94 UAL/ha.

Figure 12.4. shows the interactions between stocking rate, CC, and the annual dynamics of animal weight for female alpaca in Ulla Ulla. When the mean stocking rate of 3.01 UAL/ha, calculated for January in the UADMs (equivalent to the demand for forage), is maintained more or less constant throughout the year and is compared to the curve for CC of UADMs in Ulla Ulla estimated from forage-yield data (secondary information), an equilibrium of forage

surplus between December and April, and a shortage between May and November, becomes apparent. This forage-deficit situation during the dry season is common, and its magnitude tends to be amplified in dry years and dry periods (Alzérreca et al., 1999; Le Baron et al., 1979). Our data show a forage deficit of 1.16 UAL/ha (3.01 to 1.85) for the UADMs studied, even in the middle of the rainy season in January. The weight loss of alpaca (Villarroel, 1997) between July and October is a clear indicator of forage shortage during the dry season in Ulla Ulla. The curve for the monthly weight of female alpaca does not closely follow the curve for the CC of the UADMs ($r = 0.62$, $p = .030$), which indicates a great discrepancy between the nutritional requirements of the females and the availability of forage, further complicates the forage balance, and compromises the reproduction of the herd. When the four UADMs in this study are considered, a nonsignificant negative relationship between CC deficit and property size ($r = 0.28$, $p = .2667$), and a significant

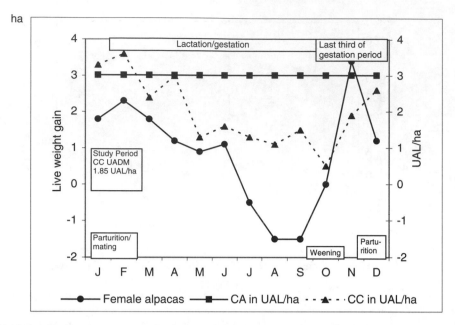

FIGURE 12.4 Monthly dynamics of supply, demand, and live weight of the alpaca in Ulla Ulla. Stocking rate (CA) in UAL/ha is the adjusted mean for the canapas of the four administrative units (UADMs), estimated for the study period in January 2001 and extrapolated up to December. Monthly carrying capacity (CC) for UADMs in Ulla Ulla was estimated from secondary information on forage yields. The curves for monthly live weight gain of adult female alpaca (in kg/month) are values for 1996. The carrying capacity estimated in this study for the month of January is 1.85 UAL/ha and was calculated as the adjusted mean of the rangeland carrying capacities in the four UADMs. The boxes indicate the stages of animal reproduction throughout the year and their interactions with CC and CA.

positive relationship between CC deficit and the number of families per UADMs ($r = 0.74$, $p = .0005$), became apparent, which suggest that the forage balance deficit is greater with smaller unit sizes and greater numbers of families. These results pose an important question in relation to the sustainability and conservation of the biodiversity in the bofedales and the adjacent prairies (considering climate as a fixed factor): Are the causes of overuse in bofedales socioeconomic rather than strictly management related? Present management practices, therefore, would be a consequence of an underlying situation (poverty) (Coppock et al., 2002; Tichit, 1995), and, therefore, the mismanagement of these bofedales may be necessary to ensure the survival of the cattle-ranching families in the face of a lack of alternative economic activities. In this respect, the better management of rangelands in the community of San José de Llanga in the central altiplano, for example, is partly related to the proximity of a

market for agricultural products such as mutton, cow's milk, and cheese, and to the access to technology and credit, factors that allow producers to sustainably manage their primary resources, soil and vegetation (Coppock, 2001). The need to reflect upon the future of the bofedale ecosystem, the diversity of forage plants, and the notable deficit of forage in the dry season is becoming urgent; even more so when one considers that this information presents a picture of management within a protected area, and that it is possible for this situation to become more critical.

RESTORATION OF BOFEDALES

Results from experiments on the possibilities of the restoration of hydromorph bofedales and similarly degraded ecosystems are few but more or less consistent in indicating the positive response of the vegetation to improved management (Alzérreca, Luna et al., 2001; Alzérreca et al., 1985a and b; Alzérreca et al., 1999; Farfán et al., 2000). Figure 12.5 shows the results of four restoration trials in the rangelands of Ulla Ulla. The increase in forage yield at the end of the second evaluation year is significant ($p < .05$) in all cases and is attributed to the interaction between recovery periods, fertilization (alpaca manure), and breaking the surface layer of compacted soil. It is thought that the alteration of the compact soil surface layer (furrows in the first and pits in the second, third, and fourth trials) facilitated the incorporation of manure into the substrate and the availability of nutrients to plants, promoted improved aeration for roots, and facilitated water infiltration. The recovery period permitted greater photosynthetic activity, resulting in the recovery of plant growth. Other trials have been established in mountain ranges in a transitional site between mountain range and prairie, and in the prairie.

The evaluation of a mesic acidic bofedal in Ulla Ulla (Alzérreca, 1998) showed significant differences in forage yield ($p = .0001$), increased from 668 to 2732 kg dry matter/ha), as well as in plant height (*Festuca dolichophylla*; $p = .0001$, increment between 9.6 to 15.6 cm) when measurements taken inside and outside a fence established in 1980 were compared. These results are consistent with the findings of Seibert (1993), who indicated that *Festuca dolichophylla* was the dominant species of the potential natural vegetation in Ulla Ulla. Our data show that the trend towards increased production is particularly notable in *F. dolichophylla*, *Festuca aff. rigescens*, and *Deyeuxia vicunarum*.

The results of the evaluation of a trial in a hydromorph bofedal in more arid regions of Sajama in the Cordillera Occidental show significant increases ($p < .05$) in phytomass yield by the third evaluation year in nongrazed (NG) treatments and treatments with controlled grazing (CG; approximate stocking rate of 1.0 UAL/ha. The animals were moved from the

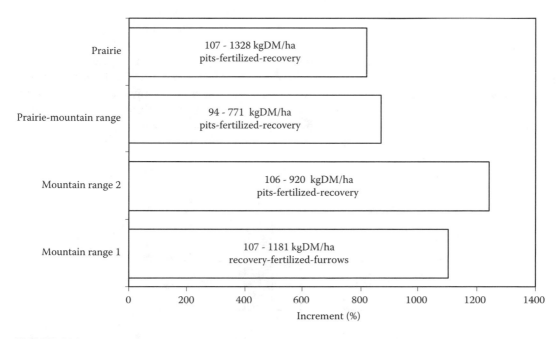

FIGURE 12.5 Results of four rangeland restoration trials, two located in a mountain range, one in a transitional site between mountain range and prairie, and one in the prairie. Only the greatest increment per trial is given. The forage yield is given for each trial in dry weight of the control and of the treatment that resulted in the greatest improvement. In the mountain range trial 1, furrows were made with two purposes, to break the compacted soil surface layer and to sow the introduced forage species, which were unable to compete with native species and disappeared rapidly.

bofedal when the utilization of key species reached approximately 50%) compared to traditional use (TU; approximate stocking rate of 2.0 UAL/ha). TU means continuous grazing for about 4 months during the transition from rainy to dry season (March to June) and 2 months of the rainy season (November and December). The rest of the year, the area is intensively grazed for short periods only. The NG and CG treatments do not differ ($p = .3099$) but both have a higher yield than TU ($p = .0014$ and $p = .0001$, respectively) (Figure 12.6).

The same trend was observed in the response of vegetation cover. However, there were no significant changes in floristic diversity, although the relative proportions of species differed between treatments and years, as can be seen in Figure 12.7, for the most common species.

The species *Eleocharis albibracteata, Werneria pygmaea, Hypochaeris taraxacoides,* and *Festuca rigescens* declined in growth in the TU plots and increased in the CG and NG sites, which classifies them as plants preferred by cattle (desirable), and their abundance indicates a rangeland with higher pastoral value. Species that increased with TU and decreased in the other treatments were *Lilaeopsis andina, Plantago tubulosa, Deyeuxia curvula, Cotula mexicana,* and *Deyeuxia rigescens*, which suggests low palatability or high grazing tolerance, but low competitive ability in restored rangelands; the same applies to *Werneria heteroloba* and *Carex* sp., although they show greater competitive ability in restored rangelands. The first five and the last species mentioned can be classified as being of medium palatability, whereas *Werneria heteroloba* can be classified as unpalatable (Troncoso, 1982a; De Carolis, 1982; Farfán et al., 2000). Species with prostrate growth forms such as *Lachemilla pinnata* and *Hypsela reniformes* have also been shown to decrease with grazing, but show no positive or even a negative response to the restoration treatments, which is not attributable to their low palatability, as they are classified as medium to very palatable, but to their low competitive ability compared to more aggressive plants. The decrease of *Scirpus deserticola* in all cases indicates that its response is independent of treatment.

Not all grazing trials in the bofedales, however, show significant increases in yields; Buttolph (1998) reported no significant increments in forage yield after a 3-year suspension of grazing in bofedales in Cosapa (Sajama Province, Oruro), but he found changes in species composition with continuous grazing. However, the indicator values with respect to grazing quality improved when the access of animals to the rangelands was permitted after a recovery period, which is possibly a consequence of the improvement in forage quality and the reduction of the stocking rate.

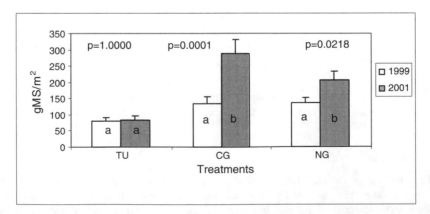

FIGURE 12.6 Traditional use (TU) shows no differences between years, unlike controlled grazing (CG) and nongrazed (NG) treatments, which show significant differences ($p < .05$) between years; both have a higher yield than traditional use in 1999 and 2001.

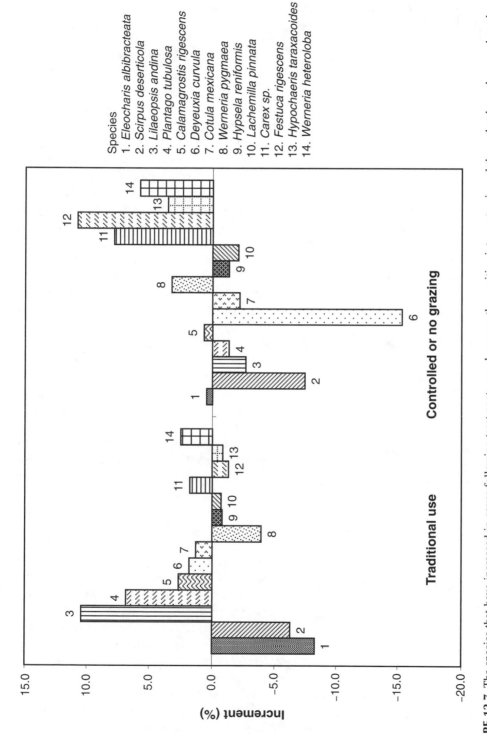

FIGURE 12.7 The species that have increased in cover following treatments are shown on the positive intercept axis and those that have decreased on the negative intercept. As no significant differences were observed between controlled grazing and nongrazed treatments, the mean of both are given. Notably decreased in grazed sites are the desirable plants *Eleocharis albibracteata*, *Werneria pygmaea*, *Festuca rigescens*, and *Hypochaeris taraxacoides*.

CONCLUSIONS

The forage balance for the study period was negative in all UADMs and varied in intensity from 0.30 UAL/ha to 7.22 UAL/ha, with higher values in the units of the cordillera (3.77 UAL/ha) than in the prairie (0.32 UAL/ha). On average, each hectare of rangeland in the study area was overstocked by 1.16 UAL/ha. Furthermore, this negative balance was a permanent state rather than a special feature of that year or evaluation period.

It is difficult to attribute the negative forage balance to a single cause, especially as our data suggest that it is the result of an interaction between variables such as climate (global warming and irregularity and high variation in the availability of water resources), management (overstocking in the dry season and the lack of alternative forage sources, among others), and socioeconomic factors (poverty).

The monthly dynamics of livestock weight show negative values during the dry season, which coincide with the low CC calculated for these months. Therefore, it is imperative, in the grazing mismanagement, that solutions be developed to cover this deficit and decrease the grazing pressure in the bofedales during dry season.

The grazing tolerance and resilience of the high-Andean hydromorph bofedales are worthy of note. In most cases, the forage vegetation responds positively to management intervention measures such as a decrease in stocking rates and the application of restoration treatments.

The differences in the abundance of key species, with respect to animal production (with good, fair, or bad fodder properties) in the floristic composition of a bofedal, is related to grazing pressure, and they can therefore be used as indicators of the quality of the grazing management practice in that particular bofedal type. For example, in hydromorph bofedales, the high proportion of *Aciachne pulvinata*, *Oxychloe andina*, and *Distichia muscoides* in the plant community indicates mismanagement, whereas the opposite is true for *Eleocharis albibracteata* and *Festuca rigescens*. The bofedales are key resources, but are not sufficient to cover the nutritional needs of cattle at the current stocking rate during the dry season and in years of drought.

SUMMARY

To study the effect of management on the vegetation of bofedales, two study sites were chosen in the high mountain range (cordillera; 4500 m asl), and two in the prairie (4340 m asl). All study sites were located in the Bolivian high Andes, northwest of Lake Titicaca, in the Ulla Ulla region near the Peruvian border. The study sites were selected based on similar ecological conditions but different socioeconomic and management characteristics between sites. In the cordillera, the stocking rate was higher than the CC, being very much higher in the Puyu Puyu (7.2 ALU/ha) than in the Kellu Punku (2.1 ALU/ha) site; significant decreases in animal populations (1996 to 2001) in both sites suggested that overstocking is common. Below-average precipitation (1996 to 1999) may also have played a role in this decrease. Rangeland condition, forage and vegetation cover, species richness, evenness, diversity, and ranch area were greater in the Kellu Punku hydromorph bofedales than in those in Puyu Puyu; in contrast, higher values for the number of proprietor families, greater dominance, and a greater decrease in animal numbers between 1996 and 2001 were detected in Puyu Puyu than in Kellu Punku. In the prairie, stocking rates were higher than carrying capacity in both sites; however, the difference was not as dramatic as in the cordillera. Cover of desirable species in the udic bofedales was higher ($t = .001$) in Jiska Joko (61.3%) than in Kellu Jahuira (29.6%); the ecological condition, total cover, evenness, and diversity were also greater in the first site. Differences between *mesic* bofedales in the prairie were minor, except for the differences in desirable species. Overall, the results suggested that mismanagement has played a major role in rangeland degradation, which threatens the biodiversity of forage species. However, there remains a potential for bofedal recovery.

ACKNOWLEDGMENTS

Part of the information is from "Estudio de la capacidad de carga en bofedales para la cría de alpacas en el Sistema Lagos Titicaca y Poopó, Río Desaguadero y Salar de Coipasa (TDPS)," commissioned by the Autoridad Binacional del Lago Titicaca (ALT) to the Asociación de Ganaderos en Camélidos de los Andes Altos (AIGA-CAA).

References

Alzérreca, H. (1998). Informe de consultaría. Asociación Integral de Ganaderos en Camélidos de los Andes Altos (AIGACAA). Informe Anual de Actividades Zona Norte. Proyecto de Mejoramiento Nutricional de las Familias de los Andes Altos a través del Aprovechamiento y Ampliación de la Crianza de Alpacas. La Paz, Bolivia.

Alzérreca, H., Cordero, R., Lara, R., and Rivero, V. (1985a). Ensayo de recuperación de la pradera nativa para Camélidos en Ulla Ulla. In Alzérreca, A.H. (Ed.), Reunión Nacional de Pastos y Forrajes, Séptima y Reunión Nacional de Ganadería, Quinta. Potosí, Bolivia, Mayo 1983. Asociación Boliviana de Producción Animal (ABOPA), Instituto Boliviano de Tecnología Agropecuaria (IBTA), Instituto Nacional de Fomento Lanero (INFOL), Banco Central de Bolivia (BCB). La Paz, Bolivia, pp. 169–184.

Alzérreca, H., Cordero, R., Lara, R., and Rivero, V. (1985b). Ensayo de recuperación de praderas nativas en serranías de Ulla Ulla. In Alzérreca, A.H. (Ed.), Reunión Nacional de Pastos y Forrajes, Séptima y Reunión Nacional de Ganadería, Quinta. Potosí, Bolivia, Mayo 1983. Asociación Boliviana de Producción Animal (ABOPA), Instituto Boliviano de Tecnología Agropecuaria (IBTA), Instituto Nacional de Fomento Lanero (INFOL), Banco Central de Bolivia (BCB). La Paz, Bolivia, pp. 157–168.

Alzérreca, H., Luna, C.D., Prieto, C.G., Cardozo, G.A., and Céspedes, E.J. (2001a). Estudio de la capacidad de carga en bofedales para la cría de alpacas en el sistema TDPS-Bolivia. Informe final de consultaría, subcontrato 21.11. Programa de las Naciones Unidas para el Desarrollo (PNUD/GEF), Autoridad Binacional del Lago Titicaca (ALT), Gerencia de Biodiversidad, Asociación de Ganaderos de Camélidos (AIGA-CAA). La Paz, Bolivia.

Alzérreca, A.H., Prieto, C.G., Laura, C.J., Luna, C.D., and Laguna, B.S. (2001b). Características y distribución de los bofedales en el ámbito Boliviano del sistema TDPS. Informe final de consultaría, subcontrato 21.12. Programa de las Naciones Unidas para el Desarrollo (PNUD/GEF), Autoridad Binacional del Lago Titicaca (ALT), Gerencia de Biodiversidad, Asociación de Ganaderos de Camélidos (AIGACAA). La Paz, Bolivia. 177 pp.

Alzérreca, H., Aquino, E., and Prieto, G. (1999). Informe de consultoría en pastos y forrajes de la zona norte de La Paz. Asociación Integral de Ganaderos de los Andes Altos (AIGACAA). La Paz, Bolivia.

Alzérreca, A.H. (1982). Recursos forrajeros nativos y la desertificación de las tierras altas de Bolivia. In Geyger, E. and Arce, C. (Eds.). *Ecología y recursos naturales en Bolivia*. 3 al 8 de Mayo de 1982, Centro Portales, Cochabamba, Bolivia. Simposio. Centro Pedagógico y Cultural Portales, Instituto de Ecología de la Universidad Mayor de San Andrés (UMSA). Imprenta Cochabamba, Cochabamba, Bolivia, pp. 23–42.

ABTEMA/ORSTOM (1988). Asociación Boliviana de Teledetección y Medio Ambiente e Instituto Francés de Investigación Científica para el Desarrollo en Cooperación. Informe Anual del Proyecto DME-SUR.

Bartels, G.B., Norton, B.E., and Perrier, G.K. (1993). An examination of the carrying capacity concept. In Behnke, R.H., Scoones, I., and Kerven, C. (Eds.), *Range Ecology at Disequilibrium: New Models of Natural Variability and Pastoral Adaptation in African Savannas*. Overseas Development Institute, London, England, pp. 89–103.

Behnke, R.H. and Scoones, I. (1993). Rethinking range ecology: implications for rangeland management in Africa. In Behnke, R.H., Scoones, I., and Kerven, C. (Eds.), *Range Ecology at Disequilibrium: New Models of Natural Variability and Pastoral Adaptation in African Savannas*. Overseas Development Institute, London, England, pp. 1–30.

Bonham, D. (1989). *Measurement for Terrestrial Vegetation*. Wiley, Interscience Publication, New York.

Bradford, P.W., Bryant, F.C., and Belaun Fraga, V. (1987). An evaluation of range condition on one range site in the Andes of Central Perú. *J Range Manage*, 40(1): 41–45.

Browman, D.L. (1974). Pastoral nomadism in the Andes. *Curr Anthr*, 15: 188–196.

Buttolph, L. (1998). Rangeland Dynamics and Pastoral Development in the High Andes: The Camelids Herders of Cosapa, Bolivia. Ph.D. dissertation, Utah State University, UT.

Buttolph, L. and Coppock, L. (2001). Project Alpaca: Intensified alpaca production leads to privatization of key grazing resources in Bolivia. *Rangelands*, 23(2): 10–13.

Caro, D. (1992). The socioeconomic and cultural context of Andean pastoralism. Constraints and potential for biological research and interventions. In Valdivia, C. (Ed.), *Sustainable Crop-Livestock Systems for the Bolivian Highlands*. Proceedings of an SR-CRSP Workshop. Published by University of Missouri-Columbia, USA, pp. 71–92.

Clements, F.E. (1916). *Plant Succession, An Analysis of the Development of Vegetation*. Carnegie Institution of Washington, Washington D.C., USA, 645 pp.

Coppock, L. (2001). Executive summary. In Coppock, D.L. and Valdivia, C. (Eds.), Sustaining agropastoralism on the Bolivian Altiplano: The Case of San José Llanga. How Culture, Livestock, Technical Innovation and Rural/urban Linkages Influence a Societies' Ability to Cope with Drought and Economic Change. Global Livestock Collaborative Research Support Programme. Department of Rangeland Resources. Utah State University, Logan, UT, pp. xxvii–xxxiii.

Coppock, L., Abaud, A., Alzérreca, H., and Desta, S. (2002). Rangeland policy perspectives from Bolivia, Ethiopia and Kenya. *Rangelands*, 24(4): 35–36.

Daget, P. and Poissonet, J. (1971). Une méthode d'analyse phytologique des praires. *Ann Agron*, 22(1): 5–41.

De Carolis, F. (1982). Caracterización de bofedales y su relación con el manejo de alpacas y llamas en el Parque Nacional Lauca. Informe de consultoría, segunda parte, Santiago, Chile.

Dodd, J.L. (1994). Desertification and degradation in sub-Saharan Africa. *Bioscience* 44(1): 28–34.

Dyksterhuis, E.J. (1958). Range conservation based on sites and condition classes. *J Soil Water Conserv*, 13: 104–115.

Ellenberg, H. (1979). Man's influence on tropical mountain ecosystems in South America. *J Ecol*, 67: 401–416.

Ellis, J.E. (1960). Influence of grazing on plant succession of rangelands. *Botanical Rev*, 26: 1–78.

Ellis, J.E. and Swift, D.M. (1988). Stability of African pastoral ecosystems: alternate paradigms and implications for development. *J Range Manage*, 41: 450–459.

Erikson, C.L. (2000). The LAKE Titicaca basin: a pre-Colombian built landscape. In Lentz, D.L. (Ed.), *Imperfect Balance: Landscape Transformations in the Pre-Colombian Americas*. Columbia University Press, New York, pp. 312–356.

Farfan, R.L., San Martin, H., and Durant, A.O. (2000). Recuperación de praderas degradadas por medio de clausuras temporales. *Revista de Investigaciones Veterinarias del Perú*, 11(1): 77–81.

Genin, D. (1997). Problemática del desarrollo sostenible de las comunidades pastoriles Andinas: El ejemplo del Altiplano Arido Boliviano. In Liberman, M. and Baied, C.A. (Eds.), *Desarrollo sostenible de ecosistemas de montaña: manejo de áreas frágiles en los Andes*. UN University–USAID: PL 480-LIDEMA-IE, UMSA, La Paz, Bolivia, pp. 141–152. (English version: Baied, C.A. and Liberman, M. (Guest Editors), 1997: Managing Fragile Ecosystems in the Andes. University of California Press, Berkeley, CA. *Mountain Research and Development*, 17(3): 1–296.)

Kent, J. (1988). El sur más antiguo: revisión de la domesticación de los camélidos andinos. In Flores Ochoa, J. (Ed.), *Llamichos y paqocheros, pastores de llamas y alpacas.* Consejo Nacional de Ciencia y Tecnología, Editorial Universitaria, UNSAAC, Cuzco, Perú, pp. 437–470.

Laycock, W.A. (1991). Stable states and thresholds of range conditions on North American rangelands: a viewpoint. *J Range Manage,* 32: 201–208.

Le Baron, A., Bond, L.K., Aitken, P.S., and Michaelsen, L. (1979). An explanation of the Bolivian highlands grazing – erosion syndrome. *J Range Manage,* 32(3): 201–208.

Liberman, M. (1987). Impacto ambiental de un proyecto de irrigación en praderas nativas del Altiplano Norte de Bolivia. In Alzérreca, H. (Ed.), *Reunión nacional en praderas nativas de Bolivia, Primera.* CORDEOR, CEE, PAC. Oruro 1987. Oruro, Bolivia, pp. 12–33.

Loza, F. (2001). Datos no publicados de ensayo de pastoreo establecido en 1999 en bofedale hidromórfico salino en Sajama, Oruro, Bolivia.

Messerli, B., Grosjean, M., and Vouille, M. (1997). Areas protegidas y recursos naturales en el Altiplano Andino desértico. In Liberman, M. and Baid, C. (Eds.), *Desarrollo sostenible de ecosistemas de montaña: manejo de áreas frágiles en los Andes.* The United Nations University, Secretaria Ejecutiva PL-480, Liga para la Defensa del Medio Ambiente e Instituto de Ecología, Imprenta Latina, La Paz, Bolivia, pp. 15–26.

Moreau, S., Le Toan, T., and Brosich, T.B. (1997). Quantification of biomass of native forages in the Northern Bolivian Altiplano through C-band SAR data. III Jornadas Euro Latinoamericanas del Espacio (ESA), México D.F., México, 1–6.

Scoones, I. (1991). Wetlands in dry lands: Key resources for agricultural and pastoral production in Africa. *Ambio,* 20: 366–371.

Seibert, P. (1993). La vegetación de la región de los Kallawaya y del Altiplano de Ulla Ulla en los Andes Bolivianos. *Ecología en Bolivia,* 20: 1–84.

Tichit, M. (1995). Diversidad de la actividad ganadera en las unidades de producción de Turco Marka. In Genin, D., Picht, H., and Lizarazu, R. (Eds.), *Waira Pampa, un sistema pastoril Camélidos-Ovinos del Altiplano Arido Boliviano.* ORSTOM-CONOPAC-IBTA-CID, La Paz, Bolivia, pp. 73–89.

Troncoso, R. (1982a). Evaluación de la Capacidad de Carga Animal del Parque Nacional Lauca. Informe de consultaría. Corporación Nacional Forestal. Región Tarapacá, Arica. Arica, Chile, p. 147.

Troncoso, R. (1982b). Caracterización Ambiental del Ecosistema Bofedal de Parinacota y su Relación con la Vegetación. Tesis Ing. Agr. Universidad de Chile, Facultad de Ciencias Agrícolas, Veterinarias y Forestales, Santiago, Chile, 1–222.

Villarroel, J. (1997). Balance Forrajero y Nutricional en Áreas de Producción de Alpacas de Ulla Ulla. Tesis Ing. Agr., Facultad de Ciencias Agrícolas y Pecuarias Martín Cárdenas, Universidad Mayor de San Simón, Cochabamba, Bolivia, 1–111.

Vuille, M., Bradley, R.S., Werner, M., and Keiming, F. (2001). Century climate change in the tropical Andes. Paper submitted to Climatic Change. Climate System Research Center, Department of Geosciences, University of Massachusetts, Amherst, MA.

Westoby, M., Walter, B., and Noy Meir, I. (1989). Opportunistic management for rangelands not at equilibrium. *J Range Manage,* 42: 266–274.

Wheeler, J. (1991). Origen, evolución y estado actual. In Fernández-Baca, S. (Ed.), *Avances y Perspectivas del Conocimiento de los Camélidos sur Americanos.* FAO, Bureau Regional América Latina y Caribe, Santiago, Chile, pp. 11–48.7

13 Functional Diversity of Wetland Vegetation in the High-Andean *Páramo*, Venezuela

Zulimar Hernández and Maximina Monasterio

INTRODUCTION

Tropical and subtropical highland areas are characterized by a high diversity per unit area (Körner 1999), which is reflected not only in species numbers but also in the functional variability of the ecosystem (Walker et al. 1999). We analyzed functional variability and architectonic models to develop an ecological interpretation of taxonomic diversity in Andean wetlands.

Plant species are often grouped according to their morphological characteristics, e.g. for temperate regions, in terms of the height of growth meristems during the unfavorable season, as proposed by Raunkier (1934). This morphological grouping, however, is not directly applicable to the plant species in high tropical mountains. Here, the widest temperature oscillations occur daily instead of seasonally, growth is continuous throughout the year, and dormancy of the growth meristems occurs during a few hours at night, when temperatures go below 0°C (simulating the latency season that lasts several winter months in extratropical regions) (Sarmiento 1986; Rundell et al. 1994).

For this reason, Hedberg (1964) proposed a classification of the Afroalpine flora according to their different adaptive strategies into five groups: caulescent rosettes, acaulescent rosettes, tussock grasses, cushion and sclerophyllous shrubs, like some forbs and grasses that are commonly temperates. Hedberg's system has been accepted as being adequately representative of the common pattern in the cold intertropics to which the diverse plant communities of the Andean páramos belong (Hedberg and Hedberg 1979; Smith and Young 1987), from the humid páramo grasslands in Colombia (Hofstede 1995) to the dry páramos in Venezuela (Monasterio 1980a).

Tropical and subtropical highland areas are characterized by a high diversity per unit area (Körner 1999), which is not only reflected in the species numbers but also in the functional variability of the ecosystem (Walker et al. 1999). From this perspective, the different lifeforms can be interpreted as architectonic models conditioned for a given function. For example, in the giant rosettes of the *Espeletia* genus, the marcescent leaves encasing the aerial stem prevents freezing during the night and allows the reestablishment of photosynthetic activities during the first hours of the day (Goldstein et al. 1984). Therefore, an analysis based on functional variability and architectonic models can be used for developing an ecological interpretation of taxonomic diversity.

Andean wetlands are located in the driest páramo of the Cordillera de Mérida, Venezuela. They occupy geomorphologic situations such as valley bottoms or microterraces, created by the deposition of fluvioglacial materials under the influence of continuous daily freeze–thaw cycles (Schubert 1979). These wetland environments are relatively more stable in terms of their temperature cycles, allowing the establishment of a grass vegetation (covering less than 10%

of land surface) made up of highly palatable forbs and grasses (80% vegetation cover), such as *Calamagrostis mulleri*, *Muehlenbergia ligularis*, *Carex albolutescens*, and *Agrostis breviculmis*, which, according to Ivlev's preference (Ramirez et al. 1996), have high protein content. In this sense, these environments are denominated as Andean grasslands. These wetland environments are dominated by Andean grasses (Molinillo 1992) with a high species richness and a high vegetation cover (80%). However, Andean wetlands occupy less than 10% of the land surface, whereas shrubland with caulescent rosettes of the *Espeletia* genus, sclerophylous shrubs, and cushions, all species with little palatable forage, dominate in the huge stretch of more than 90% of the land surface (Molinillo and Monasterio 1997a).

Andean grasses have high species richness, good stability, appropriate ground conservation and, together with other wetlands and marshes, form areas with high regional diversity (Molinillo and Monasterio 2002). However, the diversity in the Andean wetlands is seriously threatened by intensive grazing (Molinillo and Monasterio 1997a). Recently, the Andean páramos has been subjected to an accelerated process of degradation and transformation, characterized by farming intensification and continuing expansion of the agricultural frontier (Luteyn 1992; Hofstede 1995). The intensity and frequency with which the wetlands are visited by cattle are correlated with the agriculture activities (Pérez 2000). The increasing human intervention, frequently involving long fallow agriculture (Monasterio 1980b; De Robert and Monasterio 1993), led to higher stocking rates, grazing, and the formation of induced wetlands in which the dynamics are controlled by grazing patterns, especially during the dry season when the animals are gathered together in the Andean grasslands (Molinillo 2003).

During the fieldwork in 2002–2003, we observed that the cattle consumed the palatable forbs and grasses and trampled the vegetation in the Andean grasses. For this reason, the target of this study was to analyze the functional variability in species of the Andean wetlands by using ecological variables that are likely to be affected by grazing in the Andean páramos, such as the aboveground/belowground phyto-

mass rate and growth meristem's protection. We compared the species sensitive to trampling in both intensively grazed and extensively grazed wetlands. This allows us to analyze the impact that extensive grazing has on life-forms that are critical for the conservation and sustainable use of the Andean wetland. In this work, we do not study the direct effect of grazing on the studied species, but some results can be interpreted as the effect of intensive grazing (0.2–0.4 UA/ha) on Andean grass (Molinillo 1992).

STUDY AREA

The study was undertaken in the wetland of Mifafí, in the Sierra La Culata of the Cordillera de Mérida, Venezuela. The area is a dry páramo in the cold intertropic, where the annual isotherm is 2.8°C, and the average yearly rainfall is 869.3 mm (Monasterio and Reyes 1980). The precipitation regime is unimodal, with a single maximum rainfall peak and a dry season from December to March. The Ciénaga de Mifafí is an Andean grassland (Molinillo and Monasterio 1997a) dominated by highly palatable forbs and grasses, acaulescent rosettes and cushions with little palatable forage, and on the side of wetland, caulescent rosettes of the *Espeletia* genus, which come from the rosette land, where the giant species *Espeletia timotensis* and *Espeletia spicata* (Monasterio 1980a) dominate.

The study was carried out for six species; three life-forms were analyzed: acaulescent rosettes, caulescent rosettes, and cushions. The acaulescent rosettes are studied in *Plantago rigida* and *Hypochoeris setosa*, the caulescent rosettes in *Espeletia batata* and *Espeletia semiglobulata*, and the cushions in *Aciachne pulvinata* and *Azorella julianii*. Forbs and grasses were not selected for this study because, although these life-forms are preferred by cattle, we were mainly interested in measuring the impact of trampling in Andean wetlands. The species were selected depending on the following criteria: annual or perennial, low consumption, little forage, and deficient protein content.

A key case study of grazed Andean wetland in the Cordillera de Mérida (Molinillo and Monasterio 2002) demonstrated that these six species are not palatable or consumed by cattle. Acaulescent rosettes strongly benefit from grazing. In a similar way, the cushion *Aciachne pulvinata* occupies open valley bottom areas where intensive grazing facilitates its establishment (Molinillo 1992). Caulescent rosettes of the *Espeletia* genus are not very palatable because they contain toxic secondary compounds in their young leaves. Nevertheless, they may be occasionally consumed by cattle to complete the diet. Finally, it is not well known if the cushion *Azorella julianii* is consumed.

A hydrological gradient associated with superficial drainage patterns within wetlands determines plant communities in terms of the dominant life-form structure. Humid areas are dominated by acaulescent rosettes, forbs, and grasses, and in the dry areas, caulescent rosettes of the *Espeletia* genus and cushions are common (Figure 13.1).

The study area is located in the National Park of Sierra La Culata. Despite the protected status of the study area, some activities such as the livestock grazing are not controlled by the park authorities mainly due to disagreement on management plans between the state and the local community. The problem of extensive grazing has not been solved yet (Molinillo and Monasterio 1997b; Monasterio and Molinillo 2003).

METHODS

Functional variability is analyzed for those variables that respond to the micro- and mesoclimatic thermal oscillations of the Andean páramos. These variables, which allow us to understand some functional characteristics in the wetland, are: architectonic model, aboveground/belowground phytomass (AP/BP) and necromass/total phytomass (N/TP) ratios, and growth meristem's protection (a distinctive characteristic of tropical regions).

To calculate phytomass ratios, aboveground and belowground biomass are calculated on adult, reproductive individuals. Biomass was determined using the cropping method: by harvesting and separating into leaves, flowers, stems, rhizomes or belowground stems, roots, and necromass. The phytomass ratios were calculated on a dry weight basis.

Growth meristem's thermal protection for the six species was analyzed through the temperature differences inside and outside the meristems in October, November, and December of 2002. Air temperature, soil surface temperature, and humidity were measured with a Lambrecht (°K) thermohygrometer. Leaf temperatures for each species were measured using copper-constant (36 caliber) thermocouples, at 2-h intervals during 3 days.

To analyze how plant architecture is related to ecosystem functioning in páramo wetlands, soil water-holding capacity was determined in stands mainly dominated by *P. rigida*, and used as a relatively simple model system. Soil sections (of 50×50 cm surface area) were extracted at different soil depths (0–4 cm, 4–10 cm, and 0–10 cm). These sections were then saturated with water for 48 h and weighed (saturated weight) and then dried and weighed again (dry weight). The difference between saturated and dry weights indicated the percentage of water saturation and the soil water-holding capacity per unit surface area for each soil depth.

RESULTS

The results of the phytomass ratios indicated that the AP/BP ratio was the variable showing the largest difference between species, with low values for *P. rigida*, *H. setosa*, and *A. julianii* and high values for *E. batata*, *A. pulvinata*, and *E. semiglobulata* (Table 13.1). Hence, two phytomass distribution patterns are evident, with species that assign a high proportion of total phytomass in aerial structures and species that accumulate a large proportion in belowground structures (Figure 13.2).

An indicator of the importance of phytomass storage in senescent organs in páramo flora is the necromass/leaf biomass ratio. The species with the highest ratios are *P. rigida* and *E. semiglobulata*; they are also the species with the more pronounced differences in AP/BP ratios (Figure 13.3). The high aerial phytomass proportion in rosette species is largely due to the leaf necromass attached to the aerial stem (Monasterio 1986), whereas in *P. rigida*, most of the necromass is attached to the belowground stem. Even so, both species share the low ratios

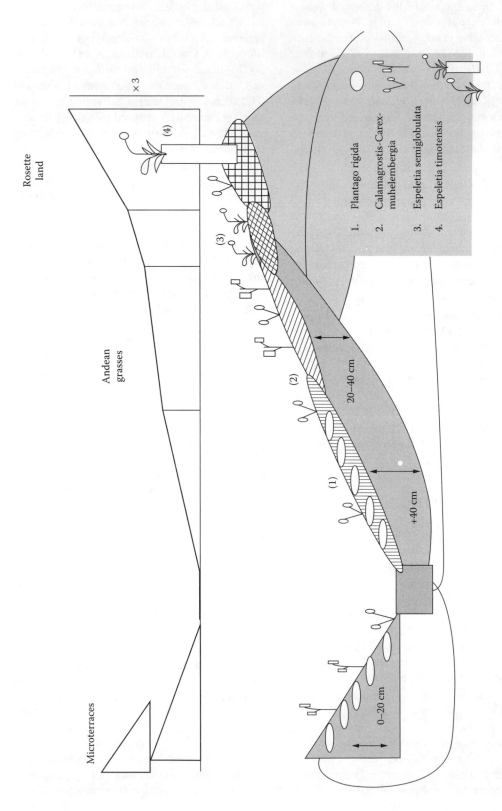

FIGURE 13.1 Horizontal spatial distribution of six species in the Ciénaga de Mifafí (4300 m), Cordillera de Mérida, Venezuela.

TABLE 13.1
Average phytomass ratios (± standard deviation) for species from Andean páramo wetlands

Species	AB/BB	Biomass Ratios				
		ALB/TB	NAB/TB	ROB/TB	RB/TB	N/TP
Plantago rigida	0.185 ± 0.074	0.130 ± 0.058	0.422 ± 0.084	0.424 ± 0.104	0.022 ± 0.028	0.704 ± 0.038
Hypochaeris setosa	0.457 ± 375	0.174 ± 0.108	0.546 ± 0.144	0.171 ± 0.108	0.108 ± 0.041	0.293 ± 0.164
Azorella julianii	0.200 ± 0.157	0.155 ± 0.101	0.537 ± 0.095	0.306 ± 0.141	0	0.281 ± 0.099
Espeletia batata	2.219 ± 1.906	0.458 ± 0.200	0.391 ± 0.203	0.032 ± 0.036	0.116 ± 0.065	0.425 ± 0.177
Espeletia semiglobulata	1484.03 ± 4405.49	0.293 ± 0.116	0.685 ± 0.125	0.021 ± 0.017	0	0.742 ± 0.092
Aciachne pulvinata	3.773 ± 1.704	0.759 ± 0.104	0	0.240 ± 0.104	0	0.528 ± 0.0857

Note: AB/BB = aboveground/belowground biomass; ALB/TB = assimilatory leaf biomass/total biomass; NAB/TB = nonassimilatory biomass (aerial and underground stems)/total biomass; ROB/TB = root biomass/total biomass; RB/TB = reproductive biomass/total biomass; N/TP = necromass/total phytomass.

Phytomass distribution (%)

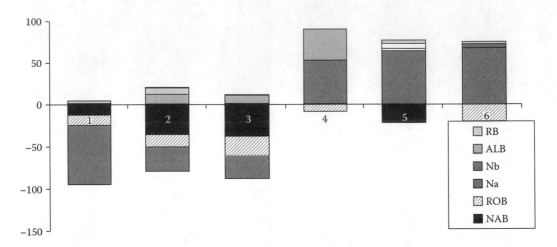

FIGURE 13.2 Vertical spatial distribution of phytomass in species of Andean wetlands. (1) *Plantago rigida*, (2) *Hipochoeris setosa*, (3) *Calandrinia acaulis*, (4) *Azorella julianii*, (5) *Espeletia batata*, (6) *Espeletia semiglobulata*, and (7) *Aciachne pulvinata*. NAB: nonassimilatory biomass, including aerial stems and rhizomes; ROB: root biomass, including primary and secondary roots; AN: aboveground leaf necromass; BN: belowground leaf necromass; LB: leaf biomass; and RB: reproductive biomass.

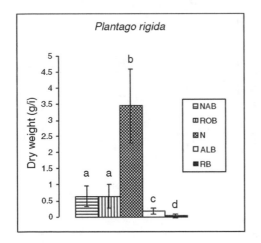
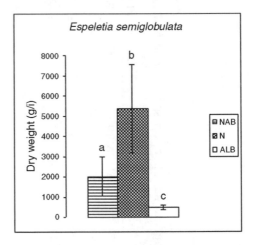

FIGURE 13.3 Average mass in different plant compartments for *P. rigida* and *E. semiglobulata* (± standard deviation). NAB: nonassimilatory biomass, in aerial stems or rhizomes; ROB: root biomass; N: leaf necromass; ALB: photosynthetic biomass; and RB: reproductive biomass. Different letters (a, b, c, d) indicate significant differences ($p = .05$).

of assimilatory leaf biomass to total phytomass, suggesting that they are slow-growing, long-lived species that store large amounts of phytomass during their life cycles.

In general terms, the six species accumulate a large proportion of phytomass as leaf necromass and show a low proportion of photosyn-

thetic biomass. As a consequence, it suggests that extensive livestock grazing may enhance the vegetation cover in the Andean wetlands because it increases the trampling of species that have a large proportion of buried leaf necromass (such as *P. rigida*, *H. setosa*, and *A. julianii*), and it decreases the low proportion of leaves in long-lived species (such as the *Espeletia* genus) that are occasionally consumed by cattle (Molinillo 1992). The root biomass ratios reported here for all species are below those typically found in alpine ecosystems (Körner 1999). Within the species studied, the large proportions of root biomass are replaced by belowground stems in species such as *H. setosa* and *A. julianii*. Hence, the density of *H. setosa* increases in areas with intensive grazing.

The Mifafí wetland showed an annual isotherm of 4.7°C (±2.1°C) and pronounced daily temperature variations, with a maximum of 12.8°C (±2.4°C) and minimum of 0.7°C (±1.3°C) for the study period. All analyzed lifeforms protect their growth meristems from night frost, and this is reflected in the higher temperatures within meristems compared to external temperatures. Depending on the number of hours that meristems stay below 0°C, three adaptive strategies of wetland vegetation can be defined: species showing no freezing temperatures, such as *P. rigida* and *H. setosa*; species staying only a few hours under freezing temperatures, such as *E. semiglobulata* and *A. pulvinata*; and species with protected mer-

istems, but which, nonetheless, spend several hours at subzero temperatures, such as *E. batata* and *A. julianii* (Table 13.2).

Moreover, the parabolic distribution of leaves (to protect growth meristems located in the center) in all species, except for *A. pulvinata* (in which more complex mechanisms are involved), contributes to the avoidance of leaf overheating during peak radiation hours (Monasterio and Sarmiento 1991). Continuous trampling by cattle can change the parabolic distribution of leaves, which protects the growth meristem from night frost, and this can explain the fast drop in temperature when the leaves of *Espeletia batata* were damaged by trampling.

Finally, the results of water saturation in vegetation stands dominated by *P. rigida* indicate that it is in the top 4 cm of the soil profile that the highest water-holding capacity is found (1640 l m^{-3}, Table 13.3). This coincides with the soil layer in which most of the leaf necromass from acaulescent rosettes are concentrated. The water capture is seriously threatened by intensive grazing and cattle trampling, which adversely affects hydrological functions in the Andean wetlands.

TABLE 13.2
Average maximum and minimum temperatures and number of hours registered with temperatures below 0°C for six species from Andean wetlands

Species	Vegetation Thermic Response					
	Daily Maximum Temperature (°C)		Daily Minimum Temperature (°C)		Number of Hours Below 0°C	
Plantago rigida	E	21 ± 6.6	E	0.2 ± 0.01	E	0
	I	18.7 ± 4.3	I	1.4 ± 0.4	I	0
Hypochoeris setosa	E	19.3 ± 6.7	E	1.7 ± 1.07	E	10
	I	15 ± 4.5	I	0.4 ± 1.4	I	0
Espeletia semiglobulata	E	17.9 ± 5.7	E	3 ± 2.05	E	10
	I	12.8 ± 3.1	I	0.3 ± 2.08	I	3
Espeletia batata	E	27.5 ± 6.3	E	4 ± 1.9	E	11
	I	30.8 ± 5.1	I	2.9 ± 1.2	I	7
Azorella julianii	E	26.1 ± 9.4	E	3.7 ± 2.5	E	12
	I	18.7 ± 7	I	1.3 ± 0.7	I	6
Aciachne pulvinata	E	34.5 ± 3.8	E	5.1 ± 1.9	E	11
	I	23.8 ± 4.5	I	0.4 ± 0.7	I	3

Note: E = external temperature; I = internal temperature.

TABLE 13.3
Water storage capacity in an Andean wetland dominated by *Plantago rigida*

		Water Storage Capacity of an Andean Wetland			
Treatment	N	Surface Area (cm²)	Saturated Weight (g)	Dry Weight (g)	Rainfall (mm)
Total soil (0–10 cm depth)	4	171.4 ± 51.9	344.3 ± 53	127.7 ± 21.2	13,445 ± 4,474
Top soil layer (0–4 cm depth)	10	47.6 ± 6.6	137.6 ± 15.5	61.3 ± 9.1	16,455 ± 3,899
Lower soil layer (4–10 cm depth)	5	77.6 ± 9.90	207.6 ± 12.7	120.4 ± 11.3	11,414 ± 1,769

DISCUSSION

It is interesting to examine how the species that occupy the wettest environment in tropical highlands distribute their resources, and to analyze if these phytomass distribution patterns constitute "successful decisions" in terms of ecosystem functioning (Monasterio 1986). Moreover, the diversity of architectonic models studied in Andean páramo wetlands has important ecological implications, as it determines the vertical spatial distribution of energy incorporated into the ecosystem.

The results presented here show two different patterns of energy distribution. On the one hand, there are species that distribute large phytomass proportions to aerial structures (more than 30 cm aboveground), with AP/BP ratios above one. This model is common in species of tropical ecosystems (Smith and Klinger 1985). On the other hand, there are abundant species in wetland ecosystems with low aerial biomass and AP/BP ratios between 0.1 and 0.001. This last model of belowground accumulation is characteristic of species of alpine, arctic, and tundra ecosystems (Smith and Klinger 1985).

In alpine regions, where the low temperatures are the main limiting factor (Aber and Melillo 1991), the species show low photosynthesis and growth rates and slow litter decomposition. Life-forms dominant in Andean wetlands show morphological and ecophysiological adaptations to low temperatures and extreme daily temperature fluctuations (Goldstein et al 1984; Monasterio and Sarmiento 1991; Rada 1993). As a result of

their adaptations to the extreme conditions of the páramo, the species show slow rates of plant growth (Rada 1993). In this sense, several authors agree that these ecosystems are fragile, showing slow rates of regeneration after disturbances such as grazing and fire (Luteyn 1992; Hofstede et al. 1995; Hofstede 2001).

The high leaf necromass proportions present in the studied species have been related to thermal insulation. In the case of giant rosettes, a cover of dead leaves isolates living tissues in aboveground stems, protecting them from nocturnal freezing and regulating their water balance (Goldstein and Meinzer 1983). This mechanism is also involved in thermal protection of leaf meristems (Smith 1974; Monasterio 1986). The stored necromass does not constitute an active energy reserve, but plays a critical role in nutrient translocation from dead leaves to active tissues (Garay et al. 1982) and might, in addition, contribute to water recharge in páramo wetland ecosystems. The same could be true of the acaulescent rosette *Plantago rigida* in our study, in which a large proportion of the leaf necromass encases the belowground stem, strongly increasing the water-holding capacity of the top few centimeters of the soil.

The effect of extensive grazing in the Andean páramos, in general, depends on the intensity, frequency, and sequence of cattle presence in the páramo grasslands (Molinillo and Monasterio 2002). A low animal intensity increases the species richness because the competitive exclusion decreases, and the fast-growing forbs are able to show explosive colonization. However, a high animal intensity decreases the diversity of species (Sarmiento et al. 2003). For example, the low frequency of grazing and fire in the west páramos decreased the tussock density and increased the fraction of forbs and grass species in the vegetation composition. However, a high animal intensity decreased the diversity of species (Sarmiento et al. 2003) and increased the fraction of less-palatable forbs (Hofstede 1995; Verweij 1995; Molinillo and Monasterio 2002).

There are some alternative management practices in the Venezuelan Andes, that emphasize the need to conserve páramo diversity (Sarmiento et al. 2003). Intensification of agriculture in some areas seems to be the best way to reduce the total area under cultivation, while maintaining production levels and improving biodiversity, given that representative natural areas are set aside for protection (Sarmiento et al. 2002). Another factor to be analyzed is the impact of grazing practice, which is likely to have a pronounced effect on the vegetation structure and diversity in Andean grasslands.

Even though the effect of extensive grazing within the wetland ecosystem is not analyzed here, the functional variability could certainly play a critical role in determining the water balance in these high-Andean páramo environments, in which the aboveground and belowground stems could act as water reservoirs, while standing leaf necromass could provide improved water capture by acting as a funnel. Therefore, the conservation of species and functional diversity for a sustainable use of the Andean wetlands necessarily implies appropriate cattle management strategies in the Venezuelan Andean region.

SUMMARY

Tropical mountain diversity is not only expressed as richness per unit area but also in terms of the functional variability of highland species. In the wetlands of the Andean páramo above 3800 m, a diverse array of plants coexist that can be grouped into acaulescent rosettes, caulescent rosettes, cushions, forbs, and grasses — the same life-forms defined by Hedberg (1964) for the Afroalpine belt. Each of these life-forms can be interpreted as an archi-

tectonic model in which phytomass distribution in aboveground and belowground structures (including senescent leaves) and thermal protection of growth meristems can provide key information on the functioning of the wetlands in the Andean páramo. The results of this study in the Venezuelan Andean wetlands show a variety of phytomass patterns, with species that accumulate phytomass in aboveground structures and species that do the same in belowground structures, particularly as buried leaf necromass. Phytomass accumulated as leaf necromass has different functions, such as protection of the growth meristems from low temperatures or water capture in the topsoil profile (e.g. an increase of water was found in wetlands dominated by the acaulescent rosette *Plantago rigida*, which has a high underground leaf necromass). Extensive grazing modifies the diversity and composition of species and, consequently, the relative abundance of the species that are not consumed by cattle (cows and horses) but are susceptible to damage by trampling. This has effects on the hydrological functioning of these ecosystems, which constitute the headwaters of important rivers draining into the Amazon catchment. Therefore, conservation of the biodiversity of the Andean wetlands necessarily implies appropriate cattle management strategies in the Venezuelan Andes.

ACKNOWLEDGMENTS

This research was supported by the Universidad de los Andes, within the project: Ecological and Social Sustainable Development of the Agricultural Production in the Cordillera de Mérida: the Flow from the Environment Services in Altiandean Páramos to the Potato Agriculture (N° CVI-PIC-C-02-01). We wish to thank Marcelo Molinillo for providing important insight to understanding some of the results in the grazed Andean wetlands.

References

Aber, J.D. and Melillo, J.M. (1991). *Terrestrial Ecosystems*. Saunders Collage Publishing, USA. 430 pp.

De Robert, P. and Monasterio, M. (1993). Practicas agrícolas campesinas en el páramo de Apure, Sierra Nevada de Mérida, Venezuela. In Rabey, M. (Ed.). *El uso de los recursos naturales en las montañas: Tradición y transformación*. UNESCO-Orcyt, Montevideo, Uruguay, pp. 37–54.

Garay, I., Sarmiento, L., and Monasterio, M. (1982). Le Parame désertique: éléments biogénes, peuplements des microarthropodes et stratégies de survie de la végétation. In Lebrun, Ph., André, H.M., De Medts, A., Grégorie-Wibo, C., and Wauthy, G. (Eds.). *Tendences Nouvelles en Biologie du Sol*. Comptes Rendus du VIII Colloque International de Zoologie du Sol. Louvain la Nueve, Belgium, pp. 127–134.

Goldstein, G. and Meinzer, M. (1983). Influence of insulating dead leaves and low temperatures on water balance in a Andean giant rosette plant. *Plant Cell and Environment* 6: 649–656.

Goldstein, G., Meinzer, F., and Monasterio, M. (1984). The role of capacitance in the water balance of Andean giant rosette species. *Plant Cell and Environment* 5: 179–186.

Hedberg, O. (1964). Afroalpine plant ecology. *Acta Phytogeographica Suecica* 49: 1–144.

Hedberg, I. and Hedberg, O. (1979). Tropical alpine life forms of vascular plants. *Oikos* 33: 297–307.

Hofstede, R. (1995). Effects of Burning and Grazing on a Colombian Paramo Ecosystem. Ph.D. thesis, University of Amsterdam, Amsterdam, The Netherlands.

Hofstede, R.G.M., Chilito, P.E.J., Evert, M., and Sandoval, S. (1995). Vegetative structure, microclimate and leaf growth of a paramo tussock grass species in undisturbed, burned and grazed conditions. In Hofstede, R. (Ed.). *Effects of Burning and Grazing on a Colombian Paramo Ecosystem*. University of Amsterdam, Amsterdam, The Netherlands, pp. 22–38.

Hofstede, R. (2001). El impacto de las actividades humanas sobre el páramo. In Mena, P.A., Medina, G., and Hofstede, R. (Eds.). *Los páramos del Ecuador. Particularidades, problemas y perspectivas*. Ed Abya Yala, Quito, Ecuador, pp. 161–185.

Körner, C. (1999). *Alpine Plant Life. Functional Plant Ecology of High Mountain Ecosystems*. Springer, Germany.

Luteyn, J.L. (1992). Paramos: why study them? In Balslev, H. and Luteyn, J.L. (Eds.). *Paramos: An Ecosystem under Human Influence*. Academic Press, London, pp. 1–14.

Molinillo, M. (1992). Pastoreo en ecosistemas de páramo: estrategias culturales e impacto sobre la vegetación en la cordillera de Mérida, Venezuela. Master thesis, Universidad de los Andes, Mérida, Venezuela.

Molinillo, M. and Monasterio, M. (1997a). Pastoralism in paramo environments; practices, forage and impact on vegetation in the Cordillera of Mérida, Venezuela. *Mountain Research and Development* 17: 197–211.

Molinillo, M. and Monasterio, M. (1997b). Pastoreo y conservación en áreas protegidas de la Cordillera de Mérida, Venezuela. In Liberman, M. and Baied, C.A. (Eds.). *Desarrollo sostenible de ecosistemas de montaña: manejo deáreas frágiles en los Andes*. UNU, Instituto de Ecologia, UMSA, La Paz, Bolivia, pp. 93–109.

Molinillo, M. and Monasterio, M. (2002). Patrones de vegetación y pastoreo en ambientes de paramo. *Ecotrópicos* 15(1): 17–32.

Molinillo, M. (2003). Patrones de vegetación y pastoreo en ecosistemas Altiandinos: una comparación de casos de estudio en páramos y punas. Ph.D. thesis, Universidad de Los Andes, Mérida, Venezuela.

Monasterio, M. (1980a). Las formaciones vegetales de los páramos de Venezuela. In Monasterio, M. (Ed.). *Estudios ecológicos de los páramos andinos*. Universidad de los Andes, Mérida, Venezuela, pp. 93–158.

Monasterio, M. (1980b). Poblamiento humano y uso de la tierra en los Altos Andes de Venezuela. In Monasterio, M. (Ed.). *Estudios ecológicos de los páramos andinos*. Universidad de los Andes, Mérida, Venezuela, pp. 170–198.

Monasterio, M. and Reyes, S. (1980). Diversidad ambiental y variación de la vegetación en los páramos de los Andes Venezolanos. In Monasterio, M. (Ed.). *Estudios ecológicos de los páramos andinos*. Universidad de los Andes, Mérida, Venezuela, pp. 47–91.

Monasterio, M. (1986). Adaptive strategies of *Espeletia* in the Andean Desert Paramo. In Vuilleuimier, F. and Monasterio, M. (Eds.). *High Altitude Tropical Biogeography*. Oxford University Press, Oxford, pp. 49–80.

Monasterio, M. and Sarmiento, L. (1991). Adaptive radiation of *Espeletia* in the cold Andean tropics. *Trends Ecol. Evol* 6(12): 387–391.

Monasterio, M. and Molinillo, M. (2003). Venezuela: El paisaje y su diversidad. In Hofstede, R., Segarra, P., and Mena, P. (Eds.). *Los páramos del mundo*. Global Peatland Initiative, NC-IUCN, EcoCiencia, Quito, Ecuador.

Perez, R. (200). Interpretacíon ecológica de la ganaderia extensiva y sus interrrclaciones con la agricultura en el piso agricola del páramo de Gaviriq, Andes Venezolanos, Tesis de maestria. Universidad de los Andes, Mérida, Venezuela.

Rada, F. (1993). Respuesta estomatica y asimilación de CO_2 en plantas de distintas formas de vida a lo largo del gradiente altitudinal en la Alta Montaña Tropical Venezolana. Ph.D. thesis, Universidad de Los Andes, Mérida, Venezuela.

Ramîrez, P., Izquierdo, F., and Paladines, O. (1996). Producción y utilización de pastizales en cinco zonas agroecológicas de Ecuador. MAG-GTZ-REPAAN, Quito, Ecuador.

Raunker, C. (1934). *The Life Forms of Plants and Statistical Plant Geography*. Oxford University Press, Oxford.

Rundel, P.W., Smith, A.P., and Meinzer, F.C. (1994). *Tropical Alpine Environments: Plant Form and Function*. Cambridge University Press, New York.

Sarmiento, G. (1986). Ecological features of climate in high tropical mountain. In Vuilleuimier, F. and Monasterio, M. (Eds.). *High Altitude Tropical Biogeography*. Oxford University Press, Oxford, pp. 11–45.

Sarmiento, L., Smith, J.K., and Monasterio, M. (2002). Balancing conservation of biodiversity and economical profit in the agriculture of the high Venezuelan Andes: are fallow systems an alternative? In Körner, C. and Spehn, E.M. (Eds.), *Mountain Biodiversity. A Global Assessment*. Parthenon Publishing, Boca Raton, FL, pp. 285–295.

Sarmiento, L., Llambi, L.D., Escalona, A., and Márquez, N. (2003). Vegetation patterns, regeneration rates and divergence in an old-field succession of the high tropical Andes. *Plant Ecology* 166: 63–74.

Schubert, C. (1979). La zona del páramo: morfología glacial y periglacial de los Andes de Venezuela In Salgado-Labuoriau (Ed.). *El medio ambiente Páramo*. UNESCO-IVIC, Caracas, Venezuela, pp. 11–27.

Smith, A.P. (1974). Bud temperature in relation to nyctinastic leaf movement in an Andean giant rosette plant. *Biotropica* 6: 263–266.

Smith, J.M.B. and Klinger, L.F. (1985). Aboveground: belowground phytomass ratios in Venezuelan paramo vegetation and their significance. *Arctic and Alpine Research* 17(2): 189–198.

Smith, A.P. and Young, T.P. (1987). Tropical alpine plant ecology. *Annual Review of Ecology and Systematics* 18: 137–158.

Verweij, P. (1995). Spatial and temporal modelling of vegetation patterns. Burning and grazing in the paramo of the Nevados National Park, Colombia. Ph.D. thesis, ITC Publication 30.

Walker, B., Kinzig, A., and Langridge, J. (1999). Plant attribute diversity, resilience and ecosystems function: the nature and significance of dominant and minor species. *Ecosystems* 2: 95–113.

14 Millennia of Grazing History in Eastern Ladakh, India, Reflected in Rangeland Vegetation

Gopal S. Rawat and Bhupendra S. Adhikari

INTRODUCTION

The Changthang Plateau in eastern Ladakh represents an important biogeographic province within the Indian trans-Himalayan region (Rodgers and Panwar 1988). This tableland forms the western extension of the Tibetan plateau and lies mostly above 4500 masl. This region, an ancient Tethyan seabed that has undergone frequent phases of glaciation and aridity, exhibits harsh climatic conditions and unique assemblages of flora and fauna. Much of the plateau comprises lake basins, sandy plains, and snow-capped rolling mountains. Internal drainage in some of the basins has led to concentrations of salt and other minerals over the millennia, making the water bodies brackish. Less than 1% of the geographical area on the plateau is cultivated, and most of the vegetated zone is used by a migratory pastoral community known as the *Changpa* for livestock (goats, sheep, yaks, and horses) grazing. Despite a sparse cover of vegetation and low-standing biomass, this area sustains a high livestock population. Whereas alpine zones of the greater Himalayas are largely humid and exhibit higher seasonal primary productivity, the cold–arid trans-Himalayan pastures have relatively low productivity (Rikhari et al. 1992, Ram et al. 1989, Miller and Schaller 1996, Jian 2002). Therefore, the effects of livestock grazing in both the zones are likely to be different, depending upon the microtopographic features and history of livestock grazing. Many questions concerning the functional aspects of rangeland vegetation in response to the pastoral system remain unanswered. There is a need for more in-depth studies on the relationships among the vegetation structure, composition, and patterns of grazing by the domestic livestock and wild herbivores.

So far, the Changpa herders of Changthang and their livestock have coexisted with wildlife in the area. Traditionally, the herders have been following a rotational and definite pattern of seasonal movement. Duration of stay at each pasture depends on the forage and water availability, weather conditions, and sociocultural events. The marsh meadows around lake banks are considered the best wintering (January to March) ranges. During heavy snowfall, male yaks are taken to nearby slopes dominated by caragana scrub. Wool is harvested during late April or early May, which marks the beginning of spring. Then, the herders shift to the middle hills and plateaus and on higher slopes during peak summer (June to August). Male yaks and horses are separated from the female yaks and calves and taken to the farthest pastures during summer. The herders collect brushwood and yak dung throughout the summer and fall for winter use. The headman (called the *gub* or *goba*) usually occupies the centrally located pastures and liaisons between the herders and the government representatives. Currently, the estimated number of sheep and goats in Changthang is about 185,700, of which goats account for 64% and sheep, 36%. There has been a steady increase in the livestock population in the area since the 1970s (Anon *Statistical Handbook* 1998) due to the influx of nomadic herders from Tibet. The Pashmina goats yield fine-quality wool, which is currently sold in the area at the rate of Rs. 400 to 500 ($9 to 11) per kg and generates a considerable revenue in the region. However, an increase in the livestock number, limited area for grazing, rapid increase in tourism, and related developmental activities have caused conflicts between herders, wildlife managers, and development agencies in the area. Despite the immense importance of conserving these rangelands as key wild habitat and as a source of livelihood for Changpa herders, no ecological studies have been conducted in the area. Systematic surveys on various components of biodiversity in Changthang have started only recently (Anon *Conserving Biodiversity in the Indian Trans-Himalaya* 2001). Kachroo et al. (1977), Chaurasia and Singh (1997), Schweinfurth (1957), Hartmann (1987), and Rawat et al. (2001) have conducted ecofloristic studies in other parts of the Indian trans-Himalayan region.

In this chapter, we describe the plant species diversity, vegetation structure, and rangeland conditions, *vis-à-vis* patterns of livestock use, in a part of the Changthang Plateau. We discuss the implications of new land use practices that might take place in the area in the near future.

STUDY AREA

This study was conducted in the Rupshu Plains of Changthang (Figure 14.1). The study area is divided into two grazing units (Samad and Kharnak), which are the two traditional subdivisions of pastures in Changthang. The intensive study area lies approximately between 33°10 and 33°30 N latitudes and 77°55 and 78°20 E longitudes, and within an altitudinal range of 4400 to 5800 masl. There are two large water bodies in the study area, i.e. Tso Kar (about 19.5 km², 4534 masl) and Tsarstapuk (about 2.5 km², 4545 masl). Tso Kar is a brackish-water lake, whereas the adjacent Tsarstapuk has fresh water, drained by two perennial streams. The northwestern and eastern banks of Tso Kar are rich in deposits of borax and common salt and hence unfavorable for plant

growth except for a few salt-tolerant species such as *Axyris amaranthoides* and *Sueda microsperma*. In other areas, soil texture varies from sandy and sandy loam to clay. Detailed meteorological data are not available for the study area. According to Dhar and Mulye (1987), mean annual precipitation for this region ranges from 15 to 20 cm, 90% of which is snow. Mean minimum and maximum temperatures during the growing season (mid-June to mid-September) ranges between 2 and 30°C. Approximate minimum and maximum temperature for the year ranges between 35 and 32°C.

The vegetation of the study area can be broadly grouped into scrub formations, desert steppe, and marsh meadows. The major plant communities include caragana–eurotia, artemisia–tanacetum, stipa–oxytropis–alyssum, and *Carex melanantha–Leymus secalinus*. Very high altitudes (>5000 m) have sparse fell-field communities with mosses, lichens, and cushion-like angiosperms, e.g. *Thylacospermum caespitosum, Arenaria bryophylla*, and *Androsace sarmentosa*. The vegetation around stream banks and marsh meadows (except areas of borax and salt deposits) is dominated by sedges such as species of carex, kobresia, and scirpus.

It is estimated that about 10,000 sheep and goats, 1200 yaks, and 188 horses graze in the intensive study area (Table 14.1). Of the total area surveyed, nearly 75% constitutes net grazing land, and the rest is barren, snowbound, and rocky. In recent years, there has been rapid fluctuation in the number of livestock, especially goats and sheep, in the study area due to mass mortality during severe winters. The wild herbivores in the study area include Tibetan wild ass (*Equus kiang*), Tibetan argali or nyan (*Ovis ammon hodgsoni*), Himalayan marmot (*Marmota bobak*), Tibetan woolly hare (*Lepus oiostolus*), Royle's vole (*Alticola roylei*), and pika (*Ochotona* sp.). The wild predators reported from the study area include snow leopard (*Uncia uncia*) and Tibetan wolf (*Canis lupus chanko*). According to Bhatnagar (2001), this area once formed the distribution range of wild yak (*Bos grunniens*) and Tibetan antelope (*Pantholops hodgsoni*). Besides, the marsh meadows on the banks of both the lakes are used by a large number of migratory birds during the summer (Singh and Jaypal 2001).

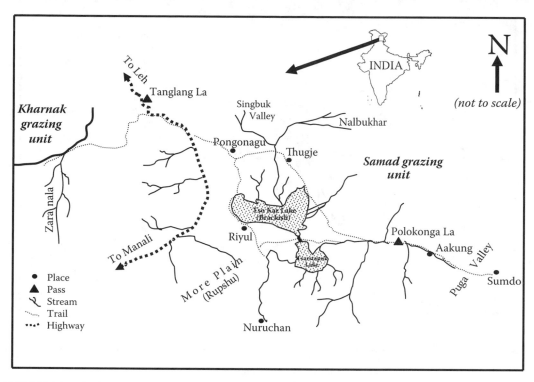

FIGURE 14.1 A diagrammatic sketch of Tso Kar basin, Western Changthang, Ladakh, India.

TABLE 14.1
Human and livestock population in the study area, eastern Ladakh (India)

Population Parameter	Samad Grazing Unit	Kharnak Grazing Unit
Human		
Total families	61	55
Males	79	112
Females	77	106
Children	142	101
Total population	298	319
Mean family size (±S.D.)	5.4 ± 2.4	5.8 ± 3.3
Livestock (LS)		
Sheep	2241	1552
Goats	2939	4344
Yaks	601	609
Horses	96	92
Total LS population	5877	6587
Mean LS holding/family (±S.D.)	106.9 ± 62.2	121.8 ± 91.6
Approximate area (km²)	500	300

Source: Block development office, Nyoma, Ladakh, India.

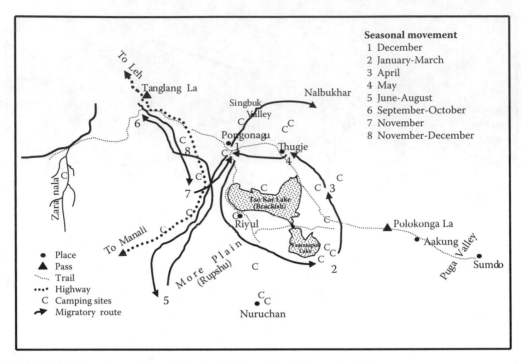

FIGURE 14.2 Migration patterns of herders in Changthang Plateau, Ladakh.

The seasonal use of various pastures and the migration pattern of the herders in the Samad area are shown in Figure 14.2. The herders arrive at Pongonagku for the New Year celebration by mid-November. By late December, at the onset of severe winter, the herders move to the southern banks of Lake Tsartsapuk. The Changpas follow the left bank of the Tso Kar, which is shorter and more vegetated, to reach the winter camps, whereas the immigrant refugee herders have to follow the drier and longer route through the right bank.

METHODS

The study was carried out during the peak growing season (July to September) of 2000. The study area was traversed on foot as well as by vehicle, wherever feasible. The plant species were collected and recorded from all the habitats and terrain types, and were categorized according to their growth habits, broadly following Raunkiaer's (1934) life-forms and habitats. We used Polunin and Stainton (1984), Aswal and Mehrotra (1994), and Murti (2001) for identification of species in the field.

The study area was classified into various habitat types for random sampling. We used random quadrats of 1 m^2 (10 at each site) following standard methods (Misra 1968). At each sample site, we recorded the altitude, geographical coordinates (using a global positioning system [GPS] for spatial analysis of vegetation, done later), aspect, terrain, soil type, and vegetation cover. The data on the abundance (number) of individual species were recorded within each quadrat along with their approximate percentage of cover. In case of graminoids (grasses and sedges), we counted the tillers irrespective of their clonal status. The number of sites sampled under each habitat type was as follows: upper slope (27), valley bottom (25), lower slope (17), marsh meadow (14), cattle camp (12), flat meadow (9), wet meadow (7), lower plateau (5), bouldery slope, higher plateau, and lateral moraine (4), and alluvial fan with pebbles (3).

The plant communities were classified using the two-way indicator species analysis (TWINSPAN; Hill 1979), a computer-based polythetic divisive clustering package. To describe the vegetation characteristics within the various habitat types, ecological parameters such as vegetation cover (percentage), density (plants m^{-2}), diversity (Shannon–Wiener index, H), and richness (Menhinick's index) as given in Magurran (1988) were calculated. Availability of forage under various categories, e.g. graminoids, forbs, and shrubs, was compared for various habitat types using the prominence value index (Dinerstein 1979) as follows:

$$\text{Prominence value} = \text{percentage of vegetation cover} \times \sqrt{\text{percentage of frequency}}$$

The Changpa herders were interviewed to get information on the patterns of rangeland use and the seasonal movement patterns. The plant specimens collected from the area (herbarium sheets) were shown to the herders to get information on the local names, uses, and preference ratings (in terms of palatability) on species preferred by their livestock. Information on human and livestock population was obtained from the block development office, Nyoma, Ladakh (Table 14.1).

RESULTS

A total of 232 species of vascular plants belonging to 101 genera were recorded from the study area. Poaceae (39 species), Asteraceae (27 species), and Cyperaceae (25 species) were the dominant families. Based on the growth forms, these species can be categorized as graminoids (66 species), dwarf shrubs and undershrubs (16 species), aquatic and semiaquatic herbs (12 species), tuberous herbs (4 species), perennial herbs with thick underground root stock (40 species), woolly and cottony herbs (15), dwarf cushionoid herbs (22), prostrate herbs (32), and annual or biennial herbs (27).

PHYSIOGNOMY AND VEGETATION CHARACTERISTICS

Based on physiognomy, four general categories of vegetation can be recognized in the study area. These are:

1. Scrub steppe (*Caragana–Eurotia*, *Artemisia*, and *Tanacetum* associations)
2. Desert steppe (*Stipa–Alyssum–Oxytropis*, *Leymus secalinus*, and *Carex melanantha* associations)
3. Wet and marsh meadows dominated by sedges under various associations
4. Fell-fields dominated by cushionoid vegetation at higher altitudes (>4800 m)

The scrub steppe, desert steppe, sedge meadows, and fell-fields formed approximately 20, 40, 5, and 10% of the land surface, respectively, in the intensive study area. The remaining 25% of the area is steep rocky slopes, scree slopes, and snowfields. The vegetation in all the physiognomic units was characteristically low (<50 cm in height) and sparse as in other parts of the trans-Himalayan region.

For the entire study area, TWINSPAN analysis identified 12 major plant communities, which were segregated into different plant associations under different habitat types. Vegetation characteristics such as number of plant associations, mean cover of ground vegetation, density of individual plants, species richness (SR), and species diversity (Shannon–Wiener index) for various habitats are given in Table 14.2. The number of plant communities was highest on the higher slopes (especially below snowbanks) and decreased at the valley bottoms (away from the lake basins). These areas exhibit sharper ecological gradients of moisture and soil depth and hence more microhabitats. In marsh meadows, species of *Carex* (*C. parva*, *C. sagaensis*, and *C. orbicularis*) formed various associations with *Kobresia*, *Ranunculus*, and *Scirpus*. Lower plateaus and alluvial fans had fewer communities (3 to 4) and moderate vegetation cover (30 to 40%). The highest vegetation cover during the peak growing season was recorded within marsh meadows (92%) and the lowest, within the alluvial fans (28%). The highest stem densities were observed in marsh meadows (544 plants m^{-2}) followed by wet meadows (507 plants m^{-2}). Such high density is largely due to dominance of tussock and tiller-forming sedges. Plant densities (m^{-2}) were low in lateral moraines (14), alluvial fans with pebbles (26), and bouldery stable slopes (29).

TABLE 14.2
Vegetation parameters across various habitat types in Tso Kar Basin

Habitat Type	Associations (Number)	Cover (%)	Density (Individuals m⁻²)	Species Richness (Number)	Shannon–Wiener Index (H)
Marsh meadow	10	91.7 ± 5.6	544.0 ± 301.0	11.4 ± 4.1	1.37 ± 0.49
Wet meadow	6	89.1 ± 11.2	507.0 ± 376.0	12.0 ± 4.2	1.55 ± 0.55
Valley bottom	18	48.8 ± 22.7	80.5 ± 110.0	6.6 ± 2.9	1.09 ± 0.57
Flat meadow	10	47.8 ± 29.4	77.6 ± 71.9	7.4 ± 3.3	1.40 ± 0.44
Lateral moraine	4	30 ± 9.1	14.3 ± 4.1	6.8 ± 2.2	1.25 ± 0.32
Lower slope	14	47.5 ± 13.6	34.4 ± 20.7	6.2 ± 1.5	1.16 ± 0.35
Lower plateau	4	69 ± 24.8	127.0 ± 115.0	6.8 ± 3.8	1.15 ± 0.67
Alluvial fans	3	28.3 ± 2.9	26.1 ± 3.1	5.3 ± 0.6	0.92 ± 0.17
Bouldery stable slope	3	39 ± 13.3	29.2 ± 8.9	11.9 ± 3.4	1.99 ± 0.51
Higher slope	25	39.6 ± 16.5	58.5 ± 132.0	9.2 ± 2.5	1.55 ± 0.40
Higher plateau	4	37.5 ± 10.4	38.1 ± 5.2	8.8 ± 1.0	1.49 ± 0.12
Cattle camp	6	42.1 ± 12.5	83.4 ± 79.8	4.8 ± 1.7	0.60 ± 0.40

Species richness was lowest around livestock camps (4.8), followed by alluvial fans (5.3) and lower slopes (6.2). Although the wet and marsh meadows had higher species richness, the bouldery slopes had greater species diversity (Shannon–Wiener index of 1.99) owing to a variety of microhabitats that supported graminoids, forbs, and shrubs. Overall cover, density, species richness, and diversity followed the gradients of soil moisture and microtopography rather than altitudinal gradient. Most of the livestock camps were characterized by the prominence of an unpalatable nitrophilous herb, *Chenopodium glaucum*, and a coarse grass, *Leymus secalinus*.

FORAGE AVAILABILITY, UTILIZATION, AND ASSESSMENT OF GRAZING LANDS

Graminoids formed the bulk of the forage available in most of the habitats, as indicated by their high prominence value (Figure 14.3). Marsh meadows had the highest prominence value of graminoids (850), followed by wet meadows (642). Dicotyledonous herbs had high prominence value (236) near livestock camps, largely due to the high abundance of unpalatable herbs such as *Chenopodium glaucum* and *Physochlaina praealta*. Valley bottoms away from the lake basins also had a high proportion of forbs (prominence value 116). Lower slopes and plateaus had higher prominence value (328.65) of

Caragana versicolor and *Eurotia ceratoides* shrubs. Of all the species in the study area, graminoids had the highest number (57%). Nearly 23% of species of dicots were dwarfs with perennial rootstock. About 35% of the species were highly aromatic, woolly, or hairy, exhibiting antiherbivory strategies.

According to the herders, about 50 to 60 species (of the 232 recorded in the study area) form the bulk of the livestock diet. All species of livestock prefer the graminoids during summer, except *Leymus secalinus* and *Carex melanantha*. Highly preferred species were pangsa (*Kobresia* spp.), sahu (*Leontopodium himalayanum*), jhak (*Elymus longe-aristaus* and *E. nutans*), havre (*Saussurea ceratocarpa*), shyot (*Stipa orientalis* and *S. sibirica*), nyargal (*Oxytropis microphylla*), semu (*Poa* spp.), and trolo (*Potentilla bifurca*). Several species that are unpalatable during summer, e.g. *Artemisia* spp., *Tanacetum* spp., *Caragana versicolor, Kochia indica, Axyris amaranthoides, Biebersteinia odora, Hyoscyamus niger, Physochlaina praealta, Scrophularia scabiosaefolia, Ranunculus* sp., and *Urtica hyperborea*, are reported to be consumed by the livestock during peak winter. This could be due to translocation of secondary compounds from shoots to roots and a higher proportion of fiber or coarse material in the shoots. Goats prefer *Eurotia ceratoides, Ephedra gerardiana, Artemisia sieversiana,* and *A. gmelinii* during the peak winter and

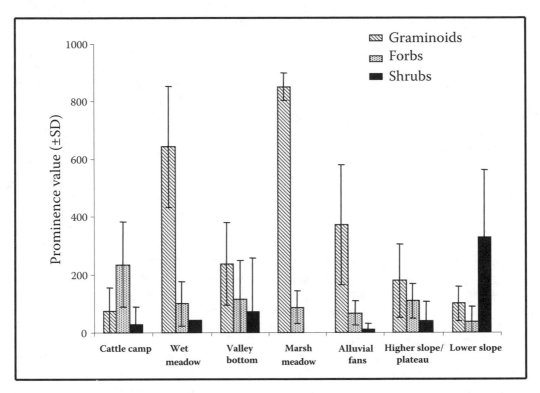

FIGURE 14.3 Prominence value of forage in major habitat types.

spring seasons. In terms of abundance among shrubs, *Eurotia ceratoides* is quite low, and only 2 to 3 clumps of *Ephedra gerardiana* were seen in the study area.

Despite the cold–arid conditions, short growing season (mid-June to mid-September), and intensive use by the livestock, as well as wild herbivores, the study area supports reasonably good vegetation cover. Although the marsh meadows around Tso Kar and Tsartsapuk Tso form only about 5% of the study area, these patches support a considerable population of livestock (nearly 5000 sheep and goats, 600 yaks, and over 80 horses) during crucial winter months. During summer, this area serves as an important breeding ground for the migratory waterfowl and black-necked crane. The abundance of graze-resistant grasses and sedges (graminoids) in the study area can be attributed to a long and evolutionary history of grazing by ungulates. The density of livestock (all species) in the study area at present is about 42 animals km^{-2}. The shortage of forage during severe winters results in the mass mortality of livestock (especially goats) and low wool pro-

duction. This is a major complaint by the herders (Konchok Stanzin, personal communication). Although most of the sandy plains and mountain slopes appear to be stable in terms of vegetation cover, a few areas show signs of degradation, especially the lower slopes close to camping sites from where caragana and eurotia bushes have been systematically removed for fuel wood. The heavily grazed areas are characterized by the abundance of *Carex melanantha*, *Leymus secalinus*, and an unpalatable herb, *Chenopodium glaucum*. The former species are less preferred during summer and may form bulk forage during the winter season. According to herders, the Tsartsapuk area has an adequate vegetation cover to support most of the livestock from the Samad area during winter whereas the lower areas of Kharnak are in poor condition. We did not notice any exotic or recently invaded species in the study area. This means that the palatable species still have a chance to regenerate, and a general cover of vegetation could increase if the area gets even slight relief from degradation and land-use practices (e.g. ill-planned developmental activities) are regulated. Based on these considerations, the overall condition of rangelands in the study area may be rated as good.

DISCUSSION

The Changthang Plateau has been reported by earlier authors as floristically poor (about 64 plant species) compared with other areas of Ladakh (e.g. Kachroo et al. 1977, Chaurasia and Singh 1997). An update on the flora, based on the present survey, reveals that it harbors at least 232 species of vascular plants. This study also shows the importance of graminoids in the flora of the region. It can be stated that, although the Tso Kar basin forms less than 1% of the geographical area of Ladakh, it contains nearly 25% of the reported flora from the region. Because the area forms a continuum with the vast Tibetan plateau, very few endemic species are expected in the study area. The species sensitive to heavy trampling, grazing, and aridity, such as orchids and ferns, were completely absent in the present study area. The following species (otherwise common in the trans-Himalayan region) were either absent or extremely rare in the study area: *Ephedra gerardiana*, *Rosa webbiana*, *Lonicera spinosa*, *Salix* spp., *Juniperus* spp., *Berberis* spp., *Hippophae tibetana*, *Cicer microphylla*, and *Cotoneaster* spp. Selective removal of some of these species for fuel wood, overgrazing through centuries, and failure of regeneration cannot be ruled out. Palynological studies in the area could reveal some more interesting facts about its phytogeography.

Habitats such as alluvial fans with pebbles and bouldery stable slopes had similar values for cover, density, and diversity that are comparable with each other. The habitat types were best classified based on the gradients of soil moisture that exhibited varied species richness (Figure 14.4). General cover, density, species richness, and diversity values were better correlated with soil moisture and microtopography than with altitude. This emphasizes the importance of microhabitats for the conservation of rarer communities. Compared to the herbaceous alpine meadows of the greater Himalayan region (e.g. Kala et al. 1998, Singh and Rawat 1999), the steppe formations in the study area

have higher proportions (both diversity and prominence value) of graminoids. Higher diversity and biomass of grazing ungulates (both wild and domestic) in the trans-Himalayan area, compared to the greater Himalayan region, suggest that the former ranges have had a much older evolutionary history of grazing that could have resulted in the prominence of graminoids. (Bock et al. 1995). Koirala et al. (2000) have also reported higher prominence of graminoids from the cold–arid pastures of the upper-Mustang area of Nepal. The aboveground phytomass in the area generally ranges from 32 to 216 g m^{-2} (unpublished data, senior author and others), which is higher than the values reported from the adjacent grazing lands of Spiti (Mishra 2001) and the Changthang Plateau of Tibet (Miller and Schaller 1996). In other moist alpine areas of the world, aboveground phytomass ranges from 200 to 3500 g m^{-2} (Körner 1999).

Are the rangelands in the study area degraded? Based on the general vegetation cover, density, and diversity (Table 14.2), as well as the high prominence values of graminoids in most of the habitat types (Figure 14.3), the overall condition of the rangelands in the Tso Kar basin (especially the Samad grazing block) could be rated as good. Miller and Schaller (1996) have reported only 10 to 16% vegetation cover (the rest is bare ground) in the eastern Chang Tang Reserve in Tibet. Similarly, from the adjacent trans-Himalayan region, i.e. Kibber Plateau of Spiti, Mishra (2001) has reported low vegetation cover compared to the present study area and stated that a majority of the rangelands in the Spiti Valley are overstocked in areas where the human impact on vegetation is intensive and widespread. Owing to more intensive agropastoral activity, it is likely that pastures on Kibber Plateau are more degraded as compared to the Tso Kar basin. Forage availability, in terms of biomass during the winter season and proportions of palatable and unpalatable species in the range, would give the true picture of rangeland conditions. Future studies in the area need to address these parameters.

Although the detailed dietary profiles of four livestock species in the study area could not be ascertained, there is, based on the Inter-

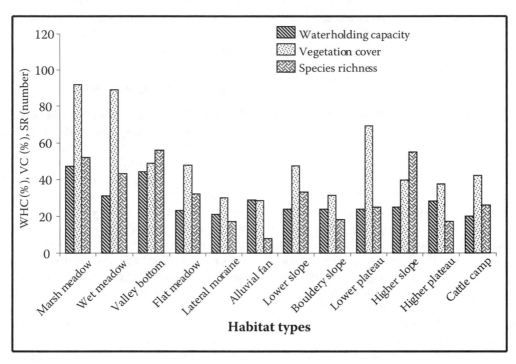

FIGURE 14.4 Soil water-holding capacity (WHC) as an index of moisture, vegetation cover (VC), and species richness (SR) in different habitat types.

views with the herders, an indication of an over 95% overlap in their diet, especially during the winter, when the species and the grazing areas are limited. There is greater choice in forage species and feeding areas during the summer. Hence, there could be less overlap in their diets. *Caragana versicolor*, a leguminous and low thorny bush, is the main source of fuel wood for the herders in the study area. Systematic removal of this species from the lower slopes, followed by soil erosion, has resulted in the desert-like conditions in such areas and reduced species diversity. The caragana, being a thorny and fairly impenetrable bush, protects several forbs and grasses from being grazed. It may also facilitate the growth of associated species by increasing nitrogen availability in the soil. Yaks feed among the caragana scrub during the peak of winter, partly on the young twigs and fallen leaf litter or on associated species. Sheep and goats are also known to suck the fallen leaves of caragana during extreme fodder scarcity and may browse on young sprouts and flowers of caragana during spring (R. Samphael, personal communication). Therefore, it is evident that, this species plays an important

role in sustaining the associated species and buffering the ecosystem from excess biotic pressure in these rangelands. Interestingly, caragana scrub cannot be classified as a distinct vegetation type in the Changtang Reserve in Tibet, although other plant communities are similar (Schaller 1998).

The present stocking densities of livestock (35 to 41 animals km^{-2}) in the study area seem to be at their peak and those of wild ungulates quite low. During a period of 4 to 5 weeks, we encountered two groups of Argali sheep (14 and 8 individuals) and about 10 to 12 groups of Tibetan wild ass (2 to 4 individuals each), in addition to the occasional marmots and Tibetan woolly hare. According to the herders, the wild animals mostly use the higher slopes to the north of Samad and Polokongkha. The data on indirect evidence collected by us (e.g. spoor, including those of the snow leopard and Tibetan wolf) corroborates this. The herders and Animal Husbandry Department (AHD) officials in recent years have begun to raise concern over the degradation of pastures and resultant shortage of forage resulting in the mass mortality of livestock during severe winters. Although cultivation of fodder in some parts of the study area has been suggested by the AHD, in Ladakh, the long-term implications of diverting water channels for this activity need to be understood. Diversion of water from the feeder streams of Tsartsapuk may change the hydrology, ecosystem, and productivity of the marsh meadows downstream.

CONCLUSIONS

The present study illustrates that a long history of wild and domestic animal grazing is not necessarily in conflict with high biological richness of high-altitude rangelands. However, grazing seems to transform vegetation, induces a clear graminoid dominance (hemicryptophytic and chamaephytic life-forms), and leads to the near or complete extinction of certain plant types such as tall forbs, subshrubs without thorns or secondary compounds, and annual herbs. Although overall rangeland conditions in the Tso Kar basin seem to be good, the conditions may degrade in the near future, because of steady increase in the livestock population and, thus, a shortage of pastures. For the sustainable use of rangelands and long-term conservation of the entire ecosystem, careful planning, based on a participatory approach backed by researched information, is essential. We, therefore, recommend the following strategies and action points for these unique rangelands: (1) establishment of long-term vegetation-monitoring plots; (2) initiation of experimental studies to understand the effect of fuel wood removal and intensive grazing on the biodiversity of the basin; (3) documentation of the process of vegetation recovery; (4) studies on the wintering ranges of livestock and wild ungulates, studies on the palatability, biomass availability, and relative forage quality (in terms of nutrients) of the major species in the region, and an analysis of the role of microhabitats in conserving rare species; and (5) giving conservation of the marsh meadows highest priority. These serve as important livestock grazing grounds in autumn and winter, and during the summer, these are the crucial breeding areas for migratory birds. Any developmental activity, such as changes in the land use around these meadows and diversion of water channels for

tourist camps and buildings, are likely to cause disturbances and threats to these areas. Hence, such activities need to be minimized.

SUMMARY

The rangelands of the Changthang Plateau represent a cold–arid ecosystem, traditionally used by the Changpa herders, a nomadic pastoral community. Although these rangelands are considered resource-poor in terms of biomass production, vegetation cover, and floristic diversity, they support a high density of domestic livestock and a variety of wild herbivores. In recent years, ecologists and conservationists have questioned the sustainability of livestock grazing in these areas and the integrity of this ecosystem, given the rapid socioeconomic changes and steady increase in the population of livestock. Baseline information on the vegetation structure and current status of these rangelands does not exist. Therefore, we investigated the floristic composition and life-form structure, forage availability, and utilization by the domestic livestock, and assessed the condition of the rangeland in a part of the Changthang Plateau.

We recorded 232 species of vascular plants within the limits of the present study area (about 500 km^2). Of these, only about 50 to 60 species formed the bulk diet of livestock. Livestock (sheep, goat, horse, and yak) prefer graminoids and avoid aromatic and woolly species during the summer months. However, most of these species are consumed along with graminoids during peak winter months. The study illustrates that a long history of wild and domestic animal grazing has given rise to hemicryptophytic (graminoid-dominant) and chamaephytic (prostrate herbs) growth forms in these rangelands. Centuries of grazing has also caused rarity among other growth forms such as tall forbs and palatable shrubs. In total, 12 distinct habitat types were identified in the study area, each representing a different set of microtopographic and soil moisture regimes. The gentle slopes at higher altitudes, especially below snowbanks, had the highest number of plant associations (25) followed by the valleys (18). These areas exhibit sharper ecological gradients of moisture and microtopography. Highest vegetation cover during peak growing season was recorded within marsh meadows (92%) and lowest within the sandy alluvial fans (28%). The marsh meadows form only about 5% of the study area, yet form crucial grazing grounds for livestock in the winter and breeding grounds for migratory birds during summer. Despite the cold–arid conditions, the short growing season (mid-June to mid-September), and its intensive use by livestock as well as wild herbivores, the study area supports reasonably good vegetation cover. Excessive pressure on the environment from tourists (camping), collection of brushwood for fuel, and diversion of water to the camping sites have led to fragmentation of the meadows and degradation of the habitats.

ACKNOWLEDGMENTS

The study was jointly funded by the International Centre for Integrated Mountain Development (ICIMOD), Kathmandu, and Wildlife Institute of India (WII), Dehra Dun. We are thankful to Camille Richard of ICIMOD for valuable suggestions and the WII for the encouragement. Our special thanks to P.L. Saklani of WII for his invaluable assistance in the field and at the headquarters. We extend our sincere thanks to T. Phuntsog, Rigzin Samphel, and Konchok Stanzin, who helped us in various stages of the study. Assistance from the late Chain Singh was also valuable.

References

Anon. (1998). *Statistical Handbook*, Government of Jammu and Kashmir, District Leh 1996–1997. Directorate of Economics and Statistics, District Statistical Evaluation Agency, LAHDC, Leh, India.

Anon. (2001). Conserving Biodiversity in the Indian Trans-Himalaya: New Initiatives of Field Conservation in Ladakh. (2001). First Annual Technical Report (1999–2000). Wildlife Institute of India, International Snow Leopard Trust and U.S. Fish and Wildlife Service.

Aswal, B.S. and Mehrotra, B.N. (1994). Flora of Lahaul-Spiti: A Cold Desert in Northwest Himalaya. Bishen Singh Mahendra Pal Singh, Dehra Dun, India, 761 pp.

Bhatnagar, Y.V. (2001). Status survey of large mammals in Eastern Ladakh and Nubra. In Anon. Conserving Biodiversity in the Indian Trans-Himalaya: New Initiatives of Field Conservation in Ladakh. Report, Wildlife Institute of India, International Snow Leopard Trust and U.S. Fish and Wildlife Service, 108–135.

Bock, J.H., Jolls, C.L., and Lewis, A.C. (1995). The effects of grazing on Alpine vegetation: a comparison of the Central Caucasus, Republic of Georgia, with Colorado Rocky Mountains, USA. *Arctic and Alpine Research*, 27(2): 130–136.

Chaurasia, O.P. and Singh, B. (1997). *Cold Desert Plants Vol. III: Changthang Valley*. Field Research Laboratory, DRDO, Leh, 84 pp.

Dhar, O.N. and Mulye, S.S. (1987). A brief appraisal of precipitation climatology of Ladakh region In Pangtey, Y.P.S. and Joshi, S.C. (Eds.), *Western Himalaya: Environment, Problems and Development*. Vol. 1. Gyanodaya Prakashan, Nainital, India, pp. 87–98.

Dinerstein, E. (1979). An ecological survey of royal Karnali-Bardia Wildlife Reserve, Nepal. Part K: vegetation, modifying factors, and successional relationship. *Biological Conservation*, 15: 127–150.

Hartmann, H. (1987). Pflanzengesellschaften trockener Standorte aus der subalpinen und alpinen Stufe im südlichen und östlichen Ladakh, *Candollea*, 42(1): 277–326.

Hill, M.O. (1979). TWINSPAN — A Fortran program for arranging multivariate data in an ordered two way table by classification of the individuals and attributes. Cornell University, Ithaca, NY.

Jian, N. (2002). Carbon storage in grasslands of China. *Journal of Arid Environments*, 50: 205–218.

Kachroo, P., Sapru, B.L., and Dhar, U. (1977). Flora of Ladakh: An Ecological and Taxonomic Appraisal. Bishen Singh Mahendra Pal Singh, Dehra Dun, India.

Kala, C.P., Rawat, G.S., and Uniyal, V.K. (1998). *Ecology and Conservation of the Valley of Flowers National Park, Garhwal Himalaya*. Wildlife Institute of India, Dehra Dun. 99 pp.

Koirala, R.A., Shrestha, R., and Wegge, P. (2000). Grasslands in the Damodar Kunda Region of Upper Mustang, Nepal. In Richards, C. et al., (Eds.) *Grassland Ecology and Management in Protected Areas of Nepal.* International Centre for Integrated Mountain Development, 53–67.

Körner, C. (1999). *Alpine Plant Life: Functional Plant Ecology of High Mountain Ecosystems.* Springer-Verlag, Berlin.

Magurran, A.E. (1988). *Ecological Diversity and Its Measurement.* Croom Helm, India. 179 pp.

Miller, D.J. and Schaller, G. (1996). Rangelands of the Chang Tang Wildlife Reserve in Tibet. *Rangelands,* 18: 91–96.

Mishra, C. (2001). High Altitude Survival: Conflicts between Pastoralism and Wildlife in the Trans-Himalaya. Doctoral thesis, Wageningen University, Amsterdam, The Netherlands.

Misra, R (1968). *Ecology Work Book.* New Delhi.

Murti, S.K. (2001). *Flora of Cold Deserts of Western Himalaya. Vol. I (Monocotyledons).* Botanical Survey of India, 452 pp.

Polunin, O. and Stainton, A. (1984). *Flowers of the Himalaya.* Oxford University Press, Delhi, India.

Ram, J., Singh, S.P., and Singh, J.S. (1989). Plant biomass, species diversity and net primary production in a central Himalayan high altitude grassland. *Journal of Ecology,* 77: 456–468.

Raunkiaer, C. (1934). *The Life Forms of Plants and Statistical Plant Geography.* Clarendon Press, Oxford.

Rawat, G.S., Adhikari, B.S., and Rana, B.S. (2001). Vegetation Surveys in the Indian Trans-Himalaya. In Anon. Conserving Biodiversity in the Indian Trans-Himalaya: New Initiatives of Field Conservation in Ladakh. Report, Wildlife Institute of India, International Snow Leopard Trust, and U.S. Fish and Wildlife Service, 7–14.

Rikhari, H.C., Negi, G.C.S., Rana, B.S., and Singh, S.P. (1992). Phytomass and primary productivity in several communities of a central Himalayan alpine meadow, India. *Arctic and Alpine Research,* 24: 344–351.

Rodgers, W.A. and Panwar, H.S. (1988). *Planning a Protected Area Network for India.* Vol. I and II. Wildlife Institute of India, Dehra Dun, India.

Schaller, G.B. (1998). Wildlife of the Tibetan steppe. University of Chicago Press, Chicago, USA.

Schweinfurth, U. (1957). Die horizontale und vertikale Verbreitung der Vegetation im Himalaya, *Bonner Geographische Abhandlungen,* Heft 20: 372 pp.

Singh, P. and Jaypal, R. (2001). A survey of breeding birds of Ladakh. In Anon. Conserving Biodiversity in the Indian Trans-Himalaya: New Initiatives of Field Conservation in Ladakh. Annual technical report. Wildlife Institute of India, International Snow Leopard Trust, and U.S. Fish and Wildlife Service, 74–104.

Singh, S.K. and Rawat, G.S. (1999). Flora of Great Himalayan National Park, Himachal Pradesh. Bishen Singh Mahendra Pal Singh, Dehra Dun, India. 304 pp.

15 Alpine Grazing in the Snowy Mountains of Australia: Degradation and Stabilization of the Ecosystem

Ken Green, Roger B. Good, Stuart W. Johnston, and Lisa A. Simpson

INTRODUCTION

The alpine ecosystems of Oceania differ from all other major continental alpine areas in lacking a long association with grazing pressure from ungulates. Ungulates are found in the mountains of Africa, Asia, Europe, and North and South America. Southeast of the Wallace Line, New Guinea and Australia evolved a fauna in which the large grazing animals were marsupials. In the isolated islands of New Zealand, the largest grazing animals were birds.

The Snowy Mountains of New South Wales (36°30 S, 148°15 E), consist of 250 km^2 of alpine and 1065 km^2 of subalpine vegetation (Costin et al., 1979). The alpine zone, extending from the treeline at about 1830 m to the highest summit, Mt. Kosciuszko at 2228 m, is characterized by a continuous snow cover for at least 4 months per year and 6 to 8 months with daily minimum temperatures below 0°C. Precipitation is in the range of 1800 to 3100 mm per year with about 60% of this falling as snow (Costin, 1957). The subalpine zone, extending from the treeline down to the winter snowline (at about 1500 m), is characterized by a continuous snow cover for 1 to 4 months per year and daily minimum temperatures below 0°C for about 6 months per year. Precipitation is in the range of 770 to 2000 mm per year (Costin, 1970). The Snowy Mountains are the headwater catchments of the three major rivers in south-eastern Australia — the Murray, Murrumbidgee, and Snowy rivers.

Costin (1955) called the Snowy Mountains "soil mountains" because of the contrast with mountains elsewhere, which are predominantly rocky or have a mantle of peat. The major rock types in the Snowy Mountains are gneisses and granites with smaller areas of metamorphic rocks and basalts (Costin, 1955). The soils are well developed, with as many as eight groups and six subgroups occurring in the alpine zone. The major soil type is alpine humus soil, commonly more than a meter deep.

In the Snowy Mountains, grazing by native vertebrates above the treeline is confined virtually to one species, the broad-toothed rat (*Mastacomys fuscus*), and this species is confined, by lack of cover, to about 27% of the alpine area (Green, 2002). The major grazing animals in the alpine zone are insects, predominantly the grasshoppers Acrididae in grasslands and Pyrgomorphidae in heaths (Green and Osborne, 1994). This lack of mammalian grazing is similar to that in the alpine areas of New Zealand and the tropicalpine grasslands of New Guinea.

In New Guinea, alpine areas occupy 5955 km^2 (Costin, 1967), but the largest of the potential alpine grazing animals, the New Guinea pademelon, *Thylogale browni*, a marsupial weighing 4.5 to 6.5 kg, is uncommon above the treeline (Green and Norment, unpublished). Alpine areas have received little usage in New Guinea (Hope and Hope, 1976), and no grazing by domestic ungulates, although many of the alpine areas were burned during hunting expeditions (Costin, 1967). The nature of the New Guinea alpine area is very much determined by the human use of fire (G.S. Hope, personal communication). The pollen data suggest that had there been no fires, the subalpine zone would consist mainly of shrub or low-forest, and the grasslands would be very constrained (Haberle et al., 2001). Of the 50,000 km² of subalpine land, an estimated 80% has been cleared by fire, although the pollen results show individual variations among different mountain areas. There are records of fire and clearance for over 10,000 years from Mt. Albert Edward, Mt. Trikora, and, to a lesser extent from Mt. Jaya; however, on Mt. Wilhelm, clearance only occurred in the past 1000 years (Haberle et al., 2001). Some mountain areas are now regenerating because of lower visitation rates (G.S. Hope, personal communication). Elsewhere, post-fire grasslands are giving way to successional communities, in some cases, with European weeds such as *Conyza* spp. making an entry (G.S. Hope, personal communication).

The alpine grasslands of New Zealand have a history of burning that predates human occupation, which only took place in the past 1000 years (Mark, 1994). New Zealand lacks native land mammals almost entirely. In the natural system, grazing animals in the alpine grasslands of New Zealand were predominantly grasshoppers (Acrididae) and herbivorous birds, the latter having suffered a decline in the face of competition from grazing mammals introduced in the 19th century (Mills et al., 1989). The effects of grazing and burning by pastoralism have been documented by Mark (1994). One hundred years of summer pasturing of sheep (mainly at subalpine altitudes) and the spread of feral animals has led to widespread deterioration of the alpine ecosystem with soil erosion problems, which are among the most difficult to rehabilitate (Costin, 1967). Himalayan thar (*Hemitragus jemlahicus*), chamois (*Rupicapra rupicapra*), and red deer (*Cervus elaphus*) were introduced into the Southern Alps of New Zealand around the beginning of the 20th century (Wilson, 1976). Batcheler (1967) reported biomass of red deer and chamois near the treeline (in the period 1965 to 1966) of 11.2 kg

ha⁻¹ in 2 localities, falling to less than 0.6 kg ha⁻¹ above the treeline at 1550 m. For grasshoppers, the figures were 10.5 kg ha⁻¹ at 1550 m, and 1.2 kg ha⁻¹ for altitudes from 1725 to 1815 m. In a similar area, a grasshopper biomass of up to 14.7 kg ha⁻¹ has been shown to consume 3% of the net annual leaf production (White, 1975).

In Australia, the biomass of grasshoppers would be about 3.24 ± 0.35 kg ha⁻¹ based on the densities of, mainly, the *Kosciuscola* spp. recorded by Green and Osborne (1981) in a high subalpine site in the Snowy Mountains. This site was mixed grassland, wet heath, and dry heath, and this figure possibly increases at higher altitudes where there may sometimes be many hundreds of grasshoppers in a square meter in optimal patches (Green and Osborne, 1994). For the same biomass, grasshoppers exert greater grazing pressure than mammals, with the large alpine grasshopper *Monistria concinna* having a daily food intake equal to its own body weight (Green and Osborne, 1994).

The use of resources in the alpine areas of the Snowy Mountains by indigenous people was restricted mainly to the harvest of Bogong moths, *Agrotis infusa*, that form dense aggregations over summer. The alpine area had little history of burning before the commencement of grazing (Hancock, 1972), and the fire frequency also was low (Figure 15.1). Helms (1896) described the high country thus: "At the highest elevations, an almost alpine vegetation covers every available spot between the rocks." He described Carruthers Peak as: "With the exception of a few protruding masses of barren rock, it is covered with vegetation." By the close of grazing, Costin et al. (1959) suggested that the area would be better described as: "Except for a few pieces of vegetation, it is covered with bare soil and stones." Such was the impact of pasturing of ungulates on an ecosystem not adapted to hard hooves and frequent firing.

THE SNOWY MOUNTAINS CASE HISTORY

The grasslands of the Snowy Mountains were first identified as potential pastures in about 1829, with the grazing of sheep and cattle com-

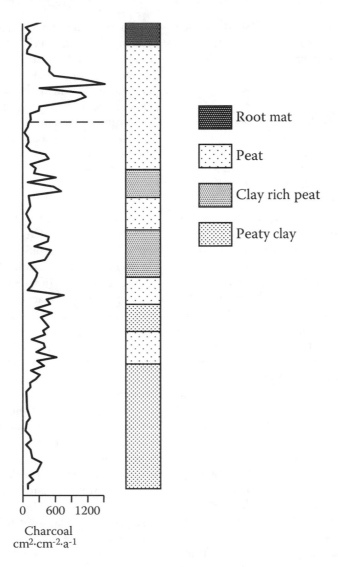

FIGURE 15.1 Fire history of over 1000 years from sediment profiles at Club Lake. (From Dodson et al. (1994.) The horizontal dashed line is at about 200 years BP.

mencing soon after (Irwin and Rogers, 1986). Grazing in alpine (>1850 m), subalpine (>1500 m), and high tableland areas (>1370 m) was initially a response to drought at lower altitudes (King, 1959). The drought of 1890–1901 led to the heaviest period of grazing (Irwin and Rogers, 1986), and cattle were brought to the Snowy Mountains from as far as 800 km away (King, 1959).

By 1920, most of the alpine and subalpine areas were being grazed (Good, 1992), and transhumance became the normal practice. The heaviest grazing was concentrated in the Main Range area surrounding Mt. Kosciuszko, where the subalpine and alpine herbfields were most favored for grazing for their succulent forbs and plentiful water supply. Mature snow grass, however, is unpalatable, and when livestock are restricted to this food, they lose body weight rapidly (Costin, 1975). The effect was that grazing was concentrated on the larger forbs and the intertussock herbs that grow as colonizing species (Costin, 1970). In terms of the area actually grazed, this increased the effective stocking rate to about 20 sheep per hectare (Costin and Wimbush, 1972).

In 1944, the Kosciuszko State Park of 5000 km² was declared, and grazing in the alpine zone was prohibited (Taylor, 1956). Advice from the Australian Academy of Science led, in 1958, to a government prohibition of all grazing above 1370 m (King, 1959).

VEGETATION RESPONSE TO GRAZING

Burning was used in the Snowy Mountains as a management tool to remove areas of heath, to remove unpalatable tussocks, and to stimulate fresh green growth (Newman, 1954; Irwin and Rogers, 1986). Pastures were burned regularly about once every 3 years (Newman, 1954) and by the commencement of the construction of the Snowy Mountains Hydroelectric Scheme in 1949, 100 to 200 fires were being lit in the higher-altitude grazing leases each autumn (Costin, 1970). Warnings of ecosystem degradation had come early. Maiden (1898) had observed fires and warned of the "great and regrettable changes" that would take place in the vegetation, and Helms (1893) warned against this "improvident practice." Snow grass (*Poa* spp.) is relatively fire tolerant, but during the grazing period, light periodic burning and the occasional severe burning resulted in a reduction of ground cover and an increased drying out of the soil and grass in dry periods (Helms, 1896; Newman, 1954). A good vegetative ground cover could return to burned areas about 12 months after a single fire (Newman, 1954), but if burned in two succeeding seasons, grazed grasslands did not recover for at least 5 years (Bryant, 1973).

Across the Tasman Sea in New Zealand, burning as a means of encouraging pasture growth also turned out to be inappropriate in the long term, resulting in extensive erosion of the mountain catchments. Although burning the tussock in spring was found to result in increased leaf growth for 2 year after fire, plant reserves were diminished, resulting in less growth for up to 14 year after burning (Payton and Mark, 1979). Three years of postfire withdrawal of grazing was required for the *Chionochloa rigida* snow-tussock to recover to a similar condition as in unburned areas (O'Connor and Powell, 1963). The low root:shoot ratio of *Chionochloa* spp. of less than one makes this genus particularly vulnerable to burning and grazing (Williams, 1977; Meurk, 1978; Lee et al., 1988). In Australia, *Chionochloa frigida*, which currently makes up >10% of alpine grasslands (Costin et al., 1979), became almost extinct due to the effects of grazing.

In the Australian Snowy Mountains, once the plant cover was reduced to 40 to 85%, continuing loss of cover occurred at the rate of about 2% per year, leading to a reduction in cover to 20 to 40% within a short period of only 20 year (Bryant, 1973). As the dominant alpine vegetation community, the tall alpine herbfield became the most extensively and severely disturbed alpine vegetation community in the Snowy Mountains (Costin, 1957; Good, 1976). This degradation consisted of selective grazing of intertussock herbaceous species, the loss of ground cover, the invasion of exotic plant species, and, most importantly, the erosion of the organic alpine humus soils that had taken centuries to develop under harsh climatic conditions (Costin, 1954, 1959; Good, 1992; Johnston and Ryan, 2000).

At subalpine altitudes, Bryant (1969) reported that grazing and burning suppressed seedling growth, limiting the life of snow gum *Eucalyptus niphophila* communities by preventing recruitment (Table 15.1). These communities were considered by Costin et al. (1961) to be valuable in optimal catchment management because of the role they play in maximizing interception and accumulation of snow, rain, fog, and cloud and in extending the period over which the spring thaw occurs, thus both increasing water accession to the mountains and regulating its release in the spring thaw (Costin, 1970).

IMPACTS ON BIODIVERSITY

By the end of grazing, important vegetation communities had been destroyed and some endemic alpine plants were near extinction (Good, 1992). At the time of the cessation of grazing in the alpine and subalpine areas, 71 of 377 alpine and subalpine plant species were recorded as only occurring in a very limited area of their total range. A number of these species being observed in these favorable refuge sites were, at that time, considered only to

TABLE 15.1
Effect of sheep grazing on regeneration of snow gum after fire along a 10-km transect

Cover Type	Ungrazed	Grazed
Snow gum	39	2
Snow grass	34	42
Shrubs	15	20
Bare ground	12	36

Note: Figures are percentage cover.
Source: From Costin et al. (1959).

occur in these sites. With the removal of grazing pressure, and the recovery and growth of these plants in other less favorable sites, it became apparent that many had a much wider distribution in the tall and short alpine herbfields, heathlands, and bog and fen communities. Of the 71 species, 25 are endemic to the alpine area and 6 are endemic to the subalpine area. All these remain under threat from fire, recreational activities, and park management activities.

Some plant species such as the spectacular buttercup, *Ranunculus anemoneus*, and the botanically interesting *Aciphylla* species and *Caltha introloba* were "almost grazed out of existence" (Costin, 1954, 1958), and despite warnings from the naturalist Stirling in 1887 that the buttercup was disappearing quickly, "owing to the inroads made into the native vegetation by stock," *R. anemoneus* is now extinct in the State of Victoria (Costin and Wimbush, 1972). In 1898, only 1 exotic plant species was recorded from the alpine zone of Mt. Kosciuszko; this number had grown to 6 by 1954 (Costin, 1954), to 37 in 1963, and is 74 today in response to the continuing postgrazing disturbance (Good and Johnston, 2004).

As early as 1893, Helms (1893) warned of soil erosion, the silting of streams and rivers, and the destruction of bogs and fens. Some groundwater communities dried out and developed herbfield-like characteristics (Good, 1992). However, few attempts were made to examine water quality until the 1960s. Simpson (2002) found that water in streams from grazed catchments had elevated nutrient levels and a higher proportion of pollution-tolerant taxa (Oligochaeta) than did water from ungrazed catchments, this being associated

with an abundant standing crop of algae. Exclusion of cattle had benefits for aquatic ecosystems, and long-term recovery in the macroinvertebrate fauna was concomitant with improvements in soil and vegetation stability within the catchment (Simpson, 2002). Using macroinvertebrate assemblages, Simpson (2002) demonstrated that streams in grazed catchments had lost between 15 and 45% of their biodiversity relative to streams from catchments ungrazed for 10 to 20 and 20 to 40 years. However, the extent of recovery in macroinvertebrate assemblages within 10 to 20 years following the removal of grazing was not significant, with recovery significantly evident after 20 to 40 years (Simpson, 2002). Other features of the aquatic ecosystem may never recover following grazing, with little natural recovery in channel morphology in many streams following 10 to 40 years of cattle exclusion (Simpson, 2002). Incising of channels is largely irreversible following grazing without intervention because the ground cover and soil protection afforded by removing grazing cannot replace the lost wetlands of subalpine valleys.

Little is known of the state of the terrestrial fauna before the onset of grazing; Helms (1896) considered that the continuous use of fire "leads directly and indirectly to the destruction of the fauna" and felt that "a great number of species must become extinct at a very early date unless the burning is prohibited by law." With very little early work by zoologists, no data are available on the impacts of grazing and fire on the terrestrial alpine fauna. However, Green and Osborne (2003) noted the absence of the herbivorous rodent *Mastacomys fuscus* (which has been declared vulnerable within the state of New South Wales) from areas where grazing has opened up the intertussock spaces. Following the major wildfires in January and February of 2003, Green and Sanecki (in press) found that although many small mammals were able to survive the fire itself, the loss of vegetation left no structural basis for the formation of subnivean space, and the mammals disappeared in the following winter. This loss of small mammals would also have been concomitant with the burning of areas of heath by graziers. Little is known of the diversity of soil animals in the alpine and subalpine zones of the Kosciuszko National Park, and even less regarding their roles in the various processes involved in soil formation, although it is thought that this role is highly significant because of the high level of activity found in the alpine soils of Kosciuszko (Costin, 1954). Previous work by Costin (1954) on the soils and vegetation in the Kosciuszko alpine area has indicated that biological factors are particularly important in the development of alpine humus soils; however, only one researcher in the past has carried out formal surveys of the soil fauna found in the alpine area of the Kosciuszko National Park (Wood, 1974). In these studies, the effects of soil fauna on the decomposition of plant litter in alpine and subalpine vegetation and the distribution of earthworms in the Kosciuszko alpine area in relation to the soils, vegetation, and altitude were investigated. Recently, however, measurements of the activity of soil biota in the area have been carried out, indicating high levels of activity for an alpine area (Johnston, unpublished). In alpine grassland areas that are greatly affected by grazing through vegetation and soil loss, the activity levels of soil biota was found to be, on average, one-tenth that found in alpine grassland areas that had a full vegetative cover and an intact soil profile (Johnston, unpublished). This loss in soil biota in disturbed areas has affected the recovery of these areas by reducing the decomposition of plant litter and nutrient cycling.

IMPACTS ON SOILS AND WATERWAYS

Helms (1893) had warned that grazing "interferes with the regular absorption, retention, and distribution of moisture." When grazing was withdrawn from all alpine and subalpine grasslands in the Snowy Mountains, soils were severely eroded, there were entrenched flow lines, and there was deep and active gully erosion (Good, 1992). Peat fires ignited by the graziers also burned for many months, destroying many of the peat beds below *Sphagnum* bogs and fens (Newman, 1954).

In 1938, the Soil Conservation Service of New South Wales declared the Snowy River catchment an area of erosion hazard (Irwin and Rogers, 1986). Most seriously affected were the bogs (6% of the alpine area), short alpine herbfields (1%), and tall alpine herbfields (approximately 64%) (Costin et al., 1979). In the bogs, the sedge *Carex gaudichaudiana* was selectively grazed by cattle, which trampled the bogs and destroyed the structure and functions of the community. They created tracks that became drainage channels, lowering the water table, and leading, in turn, to desiccation, humification, and further erosion of the peat (Costin, 1954; Costin et al., 1979). In the Kosciuszko National Park, up to 50% of the bogs were destroyed by grazing practices (Murray–Murrumbidgee Development Committee, 1955; Costin, 1959), of which less than half have recovered naturally or have been artificially restored. Bogs played an important part in maintaining catchment area function by holding water (as much as six times the mass of soil involved) and releasing it slowly over the summer to the streams and rivers (Taylor, 1956). This continuous flow of water is essential to the generation of electricity from the Snowy Mountains Hydroelectric Scheme. Simpson (2002) showed, however, that once streams are incised, the legacy of grazing affects aquatic ecosystems for many decades.

Short alpine herbfields are of restricted local occurrence found on wet drainage sites below snow patches (Bryant, 1971; Costin, 1957). The habitat consists of low-growing and mat-forming plants. This community was also heavily affected by grazing and the subsequent erosion. Costin (1954) reported serious structural vegetation degradation in many localities, and Bryant (1971) commented on the truncation of these communities by soil erosion rills and small gullies as a result of stock grazing on the highly palatable vegetation.

Surveys by the Soil Conservation Service in 1949 showed that opening up of normally closed ground vegetation had occurred over half the catchment area of the Snowy Mountains Hydroelectric Scheme (Costin, 1970). In the 1950s, further surveys by the Soil Conservation Service found 5560 ha of sheet erosion in the major alpine area (Irwin and Rogers, 1986), with the loss of 30 to 60 cm of soil over many hectares (Clothier and Condon, 1968). An estimated 1.2 million t of soil were lost from an area of approximately 200 ha during the grazing period (Morland, 1958a, b; 1959a, b). Soil cores from the sediments in Club Lake basin (one of the five glacial lakes in the Snowy Mountains) showed little disturbance for over 800 years before the onset of domestic stock grazing (apart from some minimal disturbance from infrequent fires). This changed dramatically after the advent of grazing and regular burning, with increased soil erosion and soil nutrient deposition, resulting in lake eutrophication, significant vegetation changes, and a reduction in the stability and persistence of species representation (Dodson et al., 1994). There was a 2.4-fold increase in the rate of sedimentation in the glacial lakes after the commencement of grazing and the subsequent erosion, from 0.19 $\pm$ 0.01 cm a^{-1} to 0.45 $\pm$ 0.02 cm a^{-1} (De Dekker et al., in press).

HISTORY OF RECOVERY

The Snowy Mountains Hydroelectric Scheme was a major undertaking to store and divert inland waters for irrigation and for the production of hydroelectricity. Concern was expressed over the possible silting of the dams, and summer grazing was seen as the main threat. One of the first acts of the Snowy Mountains Authority was to persuade the New South Wales State Government to cancel all grazing leases above 1370 m because, "the effect of erosion from over-grazing had already reached serious proportions" (Hudson, 1970). Above the treeline in the Snowy Mountains, up to 200 freeze–thaw cycles occur every year. Well-vegetated soil is generally protected from the erosive force of these by the vegetation cover, 10

t ha^{-1} being the minimum to afford full protection (Costin, 1966). However, on bare ground, frost heave reduces the soil to a fine powdery condition and it is then readily blown or washed away. With high-intensity rains (50 mm h^{-1}), soil infiltration, under a continuous cover of vegetation, is 50 to 75 mm h^{-1} but it is only 25 to 38 mm h^{-1} on bare ground (Costin, 1966). The only effective way to control soil erosion is by revegetation to halt the processes of frost heave and further erosion.

The simple elimination of grazing and burning had a marked effect on the vegetation and landscape. Following elimination of grazing from the alpine area in 1944, "mountain people, particularly those living at the Mt. Kosciuszko 'Chalet,' were soon impressed by the increased flowering of both forbs and snow grass which commenced in the summit area" (Costin et al., 1959). In wet areas, hygrophilous sedges and *Sphagnum* moss and associated shrubs spread rapidly (Wimbush and Costin, 1979b).

In areas where some vegetative cover was present and only minimal damage to the soil had occurred, some revegetation was possible without human intervention. Grazing land at about 1700 m altitude, at which the cover was reduced to less than 2.8 t ha^{-1}, took about 10 years to recover to the vegetative cover of 10 t ha^{-1} required for soil protection (Costin, 1970). Figures over a longer time frame (20 years) show an average increase in vegetation of 1.4% per year (Wimbush and Costin, 1979a, b). This figure is similar to the 1.3% per year over 23 years recorded for alpine heath and grassland in Tasmania (Bridle et al., 2001). Bryant (1969) gave figures for the recovery from grazing-induced loss of cover (Table 15.2) similar to those for alpine and montane pastures in New Zealand (Wraight, 1963). In New Zealand, where the red deer had a serious effect on the stability of the vegetation through browsing and trampling, eradication of the animal in some areas has resulted in "a spectacular resurgence of alpine vegetation" (Mark and Dickinson, 1997).

Johnston et al. (2003) examined the tall alpine herbfield communities that had been subjected to grazing and subsequent rehabilitation and found that disturbances produced

TABLE 15.2
Relationship between ground cover, time to revegetate, and loss of soil through erosion in grasslands of the Snowy Mountains

Cover (%)	Percentage of Total Area[a]	Projected Years to Complete Recovery[b]	Mean Cover (%)[c]	Mean Loss of Soil (t ha^{-1}) 1956–1958[c]
85–100	51.6	10	100	0
60–85	41.9	10–40	70	4.39
40–60	6.7	40–65	—	—
20–40	0.7	65–90	20	13.38
0–20	0	90	10	94.41

Source: Adapted from [a]Bryant (1973).
[b]From Bryant et al. (1969).
[c]From Costin et al. (1960).

ecosystem states different from that of the natural climax state. In areas where the impact of degradation of the ecosystem was less severe, the ecosystem retained the capacity to return slowly to its climax state. However, if the disturbance and subsequent degradation were more severe, the ecosystem did not recover, but reached, or is in the process of reaching, a different stable state (i.e. erosion-feldmark). Many of the more exposed areas of the Main Range of the Kosciuszko alpine area have, after initial stabilization and rehabilitation by the Soil Conservation Service during the 1960s, now begun to degrade again at a rate of soil loss of approximately 2 to 5 t ha^{-1} a^{-1}.

Areas that have highly degraded soil have not regenerated so easily. With no treatment, 20 years after grazing and fire had been removed from the Gungartan area, the most severely eroded areas were still not revegetated (Wimbush and Costin, 1979b), and a complete vegetation cover is still to be achieved. Soil losses from such sites exceeded manageable proportions when ground cover declined to 60% (Bryant, 1973). Wimbush and Costin (1979a) examined the response of grazed subalpine sites and suggested that succession might continue for more than 30 years before climax conditions were reached, and in severely disturbed sites, seral conditions may persist indefinitely.

A 25-year program for rehabilitation and revegetation of the alpine catchments was started in 1957 (Irwin and Rogers, 1986). In all, 450 ha of severely eroded and 1100 ha of mod-

erately eroded alpine lands were treated manually (Good, 1992). In some areas, rehabilitation caused further degradation because of a lack of understanding of the ecosystem (Johnston and Good, 1996). For example, the use of galvanized wire to hold mulch in place had a detrimental effect on the alpine ecosystems. The low levels of exchangeable zinc in natural alpine soils (<0.5 mg kg^{-1}), and the sensitivity of the alpine vegetation were not recognized, and leachate of zinc (~4 to 8 mg kg^{-1}) from the galvanized wire was enough to kill all of the native vegetation some 6 to 7 years after the initial treatment (Johnston, 1995; Johnston and Good, 1996). As a result of the zinc toxicity at these sites, erosion-feldmark-like communities have replaced the tall alpine herbfield communities (Johnston, 1995; Johnston and Good, 1996). Thus, the rehabilitation led to a mosaic of vegetation communities somewhat different from the natural state that existed before grazing had commenced (Johnston, 1995).

COSTS AND OPPORTUNITY COSTS

Helms (1893), writing on grazing and the attendant practices such as burning, noted the effects on water and erosion: "That such ignorance, and maybe greed, should be allowed to interfere so drastically in the economy of nature is pernicious and should not be tolerated." In the overall Australian economy, high-country grazing was a liability. In 1957, the value of grazing above 1370 m was less than 0.1% of the production for Victoria and New South Wales (Costin, 1959), yet the impacts on other possible uses were significant. When grazing ended and the damage was repaired, the costs of rehabilitation and revegetation exceeded the total per-hectare value of more than 100 year of agricultural production by a factor of 2 or more (Good, 1995). Figures adjusted to 2000 values by Scherrer and Pickering (2001) from Costin (1966) show that stabilization and revegetation of the alpine vegetation cost approximately A$4796 per hectare, whereas the value in terms of grazing was A$10 to A$40 per hectare per annum (Table 15.3). Even at the maximum value of grazing of A$40 per hectare per annum, the cost of revegetation was the equivalent of 120 year of grazing and, whereas the financial benefits of grazing accrued to private graziers, the cost of rehabilitation was borne by government.

The opportunity costs also were obvious; economically, the land was far too valuable for other purposes to be degraded by grazing. The value in terms of water alone, diverted to the Murrumbidgee Irrigation Areas by the Snowy Mountains Hydroelectric Scheme, was estimated to be A$54,320,000 per year or a water production value of A$143.32 per hectare (Taylor, 1956) and a combined water and hydroelectric value of up to A$225 per hectare (Costin,

1966). This figure was unlikely to be approached by grazing, in which the potential financial return in production was only A$1.25 to A$5.00 per hectare (Costin, 1959).

Helms (1893) wrote, "even from an aesthetic point of view it (grazing and its attendant practices) ought not be allowed, for what right has one section of the community to rob the other of the full enjoyment of an unsullied alpine landscape." This was before the advent of mass tourism. Currently, over 2000 people visit the alpine area per day during the peak summer holiday period (Johnston and Growcock, 2005). This summer tourism brings an estimated A$110 million into the area annually (Mules, 2004).

SUMMARY

The alpine ecosystems of Oceania evolved without the presence of ungulates. In the Australian Snowy Mountains, an inappropriate regime of grazing by hard-hoofed animals and burning of the grassland to promote growth commenced in the 1830s. Although the detrimental effects of overgrazing and burning became apparent in the late 1880s, it was not until 1944 that remedial action began. When grazing was withdrawn from alpine areas, soils had been severely eroded, there was deep and

TABLE 15.3
Costs and benefits of grazing with benefits of alternative land uses

Benefits (per year)	A$ (Millions) Total	A$ per ha	Source
Tourism	70–100	—	Pickering, personal communication
Hydroelectric power	375	220–245	Good (1995)
Irrigation water	220	130–145	Good (1995)
Grazing (to graziers)	—	10–40	Costin (1966), Scherrer and Pickering (2001)
Grazing lease value to government	—	0.9	Approximate adjustment from Taylor (1956)
Costs			
Deforestation	—	Decreases water value by 10%	Costin and Wimbush (1961)
Rehabilitation	—	4796	Costin (1966), Scherrer and Pickering (2001)
Present-day remediation	3	—	Johnston and Growcock (2005)

active gully erosion, important vegetation communities had been destroyed, and some endemic alpine plants were near extinction.

The disturbances due to grazing and burning, and the 15-years rehabilitation program itself, have produced terrestrial and aquatic ecosystem states that differ from the natural climax state. Where degradation of the ecosystem was less severe, the ecosystem retained the capacity to return to its climax state; elsewhere, the ecosystem reached a different stable state, whereas some areas are still in transition. The prohibition of grazing in the alpine catchments has resulted in benefits for aquatic ecosystems, however, recovery of aquatic features varied in both extent and rate.

Alternative uses of the alpine area as a water catchment area for the generation of hydroelectricity and for irrigation, and for increasing tourism, were also incompatible with continued grazing. The costs of rehabilitation and revegetation outweighed the financial benefits of over 100 years of grazing by a factor of more than 2 to 1.

ACKNOWLEDGMENTS

We thank Geoff Hope for his rapid response to our queries on New Guinea, and Alec Costin and Michelle Walter for their comments on the manuscript.

References

Batcheler, C.L. (1967). Preliminary observations of alpine grasshoppers in a habitat modified by deer and chamois. *Proceedings of the New Zealand Ecological Society*, 14: 15–26.

Bridle, K.L., Kirkpatrick, J.B., Cullen, P., and Shepherd, R.R. (2001). Recovery in alpine heath and grassland following burning and grazing, eastern Central Plateau, Tasmania, Australia. *Arctic, Antarctic, and Alpine Research*, 33: 348–356.

Bryant, W.G. (1969). Vegetation and ground cover trends — following the exclusion of stock at three sites in the Snowy Mountains. *New South Wales Soil Conservation Journal*, 25: 183–198.

Bryant, W.G. (1971). Deterioration of vegetation and erosion in the Guthega catchment area, Snowy Mountains. *New South Wales Soil Conservation Journal*, 27: 62–81.

Bryant, W.G. (1973). The effect of grazing and burning on a mountain grassland, Snowy Mountains, New South Wales. *New South Wales Soil Conservation Journal*, 25: 183–198.

Clothier, D.P. and Condon, R.W. (1968). Soil conservation in alpine catchments. *Soil Conservation Journal*, 24: 96–113.

Costin, A.B. (1954). *A Study of the Ecosystems of the Monaro Region of New South Wales*. Government Printer, Sydney.

Costin, A.B. (1955). Alpine soils in Australia. *Journal of Soil Science*, 6: 35–50.

Costin, A.B. (1957). The high mountain vegetation of Australia. *Australian Journal of Botany*, 5: 173–189.

Costin, A.B. (1958). The Grazing factor and the maintenance of catchment values in the Australian Alps. *CSIRO Australia Division of Plant Industry Technical paper* 10, CSIRO, Melbourne.

Costin, A.B. (1959). Vegetation of high mountains in Australia in relation to land use, biogeography, and ecology in Australia. *Monographicae Biologicae*, VIII: 427–451.

Costin, A.B. (1966). Management opportunities in Australian high mountain catchments. In Sopper, W.E. and Hull, H.W. (Eds.), *Forest hydrology*. Pergamon Press, New York, pp. 567–577.

Costin, A.B. (1967). Alpine ecosystems of the Australasian region. In Wright, H.E. and Osburn, W.H. (Eds.), *Arctic and Alpine Environments*. Indiana University Press, pp. 55–87.

Costin, A.B. (1970). Ecological hazards of the Snowy Mountains scheme. *Proceedings of the Ecological Society of Australia*, 5: 87–97.

Costin, A.B. (1975). Sub-alpine and alpine communities. In Moore, R.M. (Ed.), *Australian Grasslands*. Australian National University Press, Canberra, Australia, pp. 191–198.

Costin, A.B., Gay, L.W., Wimbush, D.J., and Kerr, D. (1961). Studies in catchment hydrology in the Australian Alps III: Preliminary snow investigations. *CSIRO Australia Division of Plant Industry Technical paper* 15, CSIRO, Melbourne.

Costin, A.B., Gray, M., Totterdell, C.J., and Wimbush, D.J. (1979). *Kosciuszko Alpine Flora*. CSIRO Collins, Melbourne/Sydney.

Costin, A.B. and Wimbush, D.J. (1972). Scientific Bases for the Exclusion of Grazing from the Kosciuszko National Park. Unpublished report to the NSW National Parks and Wildlife Service.

Costin, A.B., Wimbush, D.J., Kerr, D., and Gay, L.W. (1959). Studies in catchment hydrology in the Australian Alps I: trends in soils and vegetation *CSIRO Australia Division of Plant Industry Technical paper* 13, CSIRO, Melbourne.

Costin, A.B., Wimbush, D.J., and Kerr, D. (1960). Studies in catchment hydrology in the Australian Alps II: surface runoff and soil loss. *CSIRO Australia Division of Plant Industry Technical paper* 14, CSIRO, Melbourne.

De Dekker, P., Olley, J.M., Hancock, G., Stanely, S., Hope, G., and Roberts, R.G. (in press). Optically-stimulated luminescence dating of sediments from the Australian alpine Blue Lake; comparison with *Pinus* pollen, radiocarbon and ^{210}Pb chronologies.

Dodson, J.R., deSalis, T., Myers, C.A., and Sharp, A.J. (1994). A thousand years of environmental change and human impact in the alpine zone at Mt Kosciuszko, New South Wales. *Australian Geographer*, 25: 77–87.

Good, R.B. (1976). Contrived regeneration of alpine herbfields. In *Proceedings of the ANZAAS Congress Hobart*, pp. 1–13.

Good, R.B. (1992). *Kosciuszko Heritage*. NPWS, Sydney.

Good, R.B. (1995). Ecologically sustainable development in the Australian Alps. *Mountain Research and Development*, 15: 251–258.

Good, R. and Johnston, S. (2004). Ecological restorationof degraded alpine and subalpine ecosystems in the Alps National Parks, New South Wales, Australia. In Harmon, D. and Worboys G. (Eds.) Managing Mountain Protected Areas: Challenges and Responses for the 21st Century. Andromeda Editrice, Colledara, Italy, pp. 305–312.

Green, K. (2002). Selective predation on the broad-toothed rat *Mastacomys fuscus* (Rodentia: Muridae) by the introduced red fox *Vulpes vulpes* (Carnivora: Canidae) in the Snowy Mountains. *Austral Ecology,* 27: 353–359.

Green, K. and Osborne, W.S. (1981). The diet of foxes, *Vulpes vulpes* (L.) in relation to abundance of prey above the winter snowline in New South Wales. *Australian Wildlife Research*, 8: 349–360.

Green, K. and Osborne, W.S. (1994). *Wildlife of the Australian Snow-Country*. Reed, Sydney.

Green, K. and Osborne, W.S. (2003). The distribution and status of the broad-toothed rat *Mastacomys fuscus* (Rodentia: Muridae) in New South Wales and the Australian Capital Territory. *Australian Zoologist*, 32: 229–237.

Green, K. and Sanecki, G. (in press). Immediate and short-term responses of two faunal assemblages to a subalpine wildfire in the Snowy Mountains, Australia. *Austral Ecology,*

Haberle, S.G., Hope, G.S., and van der Kaars, S. (2001). Biomass burning in Indonesia and Papua New Guinea: natural and human-induced fire events in the fossil record. *Journal of Palaeogeography, Palaeoclimatology and Palaeoecology,* 171: 259–268.

Hancock, W.K. (1972). *Discovering Monaro: A Study of Man's Impact on his Environment.* Cambridge University Press, Cambridge.

Helms, R. (1893). Report on the grazing leases of the Mount Kosciuszko plateau. *New South Wales Agricultural Gazette*, 4: 530–531.

Helms, R. (1896). The Australian Alps, or Snowy Mountains. *Royal Geographic Society of Australia Journal*, 6: 75–96.

Hope, G.S. and Hope, J.H. (1976). Man on Mt. Jaya. In Hope, G.S., Peterson, J.A., Radok, U., and Allison, I.A.A. (Eds.), *The Equatorial Glaciers of New Guinea*. Balkema, Rotterdam, pp. 225–239.

Hudson, W. (1970). The snowy scheme. *Australian Parks-Special Congress Issue*, pp. 80–84.

Irwin, F. and Rogers, J. (1986). Above the Tree line. How the High Country was Rescued. Soil Conservation Service of NSW, Sydney.

Johnston, S.W. (1995). The Impacts of Zinc Toxicity on Short and Tall Alpine Herbfield Communities, Kosciuszko National Park. Honours thesis. School of Resource and Environmental Management, Australian National University, Canberra.

Johnston, S.W. and Good, R.B. (1996). The impact of exogenous zinc on the soils and plant communities of Carruthers Peak, Kosciuszko National Park, NSW. *Proceedings of the ASSSI and NZSSS Conference.* Vol. 3, Melbourne, pp. 117–118.

Johnston, S.W. and Ryan, M. (2000). Occurrence of arbuscular mycorrhizal fungi across a range of alpine humus soil conditions in Kosciuszko National Park, Australia. *Arctic, Antarctic, and Alpine Research*, 32: 255–261.

Johnston, S.W., Greene, R., Banks, J.G., and Good, R.B. (2003). Function and sustainability of Australian alpine ecosystems: studies in the tall alpine herbfield community, Kosciuszko National Park, NSW Australia. In Taylor, L., Martin, K., Hik, D., and Ryall, A. (Eds.), *Ecological and Earth Sciences in Mountain Areas*. Banff Centre, Canada, pp. 226–234.

Johnston, S.W. and Growcock, A.J. (2005). Visiting the Kosciuszko Alpine area. Visitor numbers, characteristics, and activitites. Technical Report: Cooperative Research Centre for Sustainable Tourism, Griffith University, Queensland, Australia.

King, H.W.H. (1959). Transhumant grazing in the snow belt of New South Wales. *Australian Geographer*, 8: 129–140.

Lee, W.G., Mills, J.A., and Lavers, R.B. (1988). Effect of artificial defoliation of mid-ribbed snow tussock *Chionochloa pallens*, in the Murchison Mountains, Fiordland, New Zealand. *New Zealand Journal of Botany*, 26: 511–523.

Maiden, J.H. (1898). A contribution towards a flora of Mount Kosciuszko. *NSW Agricultural Gazette*, 9: 720–740.

Mark, A.F. (1994). Effects of burning and grazing on sustainable utilisation of upland snow tussock (*Chionochloa* spp.) rangelands for pastoralism in South Island, New Zealand. *Australian Journal of Botany*, 42: 149–161.

Mark, A.F. and Dickinson, K.J.M. (1997). New Zealand alpine ecosystems. In Wielgolaski, F.E. (Ed.), *Ecosystems of the World 3: Polar and Alpine Tundra*. Elsevier, Amsterdam, pp. 311–345.

Meurk, C.D. (1978): Alpine phytomass and primary productivity in Central Otago, New Zealand. *New Zealand Journal of Ecology*, 1: 27–50.

Mills, J.A., Lee, W.G., and Lavers, R.B. (1989). Experimental investigations of the effects of takahe and deer grazing on *Chionochloa pallens* grassland, Fiordland, New Zealand. *Journal of Applied Ecology*, 26: 397–417.

Morland, R.T. (1958a). Erosion survey of the Hume catchment — I. *Journal of the Soil Conservation Service NSW*, 14: 191–224.

Morland, R.T. (1958b). Erosion survey of the Hume catchment — II. *Journal of the Soil Conservation Service NSW*, 14: 293–325.

Morland, R.T. (1959a). Erosion survey of the Hume catchment — III. *Journal of the Soil Conservation Service NSW*, 15: 66–99.

Morland, R.T. (1959b). Erosion survey of the Hume catchment — IV. *Journal of the Soil Conservation Service, NSW*, 15: 172–185.

Mules, T. (2004). Value of Tourism. In Independent Scientific Committee (Ed.) An Assessment of the Values of Kosciuszko National Park. NSW National Parks and Wildlife Service, Sydney, Australia, pp. 233–244.

Murray–Murrumbidgee Development Committee (1955). The Condition and Administration of the Murray–Snowy–Murrumbidgee Catchment Area. Murray–Murrumbidgee Development Committee Interim report Albury (cited in King 1959).

NPWS (1991). Kosciuszko Grazing: A History National Parks and Wildlife Service Hurstville, NSW.

Newman, J.C. (1954). Burning on sub-alpine pastures. *Soil Conservation Journal*, 10: 135–140.

O'Connor, K.F. and Powell, A.J. (1963). Studies in the management of snow-tussock grassland 1. *New Zealand Journal of Agricultural Research*, 6: 354–367.

Payton, I.J. and Mark, A.F. (1979). Long-term effects of burning on growth, flowering, and carbohydrate reserves in narrow-leaved snow tussock (*Chionochloa rigida*). *New Zealand Journal of Botany*, 17: 43–54.

Scherrer, P. and Pickering, C.M. (2001). Effects of grazing, tourism, and climate change on the alpine vegetation of Kosciuszko National Park. *Victorian Naturalist*, 118: 93–99.

Simpson, L.A. (2002). Assessment of the Effect of Cattle Exclusion on the Condition and Recovery of Sub-Alpine Streams. Honours thesis, University of Canberra.

Taylor, A.C. (1956). Snow lease management. *Soil Conservation Journal*, 12: 33–43.

Williams, P.A. (1977). Growth, biomass, and net production of tall-tussock (*Chionochloa*) grasslands, Canterbury, New Zealand. *New Zealand Journal of Botany*, 15: 399–442.

White, E.G. (1975). A survey and assessment of grasshoppers as herbivores in the South Island alpine tussock grassland of New Zealand. *New Zealand Journal of Agricultural Research*, 18: 73–85.

Wilson, H.D. (1976). Vegetation of Mount Cook National Park New Zealand. Scientific Series Number 1. National Parks Authority, Wellington.

Wimbush, D.J. and Costin, A.B. (1979a). Trends in Vegetation at Kosciuszko. II. Subalpine Range Transects, 1959–1978. *Australian Journal of Botany*, 27: 789–831.

Wimbush, D.J. and Costin, A.B. (1979b). Trends in Vegetation at Kosciuszko. III. Alpine Range Transects, 1959–1978. *Australian Journal of Botany*, 27: 833–871.

Wood, T.G. (1974). The distribution of earthworms (Megascolecidae) in relation to soils, vegetation and altitude on the slopes of Mt. Kosciuszko, Australia. *Journal of Animal Ecology*, 43: 87–105.

Wraight, M.H. (1963). The alpine and upper-montane grasslands of the Wairau River catchment, Marlborough. *New Zealand Journal of Botany*, 1: 351–376.

16 Vegetation of the Pamir (Tajikistan): Land Use and Desertification Problems[1]

Siegmar W. Breckle and Walter Wucherer

INTRODUCTION

The Pamir Mountains in Tajikistan are sometimes called a *mountain knot*, because mountain chains from various directions meet here. The various regions of the Pamir differ considerably. The eastern Pamir is a dry, arid, desert plateau, whereas other areas of the Pamir are strongly dissected and exhibit higher rainfalls. The vegetation cover in the east Pamir is very low; in the west Pamir, north Pamir, and Pamir-Alai, vegetation is more diverse and dense.

Desertification, mainly by deforestation, overgrazing, and intensive gathering, is the main environmental impact in the Pamirs. However, the various vegetation types react rather differently on these impact factors and also lead to rather different destructive processes in the various parts of the Pamirs. Loss of vegetation cover is acute in the east Pamir, leading to strong wind erosion in the west and northern Pamir. In the Alai, changes in vegetation have been observed, leading to major loss of biodiversity.

GEOGRAPHICAL SITUATION AND CLIMATE

The Pamirs have a central position in the central Asian mountain systems. They are located in the southeast corner of Tajikistan, with Afghanistan in the south, Chinese Xinxiang in the east, and the Kirgiz Tien Shan in the north.

The complex mountain system of the Pamirs is characterized by comparable geographic and climatic data shown in Table 16.1. The western Pamir chains stretch mainly east–west, so the rivers between them flow towards the west into the upper Amu-Darya (locally called *Ab-e-Panj*). The northern Pamir (Pamir-Alai) are strongly glaciated. The eastern Pamir is a high plateau, partly with endorrheic basins and hence saline lakes (Shorkul and Karakul). It resembles Tibetan conditions. It is a desert with annual precipitation below 100 mm, mainly in summer. The other regions are more humid and have a better water economy, because the precipitation is distributed more evenly over the various seasons of the year (Table 16.1). This is demonstrated by the main climatic diagram of the region (Figure 16.1).

The Pamirs have been studied in great detail for several decades by many scientists from the biological station close to Murghab and the Pamir Botanical Garden at Khorog (Ikonnikov 1963, 1979; Agakhanjanz 1965, 1978, 1985, 2002; Agakhanjanz and Jussufbekov 1975; Stanyukovich 1973; Walter and Breckle 1986; Breckle and Agakhanjanz 1994; Agakhanjanz and Breckle 1995, 2002, 2004; see also Breckle 2003; and the bibliography in Wennemann 2003).

FLORA AND VEGETATION

Parallel to the very contrasting geomorphology, the richness in species of the angiosperm flora differs conspicuously among the various mountain regions (Table 16.1). The northern Pamir and

[1] Dedicated to the memory of Professor Clas Naumann/Bonn and Eva Kleinn/Almaty.

TABLE 16.1
Geographical data of the Pamir region (within boundaries of the former USSR)

Region	Pamir-Alai	West Pamir	East Pamir
Mean latitude (N)	39°40	38°00	38°00
Elevation (masl)	1000–5610	1640–7495	3500–7134
Type of mountains	Mountain chains, deep valleys	Mountain chains, deep valleys	High plateau
Mean annual precipitation (mm a⁻¹)	440–2000	100–300	70–120
Percentage of summer rain	35	40	75
Glaciated area (km²)	8216	8400*	8400*
Permafrost soils	–	++	+++
Runoff agriculture (*lalmi*)	+++	I	–
Number of vascular plant species	4513	1524	738
Percentage of endemic species	Not known	9.2	4.6

Note: – nonexistent; + scarce; ++ common; +++ very common

* Sum of West and East Pamir

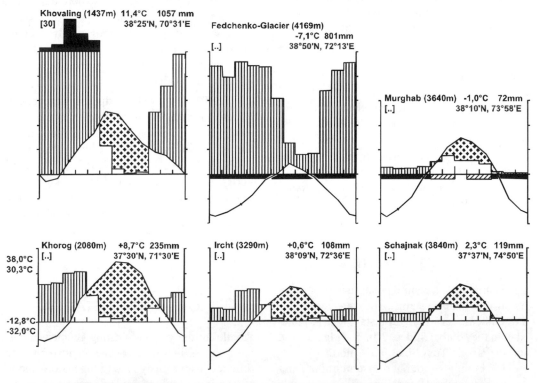

FIGURE 16.1 Climatic diagrams from the northwest (Khovaling), north (Fedchenko), southwest (Khorog), central (Ircht), east (Murghab), and southeast Pamir (Schajnak), indicating a very contrasting climatic pattern (humid and arid, winter and summer rains). (From Walter et al. 1975.)

Alai ranges exhibit a very high biodiversity distinct from the eastern Pamir Plateau, where only about 700 species have been recorded. However, usually only a few species contribute to the dominant vegetation types. Between the valley bottoms (2200 to 2500 masl) and the higher plateaus where the creeks and ravines start (about 3800 to 4200 masl), a considerable number of shrub species, with genera such as *Amygdalus, Atraphaxis, Berberis, Caragana, Cerasus, Colutea, Crataegus, Lonicera, Rhamnus, Ribes, Rosa, Rubus, Sorbus,* and *Zygophyllum* are present. There are almost a dozen species of *Rosa,* but many other genera from the Rosaceae family are present in remarkable numbers, in addition to several geophytes. Because most areas of the Pamirs are semiarid or arid, typical forest belts cannot be distinguished. Only forest patches have been known, many of which have been cut down. They were very rich in species. Wild progenitors of fruit trees in the lower-montane belt of the west and northern Pamirs are a very valuable genetic stock, examples of which are *Juglans regia, Malus sieversii, M. kirghisorum, Prunus sogdiana, Pyrus communis, P. korshinskyi, P. regelii, Cerasus tienshanica, C. mahaleb, Berberis oblonga, Amygdalus communis, Pistacia vera, Crataegus songorica, C. turkestanica,* and many others. This montane, rather an open shrubby vegetation in remote locations, is mixed with the isolated *Juniperus* in the upper-montane belt, where *Juniperus seravschanica* and *J. turkestanicus* can even be a small tree. Their upper altitude limit is about 2800 to 3200 masl (Agakhanjanz and Breckle, 2004). Centuries ago, open forest patches of *Juniperus* were more widespread, but they have been cut down, and erosion has taken place. *Salix, Betula,* and *Populus* can be found in much higher altitudes and, theoretically, the treeline would be about 3600 to 4000 masl.

In the upper vegetation belts, mainly above 3000 masl, dwarf shrubs dominate, and the number of endemics is conspicuous. Among the many endemics, species of *Acantholimon, Artemisia, Astragalus, Corispermum, Cousinia, Oxytropis, Poa, Stipa,* and *Suaeda* play particularly important roles.

According to Ikonnikov (1979), the Asteraceae (70 genera, 213 species) is the most common Angiosperm family in the western Pamir, followed by Poaceae (60 genera, 200 species), Fabaceae (26, 124), Brassicaceae (59, 110), Caryophyllaceae (24, 61), Lamiaceae (23, 56), Boraginaceae (19, 49), Rosaceae (11, 48), Scrophulariaceae (11, 47), and Cyperaceae (9, 46).

In the high plateau desert of the eastern Pamir, the α-diversity is often rather low, but the ß-diversity is still high because of the variability of sites.

There are many vegetation types described from the various mountain parts and valleys, but often only a few vegetation units are dominant over vast parts of the landscape. The percentages of areas of different vegetation types are shown in Table 16.2.

LAND USE AND DESERTIFICATION

After the breakdown of the Soviet Union and the independence of the state of Tajikistan, the food and energy supply of the people were altered completely. The grazing pressure increased despite the civil war, and the former supply source of coal from Moscow had to be replaced; thus, many trees and shrubs were cut and woody dwarf shrubs collected. Lack of fuel and subsequent depletion of vegetation have led to increased erosion by wind and water around the villages.

Desertification with all its aspects became prominent. Deforestation and overgrazing lead to gully erosion and to more frequent dust and sandstorms. With increased erosion, accumulation of sand and mud also increased. Both led to a loss of productivity and biodiversity. Disastrous events such as mud streams and huge avalanches became a threat to the villages in the steep valleys. Increasing salinity in irrigation fields became a problem only in some parts, mainly in the eastern Pamir. Eutrophication became a local problem, where sheep, goats, cows, and camels are regularly rested. Gradual differences appeared as changes of land use in the eastern and western Pamir, as indicated in Table 16.3.

TABLE 16.2
Percentages of areas of land cover and vegetation types in the Eastern Pamir

Land Formation/Vegetation Type	Elevation (m)	Area (ha)	Percentage
Mountain deserts,		73.210	33.20
mainly *Ceratoides*,	3500–4200		28.49
mainly *Artemisia*, *Ajania*, etc.	3500–4600		4.71
Mountain steppes,	3500–4000	21.000	9.54
mainly *Stipa*			
Mountain xerophytes,	3500–4100	950	0.44
thorny cushions			
Mountain meadows,	3500–4400	11.120	5.06
riverbanks, floodplains, *tugai*,			
mainly *Leymus*, *Kobresia*			
Bare open scree, rocks, glaciers,	(3500)–7000	114.220	51.8
including kryophytes			
Total		220.500	100

Source: From previously unpublished material, mainly vegetation maps from Agakhanjanz, cf. also Agakhanjanz, O.E. and Jussufbekov, C.J. (1975).

TABLE 16.3
Land use practices, effects on desertification processes, and ecosystem functions in the Pamir

Land Use	Grazing		Gathering		Deforestation		Irrigation (Agriculture)	
Desertification Symptoms	W	E	W	E	W	E	W	E
Wind erosion	4	4	3	4	3	4	3	4
Fluvial erosion	4	4	3	0	4	3	4	4
Land slides	4	2	3	0	4	2	3	1
Salinization	0	0	0	0	0	0	3	4
Eutrophication	3	4	0	0	0	0	4	4
Biodiversity losses	4	4	4	4	4	4	4	4
Loss of productivity	3	4	1	4	4	4	0	0

Note: W = west Pamir; E = east Pamir; effects: 4 = strong, 3 = distinct, 2 = moderate, 1 = slight to negligible, and 0 = zero.

DEFORESTATION OF JUNIPER WOODS

Three plant species form the basis of this vegetation type: *Juniperus seravschanica*, *J. semiglobosa*, and *J. schugnanica*. The corresponding plant communities have a limited distribution in the west Pamir. They are lacking in the east Pamir, where the climatic conditions for the juniperus species are unfavorable. This is proven by very slow annual growth and lack of *Juniperus* seedlings (Agakhanjanz 1975). Most of the *Juniperus* were cut down in the last few decades. The clearing started in the southern part of the Pamirs (Rushan, Schugnan, and Vakhan). It is estimated that only 0.1% of the juniper woods remain. Without special protective measures, the *Juniperus* plant communities will disappear.

DEFORESTATION OF TUGAI FORESTS

The azonal forest stands in river valleys in Central Asia are called Tugai forests. *Betula pamirica* plant communities are common in the west Pamir but cover small areas. Their upper altitude limit is about 3700 m asl, thus also reaching small parts of the east Pamir. Their average height is 15 to 18 m. The accompanying vegetation is rich in grasses and herbs. *Populus pamirica* forms communities in small patches in the side valleys but is normally mixed with *Betula* and *Salix*. The *Salix*–Tugai forests comprise several species: *S. pycnostachya*, *S. turanica*, *S. schugnanica*, and *S. wilhelmsiana*. They reach the valley bottoms of the eastern Pamir up to almost 3900 masl. The Tugai forests are used intensively. Felling of adult trees, overgrazing, and increased fluvial erosion of the river terraces have strong impacts, causing not only the loss of biomass and of rich riverine forests but also the loss of fertile alluvial soils. Tugai forests play an important role in water regulation, flood control in the valleys, and improvement of the microclimate.

GRAZING AND DESERT PASTURES OF THE WEST PAMIR

The mountain meadows, the high-mountain deserts, the high-mountain steppes, and the xerophytic plant communities in the Pamir represent sufficient food potential for cattle. They are used as summer pastures. Only small areas with *Artemisia* pastures with a high share of Ephemeroides (*Poa bulbosa*) can be used as spring pastures. However, the arid natural pastures of the west Pamirs have little productivity because of the high share of *Acantholimon* species and cushion life-forms, their low-cover degree (less than 20%), and the high percentage of open rock areas. The production of biomass in these pastures is 0.03 to 0.3 t ha^{-1}. Grazing under these conditions can always be categorized as overgrazing, leading to the degradation of the pastures. The percentage of biomass of the ruderal plant species in the mountain meadows can reach 50 to 70% as a result of overgrazing. The species of *Acantholimon* have spread in the primary *Artemisia* plant communities of the lower mountain belts and become dominant. Observations show that the restoration of *Artemisia* pastures takes about 20 to 30 years after the disturbance. The natural regeneration can last even longer, if impeded by invasive thorny cushions such as *Acantholimon* and *Cousinia*. The summer pastures of the west Pamir are greatly degraded (see Table 16.4) with the exception of the high-alpine pastures whose share, however, is less than 10%.

The natural vegetation suffers in most of the densely populated valleys of the west Pamir, it differs in various valleys because of their different geomorphological structure. In the Bartang Valley, the percentage of areas of rocks and scree is about 76.2, it is 49.8 in the Schachdara region, and only 29.2% in the Gunt Valley.

Herbs and subshrubs are collected in great masses from the slopes (Photo 16.1) to feed cattle and other animals, but the very selective collecting of herbal medicinal plants has also

TABLE 16.4
Degradation of vegetation in the Gunt Valley (Western Pamir)

	Moderately Degraded Area (%)	Strongly Degraded Area (%)	Extremely Degraded Area (%)
Mountain tugai forest	10	20	70
Mountain deserts	20	25	55

PHOTO 16.1 Intensive gathering of fodder material from the steep slopes in the western Pamir, brought home by crossing the river (Schachdara Valley). Photo by Clas Naumann, August 2002.

been greatly increased. The effects on the flora can be only roughly estimated. The conversion of natural habitats to low productive fields, unregulated deforestation, overgrazing, and invasion of weedy species, uncontrolled fires, and illegal poaching are major threats to species biodiversity.

GRAZING AND DESERT PASTURES OF THE EAST PAMIR

A more arid climate than the other areas is a characteristic of the east Pamir. In winter, there is often no snow cover, but severe frost. Pastures, therefore, can be used as winter. The natural plant communities are very open (vegetation covers 5 to 15%). There are only a few summer pastures with the dominance of *Artemisia* and *Festuca* plant species. The productivity of *Artemisia* pastures is 0.3 to 0.4 t ha^{-1}, and the *Festuca* pastures, 0.8 to 1.2 t ha^{-1}. These have been the best pastures of the east Pamir. Thus, grazing pressure is very high, and so is the degradation of pasture.

Pastoralism in the Pamirs is extensive but technically still underdeveloped and unstable. Dry years, or cold and long winters cause a strong reduction of livestock. Grazing has led to a major degradation of the natural vegetation and the soil. The transformation process from state farming to private farming and agriculture has caused a very uneven use of the pastures. The size of the livestock and grazing intensity, seasonal rotation of grazing, better land management, and seeding of more productive plants for a secure food base have to be adjusted in the future. The goal is to reduce the anthropogenic pressure on the natural ecosystem and thus establish protected areas for the conservation of biodiversity.

THE TERESKEN SYNDROME IN EASTERN PAMIR

The desert vegetation is dominated by teresken (*Ceratoides papposa*, Figure 16.2). Its production of biomass is relatively low at 30 to 70 kgt ha^{-1} (up to 0.15 t ha^{-1}). The teresken also forms plant communities in the west Pamir in altitudes between 2000 and 3400 masl, but is dominant in the east Pamir between 3500 and 4200 masl (Table 16.2). The wide altitudinal and ecological range of the teresken causes a very high diversity of teresken vegetation types.

These vast stretches became subjected to heavy collection of teresken (Photo 16.1 to Photo 16.5). This may be looked upon as a strong form of desertification, as teresken is not only used as an important energy source for heating and cooking (despite the fact that one

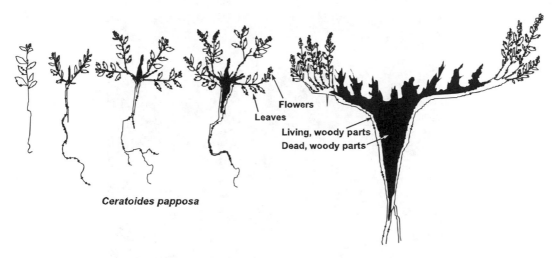

FIGURE 16.2 Development of *Ceratoides papposa*. From left to right: 1-year, 2-years, 3- to 5-years, 8- to 12-years, and more than 100-years old. An age of 250 to 300 years has been determined for some plants, and even this may not be the maximum. (From Steshenko 1956.)

PHOTO 16.2 *Ceratoides papposa* from high mountains deserts in eastern Pamir, 4000 masl. *Ceratoides* dwarf shrubs are torn out with the rootstocks, gathered, and piled up. Photo 16.2 to Photo 16.5 by S.W. Breckle, August 2002.

tereskem has little woody biomass — especially when young [see Figure 16.2]), but it is also the main source for feeding animals (sheep, goats, cows, camels, horses, and donkeys), as well as for wildlife grazing (wild goats and Marco Polo sheep).

The depletion of the tereskem pastures by grazing is of second priority. The main problem is collection of the woody rootstock (Figure 16.2) as an important energy source for house-holds (Photo 16.5). The intensive depletion of tereskem has considerably reduced the value of tereskem pastures. As it plays such a dominant role in the east Pamir, its degradation is called the *tereskem syndrome*.

The degraded tereskem deserts are very susceptible to wind erosion, and all the pits left after harvesting can be sources for gully erosion, too. Biodiversity and productivity of these

PHOTO 16.3 A huge pile of *Ceratoides papposa* dwarf shrubs, each plant is about 25 to 40 years old. This drought-resistant desert plant exhibits its main biomass in the rootstock.

PHOTO 16.4 The collected *Ceratoides papposa* are brought to a summer village. *Ceratoides* is used as fuel and fodder.

deserts are declining; many species show a degressive behavior (Table 16.5), with only a few thriving.

Around many villages, there are circles of several kilometers in diameter, where teresken has been almost eliminated, and its regeneration under such harsh climatic conditions occurs probably only once in a decade. Wind erosion has blown away all fine soil; thus, the damage and loss of other (widely dispersed)

herbal species is huge. Soil surface often is a desertic stone pavement. Particularly during harsh winters, livestock has become more vulnerable, as a result of depleting land resources and the cessation of winter fodder imports (Breu et al. 2003).

Dependance on only one life-form of plants in extreme climatic conditions is common. A striking similarity to the east Pamir is found in Bolivia. For example, in the arid parts of the

PHOTO 16.5 Summer village of the Kirgis people in the eastern Pamir, with yurts and small stone houses (4250 masl) adjacent to a small well. The energy supply is based on dried dung (foreground) and teresken (*Ceratoides papposa*; pile in the middle). The latter is also used for fodder.

altiplano in Bolivia, the *tola* (*Parastrephia lucida* and other species of *Parastrephia*, as well as similar dwarf shrubs such as *Baccharis*, *Fabiana*, and *Adesmia* from the *tolar* vegetation belt) seem to have a very similar role as teresken. It is an important source of cooking fuel, animal fodder, and grazing, mainly for llamas and alpacas (the competing uses resulting in what is known as tola syndrome). The climate of the altiplano is more favorable than the Pamir, however; the vegetation density is higher, and the tola seems to regenerate more frequently.

DESERTIFICATION AND LOSS OF SPECIES

Improper land use by grazing, clearing, gathering, and overuse of soils, in combination with the dry climate of the Pamir, leads to the destruction of the vegetation, reduction of the vegetation cover, change of species composition, loss of biological diversity and productivity, and erosion of soil (Table 16.3). The desertification or land degradation in the Pamir is very distinctive. It has become a major problem in the last few decades. The high degree of desertification is known only to some parts of

the area (Table 16.4). Most of the areas need a strong restoration management system and require investment and much time for restoration. Conservation of biodiversity and life-forms will need different strategies in the west and the east Pamir regions, due to the differences of landscape structure and land use. Biodiversity has been documented quite well by the various studies of the region. For all vegetation types, land use and desertification factors are strongly interdependent and usually are a threat to richness in species (Table 16.5).

In the Red Data Lists of the Soviet Union, some plants are mentioned as threatened species of the Pamir area: *Androsace bryomorpha*, *Ficus carica*, *Fragaria bucharica*, *Fragaria nubigena*, *Halimiphyllum darvasicum*, *Malus niedzwetzkiana*, *Platanus orientalis*, *Primula flexuosa*, *Punica granatum*, *Pyrus cajon*, *Sorbus turkestanica*, and *Vitis vinifera*. Other species that have been recorded as very rare include *Acantholimon alexeenkoanum*, *Alajja rhomboidea*, *Allium darwasicum*, *A. elatum*, *Amygdalus bucharica*, *Arum korolkowii*, *Betula murgabica*, *Biebersteinia multifida*, *Botrychium lunaria*, *Celtis caucasica*, *Cercis griffithii*, *Clematis saresica*, *Clementsia semenovii*, *Cryptogramma stelleri*, *Delphinium minjanse*, *Epipactis helleborine*, *Fraxinus raibocarpa*,

TABLE 16.5
Progressive or degressive spreading of plant species in the Pamir as a consequence of land use and desertification factors

	Degressive Behavior	Progressive Spreading
Mountain forests		
Juniperus schugnanica	D	
Juniperus semiglobosa	D	
Juniperus seravschanica	D	
Forests of the flood plains (tugai)		
Betula pamirica	D	
Populus pamirica	D	
Salix spp.		D
Hippophaë rhamnoides		D
Mountain meadows		
Agrostis spp.	I	
Trifolium spp.	I	
Kobresia spp.	Z	
Mountain deserts		
Artemisia korshinskyi	Z	
Artemisia vakhanica	Z	
Artemisia rhodantha	Z	
Artemisia rutifolia	Z	
Ceratoides papposa	C+Z	
Ephedra tibetica	C+Z	
Ephedra gerardiana	C+Z	
Ajania tibetica	C+Z	
Mountain xerophytes		
Acantholimon diapensioides	Z	
Acantholimon parviflorum		Z
Acantholimon pamiricum		Z
Astragalus roschanicus		Z
Mountain steppes		
Stipa spp.	Z	
Festuca sulcata	Z	
Cousinia rubiginosa		Z
Nepeta podostachys		Z
Alpine vegetation (kryophytes)		
Potentilla pamirica	Z	
Sibbaldia tetrandra	Z	
Primula macrophylla	Z	
Primula turkestanica	Z	
Leontopodium ochroleucum	Z	
Ephedra fedtschenkoi	D+Z	
Smelovskia calicyna	Z	
Saxifraga hirculus	Z	
Androsace akbaitalensis	Z	
Oxyria digyna	Z	
Oxytropis immersa		Z
Dracocephalum paulsenii	Z	

Note: C = collecting, D = deforestation, I = irrigation, and Z = grazing.

Ixiolirion karateginum, Lagochilus macrodontus, Lonicera nummulariifolia, Myricaria elegans, Neottia camtschatea, Pentaphylloides dryadanthoides, Pistacia vera, Populus pruinosa, Pyrola tianschanica, and *Rhamnus minuta.* This list is certainly not complete. Most of these species are severely threatened by the intensive land use or may already be extinct.

MEANS OF RESTORATION

The poverty of the whole region is striking. Only with the considerable help from the Aga Khan Foundation in recent years have famines been prevented. In the 4 years since the end of civil war in the region, a positive development has begun, and the village economies have grown. The means for nature conservation and the consciousness for its necessity have also been enhanced.

There are many attempts by international organizations, as well as by other countries (e.g. Switzerland [Breu et al. 2003]) to improve the economic situation and minimize ecological risks. It is, however, doubtful that a 2-week biodiversity assessment in each of the Central Asian states (except Tajikistan, which in 2000 was considered inaccessible due to security issues) can reveal enough reliable results (USAID 2001). Desk-study approaches are necessary but insufficient. Despite the fact that the Pamir Mountains have been investigated for decades, there is not only a strong need for scientific projects and a sound scientific basis for the application of developmental projects but also an imperative to stop further desertification (Photo 16.6).

The land use in various parts of the Pamirs and drastic changes in land use in recent years have had very different effects on changes in vegetation and surface cover, and thus on mountain biodiversity. In the past decades, mainly productivity, in general, and the quantitative effect of land use by grazing and fuel collecting were estimated (Agakhanjanz, 1975; Breckle and Agakhanjanz 1994). The qualitative richness of the flora was recorded thoroughly but without establishing a means of protection. The consciousness of the people is sufficient to create and maintain protected areas and to introduce new methods for energy supply. There is

a large potential for ecotourism (Photo 16.7) and for handicrafts, which could be a part of their income, thereby improving living conditions. This would certainly protect the richness of the vegetation cover and restore parts of the open shrub vegetation on the steep slopes of the western Pamir, thus eliminating the high risk of mud avalanches and landslides. In the eastern Pamir, the "teresken problem" has to be solved, and another diversified energy supply has to be introduced, as well as new methods of intensive fodder production in small areas to prevent further destruction of the open vegetation in the plateau desert.

SUMMARY

The eastern Pamir is characterized by high-alpine plateau deserts (often above 4000 masl). In contrast, the western and northern Pamir and Pamir-Alai region features deep river valleys with steep slopes, showing a very high relief energy. The western and northern Pamir are semihumid to semiarid. Their flora is very rich, many different vegetation types having developed according to the high geomorphological diversity. The eastern Pamir is arid, the dominant desert vegetation characterized by teresken (*Ceratoides papposa*; Chenopodiaceae), a slow-growing dwarf shrub (which can live very long). Its woody rootstock is now the main source of energy for the people. Harvesting the shrub has led to a severe decrease in plant cover and loss of old plants. In addition, uprooting teresken in large numbers has led to a severe decrease in grazing lands, as it is the main feeding source for cattle, goats, sheep, and camels (teresken syndrome). Desertification by overgrazing, gathering, deforestation, and inappropriate agriculture has led to severe erosion problems, loss in productivity, and is a strong threat to biodiversity. Sustainable land use is possible only by drastic changes in the energy supply and grazing habits. An improved economic base for the people must be found.

ACKNOWLEDGMENTS

The help of the GTZ CCD office (Bonn) and of the DAAD (Bonn) is greatly appreciated.

PHOTO 16.6 Grazing by yaks (*Bos grunniens*) in the high plains at Sorkol Lake, eastern Pamir. Photo by Clas Naumann, August 2002.

PHOTO 16.7 Intact Tugai forests in the very dynamic Pamir riverbed, shortly before its joining the Wakhan River and thus forming the Ab-i-Panj, the upper Amudarya, the border between Tajikistan and Afghanistan. Photo by Clas Naumann, August 2002.

The help and hospitality of the counterparts of these offices in Tajik (Ministry of Environment, Dushanbe, Tajikistan), the Pamir Botanical Garden (Khorog; Chekchekti), and other institutions in the Badakhshan Autonomous Province, from numerous people in the villages, as well as from Markus Hauser, Eva Kleinn, and Surat Toimastov, are greatly appreciated. Thanks also to the ZiF (Center for Interdisciplinary Studies) at the University of Bielefeld, Germany, for the facilities provided to organize the Pamir Symposium in January 2003.

References

Agakhanjanz, O.E. (1965). *Die hauptsächlichen Probleme der physischen Geographie des Pamir.* Band 1, Dushanbe, Tajikistan, 240 pp. (Russian).

Agakhanjanz, O.E. (1975). Above-ground phytomass of the Pamir pastures. *Izv Akad Nauk SSR Ser Geogr* 43–50 (Russian).

Agakhanjanz, O.E, (1978). Ecological scheme of the elevational belts of the vegetation of the Pamiro-Alai mountain system. *Ekologia (Nauka, Leningrad)*, 5: 18–24 (Russian).

Agakhanjanz, O.E. (1985). Ein ökologischer Ansatz zur Höhenstufengliederung des Pamir-Alai. Peterm. *Geogr Mitteil*, Heft 1: 17–24.

Agakhanjanz, O.E. (2002). *Der Wind, der heißt Afghane. Forschungen auf dem Pamir im Jahr der Schlange.* Shaker, Aachen, Germany, 238 pp.

Agakhanjanz, O.E. and Breckle, S.-W. (1995). Origin and evolution of the mountain flora in Middle Asia and neighboring mountain regions. *Ecol Studies* 113: 3–80.

Agakhanjanz, O.E. and Breckle, S.-W. (2002). Plant diversity and endemism in high mountains of Central Asia, the Caucasus and Siberia. In Körner, C. and Spehn, E. (Eds.), *Mountain Biodiversity — A Global Assessment.* Parthenon, Boca Raton FL, pp. 117–128.

Agakhanjanz, O.E. and Breckle, S.-W. (2004). Pamir. In Burga, C.A., Klötzli, F., and Grabherr, G. (Eds.), *Gebirge der Erde — Landschaft, Klima, Pflanzenwelt.* Ulmer, Stuttgart, Germany, 151–157.

Agakhanjanz, O.E. and Jussufbekov, C.J. (1975). *The Vegetation of the West Pamir and an Attempt of Reconstruction.* Dushanbe, Tajikistan, 310 pp. (Russian).

Breckle, S.W. (Ed.). (2003). Natur und Landnutzung im Pamir. Wie sind Erhalt der Biodiversität, Naturschutz und nachhaltige Landnutzung im Pamirgebirge in Einklang zu bringen? Pamir-Symposium ZiF, Center for Interdisciplinary Studies, 2003. Bielefelder Ökologische Beiträge (BÖB) 18.

Breckle, S.W. and Agakhanjanz, O.E. (1994). *Ökologie der Erde.* Band 3: Spezielle Ökologie der gemässigten und arktischen Zonen Nord-Eurasiens. 2nd ed., Fischer, Stuttgart, Germany, 726 pp.

Breu, T., Hurni, H., and Wirth Stucki, A. (Eds.). (2003). *The Tajik Pamirs. Challenges of Sustainable Development in an Isolated Region.* Center for Development and Environment CDE, University of Bern, Switzerland, 80 pp.

Ikonnikov, S.S. (1963). *Flora of the Plants of the Pamirs.* Dushanbe, Tajikistan, 282 pp. (Russian).

Ikonnikov, S.S. (1979). *Flora and Fieldguide of Higher Plants of Badakhshan.* Nauka, Leningrad, 400 pp.

Stanyukovich, K.V. (1973). *Mountain Vegetation of the USSR.* Akad. Nauk Natshik SSR, Dushanbe, Tajikistan, 31 pp. (Russian).

Steshenko, A.P. (1956). Formation of the semi-shrub structure in the high mountains of Pamir. *Tr. Akad. Nauk Tajik SSR* 50: 162 (Russian).

USAID (2001). *Biodiversity assessment for Tajikistan.* Chemionics Intern. Inc., Washington, USA, 36 pp.

Walter, H. and Breckle, S.-W. (1986). *Ökologie der Erde.* Band 3: Spezielle Ökologie der gemäßigten und arktischen Zonen Nord-Eurasiens. Fischer, Stuttgart, Germany, 587 pp.

Walter, H., Harnickel, E., and Mueller-Dombois, D. (1975). *Klimadiagramm-Karten der einzelnen Kontinente und die ökologische Klimagliederung der Erde.* Fischer, Stuttgart, Germany.

Wennemann, M. (Ed.). (2003). *Pamir-Expedition 2002* (Bielefeld-Bonn-Dushanbe 27.7.-24.8.2002).

17 Effects of Grazing on Biodiversity, Productivity, and Soil Erosion of Alpine Pastures in Tajik Mountains

Khukmatullo M. Akhmadov, Siegmar W. Breckle, and Uta Breckle

INTRODUCTION

Tajikistan is a typical mountainous country. Mountains make up 93% of its territory. The highest mountain systems of Central Asia are the Tyan-Shan and the Pamir-Alai. The maximum elevation is 7495 masl. More than 60% of the territory of Tajikistan is located at or above 2500 masl. This region is mostly used as summer pastures. In the Pamir-Alai mountain system, it consists of Darvaz, Academy Sciences, Peter the First, Alai, Zaalai, Karategin, Hissar, Zerafshan, and Turkestan mountain ranges.

The high-mountain areas (subalpine and alpine zone) exhibit a strong continental climate. The severe long winter is followed by a rather short and cool summer. The average annual temperature is about 0.2 to 1.6°C. The warmest months are July to August (maximum temperature: 22°C). The coldest month is January (absolute minimum: −36°C). The annual period without frosts lasts about 88 to 101 d. The annual precipitation is very varied: in the East Pamir it is only about 72 to 200 mm, in the West Pamir it can reach more than 500 mm, and in central Tajikistan (Hissar Mountains) it is about 600 to 1200 mm (Narzikulova, 1982). About 50% of the precipitation falls during the spring months. In the other seasons of the year, the scarce precipitation is distributed rather equally. Precipitation in autumn, winter, and, partially in spring is as snowfall; snow cover can last until early summer. The climatic conditions are demonstrated by climatic diagrams (see Breckle and Wucherer, this volume; Walter and Breckle, 1986a, b, 1994; Breckle and Agakhanjanz, 2004).

PASTURES AND GRAZING AREAS

The conservation of biodiversity in the high mountains that have been used intensively as summer pastures has become an important problem. Until 1992, high pasturelands were used for 2 to 3 months, and in the autumn–winter period the cattle were moved to winter pastures. The exploitation of the summer pastures had decreased, and this had presented an opportunity for the restoration of pasturelands. However, high pasturelands are now used all year round, and the grazing intensity has increased manyfold, with the result that severe erosion processes and degradation of grassy vegetation have taken place. Therefore, there are two problems in the biodiversity conservation of high pasturelands: (1) protection of the unique highland grassy vegetation, which has many species that are included in the Red Book of NIS and Tajikistan (e.g. *Taraxacum*, *Rosularia*, *Desideria*, and *Tulipa*) and (2) conservation of different types of vegetation formations through protection of pastures from weeds that can turn land into marginal deserts.

Natural pastures and the haymaking areas of Tajikistan occupy more than 3.5 million ha and are major sources of high-grade and other forages for the livestock sector. The stock of forages on pastures, according to expert data, is more than 1.6 million t of dry mass per year. The livestock sector is economically profitable — it has the lowest production costs, much below the cost of production of the stalls necessary for the maintenance of cattle in the winter period. The present conditions of pastures in the republic are characterized by an accelerated decrease in their fodder efficiency due to anthropogenous factors. The productivity of the pastures has changed greatly over the years, and the animal population is rarely provided with a steady supply of forage. For the maintenance of stable livestock populations during periods with very low fodder availability (winter), additional feeding of animals is required. Therefore, the strategy for the use of pastures and keeping cattle should reflect the specific conditions of the region.

DEGRADATION OF PASTURES AND BIODIVERSITY CHANGES

Extensive use of pastures and periods of drought have caused severe degradation in many parts of the subalpine and alpine zones. The basic indicator of such degradation is the change in vegetation. Under excessive grazing, there is a significant change in the species composition in the bluegrass–sedge pastures. Poisonous, harmful, and unpalatable plant species (e.g. *Hordeum leporinum*, *Centaurea squarrosa*, and *Alyssum desertorum*) start to develop; production of herbage decreases five- to tenfold; and biodiversity changes from highly varied vegetation patches to monotonous overgrazed areas. The large number of species (Ikonnikov, 1979; Agakhanjanz and Breckle, 1995, 2002) is threatened by the spread of toxic and unpalatable weeds. The effects differ in the various pasture types, but in all types, a reduction of 20 to 60% in the number of species can be postulated. In addition, the replacement of long-term fodder plants (with strong taproots) by fast-growing annuals (with a superficial root system) has resulted in increased water and wind erosion on pastures.

During recent years, the grazing pressure on the winter–spring pastures has increased considerably. All-year-round grazing of cattle has become common. Not only overgrazing of vegetation and pastures but also the felling of trees and cutting of bushes and semishrubs for fuel has had very negative effects. Production of herbage has decreased, and pastures have become seasonally narrow. This has resulted in a significantly higher seasonal and annual variability in available fodder. Unlimited grazing has not only influenced the plant composition of the pasture but also appreciably changed ecological conditions, especially in localities in which the forest vegetation was destroyed and the area then was transformed into pastures. A marked reduction of vegetation density and destruction of the soil cover can be observed, as well as the formation of numerous sheep paths (Table 17.1). Water permeability of the soil is reduced by overgrazing, leading to enhanced erosion and drought. The herbaceous vegetation significantly lost soil stability because of loss of protective vegetation cover. Overgrazing of a meadow coenosis with *Alopecurus seravschanicus* develops by pasture degradation to an open vegetation dominated by *Adonis turkestanicus* or by *Scorzonera acanthoclada* and *Lagotis korolkovii*. Additionally, a weedy, tall herbaceous vegetation develops with *Artemisia dracunculus*, *Cousinia franchetii*, *C. splendida*, and other unpalatable short-grass meadow species (Akhmadov, 2003a). In general, desertification of the various pastures leads to an invasion by considerable numbers of toxic and harmful species and a great loss of biodiversity due to the disappearance of many high grasses (cereals), sedges, legumes, valuable forbs, and associates (as described in the following text).

SURVEY OF PASTURE TYPES

In the subalpine and alpine zones of the Pamir-Alai, the following types of pastures can be distinguished (Ovchinikova, 1977): (1) summer cryophilic (alpine) pastures and heath; (2) prickly-grass (tragacanth) summer pastures; (3) summer steppe (mountain and high-mountain); (4) swamps and meadow summer pastures, (5) long-grass mountain steppe;

TABLE 17.1
Density of sheep paths, soil washout, and steep slopes in Tajikistan

Inclination of Slope (in Degrees)	Quantity of the "Sheep Paths" (1000 Units km⁻²)	Soil Washout (t ha⁻¹)	Density of Gullies (Units km⁻²)	Length of Gullies (km km⁻²)
5–7	less than 1.0	1.2–21.1	0.1–0.2	1–2
10–12	1.4–3.2	35–72	0.5–0.7	5–7
15–17	2.3–6.8	64–400	0.9–2.4	7–20
20–22	4.5–8.7	250–1200	1.4–3.2	16–32
25–27	6.7–9.6	470–1800	1.7–4.7	24–47
30–32	9.1–11.2	800–2700	2.4–7.4	36–62
35–37	12.5–14.7	1300–3800	1.7–6.2	16–74
More than 40	14.2–17.4	2700–5200	1.9–8.2	20–84

Source: From Akhmadov (1997).

(6) autumn–winter desert pastures (with *Artemisia* and *Ceratoides*); and, in the lower areas (7) short-grass mountain steppe as winter pastures; and (8) winter–spring pastures.

Some basic characteristics, such as the yield, the degree of erosion, and area subjected to desertification of these different types of high-mountain pastures, are shown in Table 17.2. Table 17.3 gives estimates of the biodiversity of the different pasture types.

Summer Cryophilic (Alpine) Pastures and Heath

Summer cryophilic or heath pastures are often called *alpine meadows*. They are located below the nival zone, at an altitude above 3300 to 3500 masl. Heterogeneous climatic, geomorphological, and botanical characteristics do not allow an overall characteristic to be given. Thus, in the valleys of the West Pamir, more hygrophilous pastures with short-grass heaths, high meadows, and steppes are prevalent. The alpine formations on the West Pamirs have the characteristics of deserts. Xerophilous vegetation is widespread in the alpine zone but does not have a continuous distribution because of various mounds, glaciers, and snow patches. Thus, it is very often represented only by many small separate fragments. The common trait of all heath vegetation is its suitability to a short and cold vegetative season and the adaptation of the vegetative organs and buds (which are very close to the soil surface) to a long harsh winter, as

well as its ability to endure considerable frosts during the summer. Xerophilous species are very unequally distributed in pastures, depending on their specific structure and on the grass cover density.

Intensive grazing has caused a lack of regeneration of the grasses. Shoots are completely absent, morphological structure has changed, plants have become stocky, the aboveground system of shoots exhibits a partially rosettic shape, leaf size has decreased 2 to 3 times, the height of the grass stands has been reduced drastically (from 40–50 cm to 3–5 cm in low-herb meadow pastures), and aboveground mass of plants is concentrated in the lowermost layer. Valuable fodder and grass species have disappeared, the first to go being *Poa alpina*, *P. bucharica*, *P. litvinovii*, *Alopecurus himalaicus*, *Festuca alaica*, and *Allium fedtschenkoanum*. Numerous field experiments carried out by Akhmadov and coworkers in the basic types of pastures and hayfields have shown that intensive grazing leads to a decrease in soil fertility and a sharp decrease in productivity, resulting in deterioration of the quantitative structure of herbage on those pastures.

Overgrazing of cryophilic pastures in summer favors the growth of *Cousinia franchetii*, *C. pannosa*, *Scorzonera acanthoclada*, and *Lagotis korolkovii*. Consequently, these pastures further lose their economic value, becoming unsuitable for grazing. Observations (Akhmadov, 1999) have shown that the soil

TABLE 17.2
Area, yield, and degree of desertification of high-mountain pasturelands in Tajikistan

Type of Pasture	Area (x1000 ha)	Altitude (masl)	Yield of Dry Mass (t ha⁻¹)	Total Grazed (t ha⁻¹)	Degree of Land Erosion (Percentage of Total Area)	Area Subjected to Desertification (Percentage of Total Area)
Summer cryophilic pastures	100	3300–4800	0.05–0.63	0.03–0.56	78–96	97
Prickly-grass pastures	400	2400–3200 (3800)	0.05–2.5	0.03–0.57	90–95	100
Summer steppe pastures	420	1800–3500	0.35–1.3	0.32–0.62	78–91	96
Swamps and meadow pastures	170	1500–2900 (3200)	0.48–3.2	0.30–3.0	57.4–72	80
Long-grass mountain steppe	600	1000–3300	0.36–3.6	0.25–3.2	88.4–92	95
Autumn– winter desert pastures	700	400–800 (1200) and 3000–4700	0.005–1.0	0.002–0.8	96–100	100
Short-grass mountain steppe and winter–spring pastures	500	300–700 (1100)	0.15–2.0	0.06–1.7	86–94	95

Note: Numbers in brackets denote extreme values or highest altitude.

Source: From Akhmadov and Gulmakhmadov (1999).

protection provided by a covering of 60 to 80% by *Carex* and *Kobresia* is lost when the cover percentage reaches less than 35 to 55%.

PRICKLY-GRASS SUMMER PASTURES

Prickly-grass (tragacanth) summer pastures are very common in many mountain regions of Tajikistan. They are constituted by nonpalatable grasses, prickly subshrubs, and undershrubs. The tragacanth growth form is common in *Astragalus, Onobrychis*, and some other genera. With few exceptions, they are unpalatable, woody, and prickly plants, with a somewhat hedgehog shape. Only some *Cousinia* species can be grazed, mainly very late in the season, after their germination. Prickly and spiny species increasingly persist on the pastures and, thus, gradually replace valuable fodder plants.

Among the prickly grasses important for pastures there are different Gramineae, such as the meadow species; *Poa bucharica* and *P. zaprjagajevii*, and the steppe species; *Poa relaxa, Festuca sulcata, Stipa, Leucopoa karatavica*, and others. This zone often suffers from intensive erosion processes because the vegetation here does not have a good density of sward (large intertussock space).

SUMMER STEPPE (MOUNTAIN AND HIGH-MOUNTAIN PASTURES)

The summer steppe, mountain, and high-mountain pastures are the most common types of summer pastures. The prevailing constituent grasses are *Festuca sulcata, Poa relaxa, Leucopoa olgae, L. karatavica*, and a few others. Additionally, there are very palatable cereals:

TABLE 17.3
Plant species richness of different high-mountain pasturelands in Tajikistan (approximate number of species)

Type of Pasture	Total Species	Gramineae	Cyperaceae	Leguminoseae	Forbs Grazed	Associated Species	Toxic and Harmful
Summer cryophilic pastures	58	12	5	2	15	20	4
Prickly-grass pastures	88	25	3	5	10	41	4
Summer steppe pastures	145	36	1	12	23	63	10
Swamps and meadow pastures	54	17	8	6	8	11	4
Long-grass mountain steppe	82	23	4	10	12	27	6
Autumn pastures	86	13	2	4	27	35	5
Short-grass mountain steppe and winter–spring pastures	108	25	1	9	33	34	6

Piptatherum sogdianum, P. pamiroalaicum, Zerna angrenica, Alopecurus seravschanicus, and *Roegneria ugamica.* This vegetation is the best grazing area for sheep and, to some extent, for domestic cattle and horses. Because of overgrazing and prolonged unsystematic use, steppe pastures are greatly degraded, and on trampled pastures, there has been a sharp decrease of productivity. Because of overgrazing, many valuable grasses have disappeared and *Artemisia,* Polygonaceae, *Scorzonera,* and *Cousinia* have taken over. The steppes of the Pamir-Alai extend from the zone of the thermophilic juniper slopes in the valleys up to the subalpine region. The most complete steppes are found on high-mountain plateaus (dashts) and in dry valleys. Everywhere in the Pamir, the main belt of their distribution is the subalpine zone, normally above the timberline. Usually the subalpine zone is characterized by short dry summers and long, inclement, snowy winters. Therefore, winter grazing is not possible. The intensive grazing of the mountain steppe summer pastures by cattle results in the loss of many valuable plants. Trampled pastures are subject to erosion and degradation of the soil. The productivity can reach up to 3 t ha⁻¹ of dry mass;

the palatable parts, on average, reach only 0.2 t ha⁻¹. But, taking into account that up to 90% of plant dry mass is concentrated in the region of 0 to 2 cm above the surface of soil (below the level of grazing by sheep), it means that fodder amounts to only about 0.04 t ha⁻¹. Thus, the actual used plant dry mass on sites with extensive pasture grazing is only about 5 to 8% compared with the total herbage mass.

SWAMP AND MEADOW SUMMER PASTURES

Swamps (habitats with a high and permanent water table; also called *saza*) and meadow summer pastures are not widespread. By the character of the dominant plant functional types, they can be subdivided into two groups: forbs and meadow grasses. On meadow summer pastures with prolonged overgrazing, there is a change from productive and palatable plants to low and unpalatable grasses. These pastures additionally become weedy with harmful, poisonous plants, e.g. *Thermopsis, Trichodesma, Heliotropium,* and others. Meadow summer pastures are rather widespread, mainly in montane (with moderate and tall herbs) and in alpine mountain zones (with low herbs). On montane meadows, cereal grasses (*Zerna turkestanica, Dactylis glomerata, Roegneria ugamica, Poa bucharica, Hordeum turkestanicum, Alopecurus seravschanicus,* and *Agrostis alba*) and some leguminous species (e.g. *Vicia tenuifolia*) are dominant. These meadows are used for haymaking. The productivity is high, with 1.5 to 2.5 t ha^{-1} of dry mass. As a result of the prolonged uncontrolled use of the tall-grass meadows of the subalpine zone for grazing, most areas are degraded and contain many weeds, as well as unpalatable prickly grasses. Controlled areas with natural borders (Ziddy region, Hissar mountain range), which are isolated from pastures and are only used for haymaking, give 2.0 to 2.5 t ha^{-1} of high-quality hay, whereas in intensively grazed sites, the edible part of herbage only makes up 0.25 to 0.3 t ha^{-1}, with 50 to 70% being unpalatable, mostly prickly *Cousinia*. The percentage of the area of these degraded (and now prickly-grass pastures) is more than 30% of the entire summer pastures of Tajikistan. The percentage of palatable fodder on some sites does not exceed 10%.

Long-Grass Mountain Steppe: Summer Pastures

These are tall-herb and long-grass mountain steppes used as summer pastures, which are characterized by *Ferula ovina, F. jaeschkeana, F. karatavica, F. kokanica, Prangos pabularia, Alcea nudiflora, Crambe kotschyana, Inula grandis,* and other large herbs. These large or giant herb vegetation types are very conspicuous in the different mountain zones; they belong to many different vegetation types and are always different in each site. The main dominants are ephemeroids. The various species of *Ferula* are not only characterized by their adaptation to a short vegetation period but also by their monocarpic (hapaxanthic) behavior. Such dominant species as *Ferula* and *Inula grandis* differ not only in their size and the roughness of their tissue but also in their vegetation mosaic. Tall-herb and long-grass steppe summer pastures are located from 1000 to 3300 m asl. The vegetation is basically made up by tall cereal grasses: *Hordeum bulbosum, Elytrigia trichophora,* and many ephemers and

ephemeroids. The productivity of *Hordeum bulbosum* and *Elytrigia trichophora* used for haymaking can reach, in some rangelands, up to 2.16 t ha^{-1} of hay. Fluctuations of crop productivity from year to year are between 1.4 and 3.51 t ha^{-1}. The cereal grasses *Piptatherum sogdianum* and *Roegneria ugamica* deliver the basic fodder value. In overgrazed areas, there is a change to unpalatable small grasses and an invasion by harmful, poisonous weeds, such as species of *Thermopsis, Trichodesma, Heliotropium, Cousinia,* and *Origanum tyttanthum,* etc. Additionally, overgrazing results in higher proportions of *Artemisia,* indicating a shift to semi-desert-like conditions (Akhmadov, 2003b). One of the widespread tall-herb formations is characterized by *Prangos pabularia,* which reaches as high as 3200 masl, the subalpine zone. This characteristic association with *Prangos pabularia* contains many ephemers (up to 1500 m asl), many forbs (between 1500 and 2200 m asl), mainly *Polygonum coriarium* (between 2500 and 3200 m asl) and *Ferula jaeschkeana* (between 1600 and 3200 m asl). Common species in all associations are *Hypericum scabrum, Artemisia persica, Ziziphora pamiroalaica,* and *Dactylis glomerata,* etc. The widespread *Ferula jaeschkeana* and *Prangos pabularia* sometimes displace other vegetation types and depreciate the pastures. *Ferula jaeschkeana* and *Prangos pabularia* contain essential oils and strong rough fibers, forming a hard straw, and therefore they are not eaten by cattle.

Autumn–Winter Desert Pastures with *Artemisia* and *Ceratoides* in High Mountains

Autumn–winter desert pastures develop in peneplain and low-mountain zones and, especially, in the high mountains of the East Pamirs. The insignificant snow cover and a relatively dependable availability of dry stems and some leaves and fruits make the alpine deserts suitable for winter grazing. The prevailing plants here are almost unpalatable in spring or in summer during their vegetative conditions. They become edible only in the dry conditions in winter or in late autumn. In the autumn–winter period, this grazing is synchronous in the river valleys with partly halophilous vegetation and on saline meadows.

Jungles, which develop on the sands of creeks, also are included in the winter pasture cycle.

Teresken or *Eurotia* (*Krascheninnikovia ceratoides* syn. *Eurotia ceratoides*, now called *Ceratoides papposa*) is the most common species in the East Pamir (see Breckle and Wucherer, Chapter 16). The teresken high-mountain autumn–winter desert pastures prevail in the West Pamirs only in the wide subalpine zone (3500 to 4200 masl, and rarely, up to 4500 masl). They prevail along wide and straight valley bottoms, on gentle slopes, on debris cones, and on the smoothed hills of ancient moraines. In the lower parts, the bigger *Ceratoides ewersmanniana* is also present. All soils in which teresken is found are slightly salty. They are pastures of very low productivity (0.05 to 0.2 t ha^{-1}). Vegetation cover is only 5 to 15%. In recent years, because of shortage or lack of fuel, widespread uprooting of bushes and half-bushes took place and, thus, teresken became the basic fodder source, as well as fuel, in winter pastures.

High-mountain autumn–winter desert pastures with *Artemisia* are also widely distributed in the lower mountains, in middle mountains, and in the high-mountain zones of Tajikistan up to an altitude of 4300 masl. The productivity of these pastures reaches 1.45 t ha^{-1}, and the palatable mass makes up 0.9 t ha^{-1}. The plant cover is 15 to 40%.

Alpine high-mountain autumn–winter desert pastures are distributed widely but in small patches in the alpine zone of the East Pamirs between the heights of 4300 to 4700 masl. They are represented by formations of the xerophytic dwarf semishrub *Ajania tibetica*. They are found along gentle slopes with low snow cover and on debris cones with desert skeletal soils. The productivity of these pastures reaches up to 0.18 to 0.25 t ha^{-1}, and the palatable portion amounts to 0.9 t ha^{-1}. Plant cover is 10 to 15%, and rarely, 25 to 30%.

CONCLUSIONS AND FUTURE ASPECTS

There is widespread animal husbandry in mountainous Tajikistan, and livestock keeping is largely determined by environmental condi-

tions. The various types of pastures are mainly distinguished by the seasons of their main use. The very reduced areas of pastures in the mountains and their remoteness from areas with good summer pastures have been responsible for the creation of an all-year-round grazing system, which has led to a strong reduction in biodiversity and a change of pasture type.

Overgrazing over centuries has led to a substantial change of the vegetation from its natural species composition. Pastures are contaminated by unpalatable plants — mainly *Cousinia* and *Acantholimon*, but also many other harmful and poisonous herbs and nonproductive grasses. Not only range degradation, but also the loss of biodiversity and accelerated soil erosion are consequences of a prolonged unsystematic pasturing. Development of methods for sustainable use and for the restoration of natural pastures, and creation of highly productive cultural pastures is the most effective and reliable way to combat the degression of pastures to maintain pastoral forages for cattle and other livestock.

For nature conservation and for maintaining a high biodiversity in vegetation and pastures, it is necessary (1) systemize the pasture of cattle (create good management plans); (2) use the same territory only once in 3 years; (3) apply, once in 2 to 3 years, small doses of mineral fertilizers for the improvement of quality and biomass of the plants; (4) apply meliorative measures for the improvement of the pasturelands; (5) get local communities to remove (by hand, because all summer pasturelands are located on steep slopes) the poisonous and unpalatable plants brought in by cattle as manure from the winter pastures; and (6) demonstrate (on experimental plots) to the local communities and farmers, progressive technologies for efficient conservation and improvement of high-mountain pasturelands, e.g. management of water and soil in the pasture zones; the regulation of cattle grazing; the restoration of forests, where possible; and the use of crops as an antierosion measure and to promote a species-rich grass cover.

Last but not least, it will be very important for the future development of the region to use alternative energy sources to achieve independence from organic fuels. This would be the best way to preserve the unique pasture

vegetation that protects the soil from degradation. Use of wind-power generators is a
prospect in many regions of Tajikistan, especially on the high Pamirs, where pastures are
used all year round. In addition, Tajikistan, due
to its geography and natural climatic conditions, is a very suitable region for the widespread use of solar radiation. The number of
sunny days is from 250 d (Fedchenko Glacier)
to 330 d (Murgab, East Pamirs) per year, providing 2000 to 3000 h of radiation per year. The
intensity of the solar radiation reaches up to
1 kW m^{-2} (on average, 500 to 700 W m^{-2}). Such
high-potential power from solar energy
resources is not used at all. Thus, in the future,
the use of solar energy could become an important step for biodiversity conservation in the
high-mountain pastures in the Pamir-Alai.

SUMMARY

Natural pastures and haymaking areas occupy
more than 3.5 million ha in Tajikistan. They are
the major sources of various high-grade forages
for livestock. The reserves of forages on pastures comprise more than 1.6 million t a^{-1} of
dry mass. Fodder productivity varies from year
to year, and therefore, does not ensure a stable
source for livestock. In recent years, fodder productivity of pastures has decreased due to
anthropogenic reasons. There are six different
types of pastures, depending on vegetation and
altitude and four types depending on land use.
Each type is characterized by the composition
of plants, productivity of the pasture, function,
use, and other features. Grasses are almost completely deprived of regeneration by intensive
grazing, and valuable fodder grasses are the first
to disappear, e.g. *Poa bucharica*, *P. bulbosa*,
Dactylis glomerata, *Helictotrichon asiaticum*,
H. hissaricum, *Festuca pratensis*, and *Allium
varsobicum*. Pasturable areas have been transformed by prolonged and excessive grazing into
inconvenient or marginal soils. Nowadays,
unpalatable grasses make up 75 to 90% of the
herbage. In total, the production of fodder mass
has decreased to 20%, or possibly, even 10%.
Better methods for the sustainable use and restoration of natural pastures and the creation of
cultural pastures are urgently needed (Breckle
et al., 2001; Breckle, 2003) to prevent further
pasture degradation and to provide livestock
with pasturable forages.

References

Agakhanjanz, O.E. and Breckle, S.-W. (1995). Origin and evolution of the mountain flora in
Middle Asia and neighbouring mountain
regions. *Ecol Studies* 113: 3–80, Springer-
Verlag, Berlin.

Agakhanjanz, O.E. and Breckle, S.-W. (2002). Plant
diversity and endemism in high mountains
of Central Asia, the Caucasus and Siberia.
In Körner, C. and Spehn, E.M. (Eds.), *Mountain Biodiversity — A Global Assessment*,
Parthenon, Boca Raton, FL, pp. 117–128.

Akhmadov, K.M. (1997). *Development of the Erosion Process in Tajikistan*. Dushanbe, p. 49
(Russian).

Akhmadov, K.M. (1999). Biodiversity and Dynamic
Processes on the Rangelands in the Different
Usage Periods. VI International Rangeland
Congress, August 1999, Sydney, Australia,
pp. 145–147.

Akhmadov KM (2003a). Biodiversity Conservation
in High-Mountain Rangeland: Problems and
Ways of a Solution. VII International Rangeland Congress, July 28–August 1, 2003,
Durban, South Africa.

Akhmadov, K.M. (2003b). Mountain Rangeland
Resources and Its Rational Use. VII International Rangeland Congress, July 28–August
1, 2003, Durban, South Africa.

Akhmadov, K.M. and Gulmakhmadov, D.K. (1999).
*Social–Economic Consequences of Deserti-
fication in Tajikistan*. Dushanbe: Donish. p.
64 (Russian).

Breckle, S.-W. (Ed.). (2003). Natur und Landnutzung
im Pamir. Wie sind Erhalt der Biodiversität,
Naturschutz und nachhaltige Landnutzung
im Pamirgebirge in Einklang zu bringen?
Bielefelder Ökologische Beiträge, 18, 104
pp.

Breckle, S.-W. and Agakhanjanz, O.E. (2004): Pamir.
In Burga et al. (Eds.). *Die Hochgebirge der
Erde* (Ulmer/Stgt.). pp. 151–157.

Breckle, S.-W., Veste, M., and Wucherer, W. (Eds.).
(2001). *Sustainable Land-Use in Deserts*.
Springer-Verlag, Berlin, 465 pp.

Ikonnikov, S.S. (1979). *Flora and Fieldguide of
Higher Plants of Badakhshan*. Nauka/Lenin-
grad, 400 pp.

Narzikulova, A. (ed.) (1982). Nature and Natural
Resources. Dushanbe: Danish (Russian).

Ovchinnokova, P. (ed.) (1977). The Pastures and the Haymaking Areas of Tajikistan. Dushanbe: Donish p. 305 (Russian).

Walter, H. and Breckle, S.-W. (1986a, 1994) *Ökologie der Erde* (vol. 3) *Spezielle Ökologie der gemäßigten und arktischen Zonen Euro-Nordasiens*. UTB Große Reihe, Fischer, Stuttgart, 587 pp. (2nd ed., 726 pp.)

Walter, H. and Breckle, S.-W. (1986b). *Ecological Systems of the Geobiosphere (Vol. 3) Temperate and Polar Zonobiomes of Northern Eurasia*. Springer-Verlag, Berlin, Germany.

18 Plant Species Diversity, Forest Structure, and Tree Regeneration in Subalpine Wood Pastures

Andrea C. Mayer, Christine Huovinen, Veronika Stoeckli, and Michael Kreuzer

INTRODUCTION

Mountain forests in the Alps traditionally offer several independent benefits, such as timber production, protection against natural hazards (e.g. avalanches and rockfall), recreation area, and habitat for diverse vegetation and wildlife. Wild animals, including ungulates such as red deer, use the mountain forests all the year round. In summer, 15% of the Swiss mountain forests are additionally grazed by domestic animals, mainly cattle (Brassel and Brändli, 1999). Farmers graze their cattle in the mosaics of coniferous forests, open pastures (with diverse herbaceous vegetation), and half-open pastures (with dwarf shrubs and young trees). Foresters fear that such grazing hinders tree regeneration and reduces the protective function of mountain forests (Delucchi, 1993). Others claim that forest grazing provides the opportunity for extensive agricultural and forest production, and that wood pastures have a high structural diversity, thus positively influencing the landscape amenity and biodiversity (Ten Klooster, 2000).

In this study, plant species diversity, forest structure, and tree regeneration in subalpine wood pastures were investigated, and conclusions were drawn concerning the value of subalpine forest grazing as a management strategy and its influence on biodiversity.

MATERIAL AND METHODS

STUDY SITE

The study was carried out on wood pastures traditionally stocked with cattle, located in the Dischma Valley (46°46 N, 9°53 E) between 1560 and 2000 masl at Davos, Canton of Grisons, Switzerland. The soil in this area is a humus podzol (Krause and Peyer, 1986), derived from crystalline rocks (Bosshard, 1986). The climate has aspects of both continental and oceanic character, as the region is located at the transition between the central and northern part of the Alps. Compared to other alpine regions, the annual precipitation from summer rain and winter snowfall is relatively low, reaching approximately 1050 mm per year (Günter, 1986b). The maximum snow depth is approximately 2 m (Schönenberger and Frey, 1988). The mean temperature ranges from −7°C in the coldest month (January) to 12°C in the warmest month (July). The growing season starts in May and ends in October. Thus, grazing is possible only in this period. During winter (November to April), avalanches are common natural hazards in the Dischma Valley (Brugger, 2003), responsible for the treeless vertical stripes through the forest belt.

The forests in the Dischma Valley are dominated by *Picea abies* (L.) Karst (Norway Spruce) and *Larix decidua* Miller (European larch); less frequently found tree species are

Pinus cembra (Swiss stone pine) and *Sorbus aucuparia* (rowan) (Bosshard, 1986). The ground vegetation represents a mosaic of different vegetation types, including parts of nardion for the open areas, *Rhododendron vaccinietum* and *Alnetum viridis* in the avalanche tracks and *Vaccinio piceion* in the woodland (Landolt et al., 1986). The forest is mainly owned by farmers and has a selective logging history since settlements started in the 13th century (Laely, 1984). In 1873, the cantonal forest law was renewed and the utilization of wood pastures by goats was interdicted. Since then, the wood pastures have mainly been used for grazing cattle. The intensity of agricultural activities in the valley decreased greatly during the 20th century. In the beginning of the 20th century, 35% of the forest area of Davos was grazed by cattle (Günter, 1986a). During those times, the wood pastures were much more open than now (Figure 18.1). Today, 17% of the forest area is used for grazing cattle, usually for a few weeks per year, before and after grazing alpine pastures above the treeline (Günter, 1984).

STRUCTURE AND SPECIES COMPOSITION OF GRAZED AND UNGRAZED FORESTS

In autumn 2001, 30 225-m² plots were located on a south-facing slope of the Dischma Valley, ranging from an altitude of approximately 1600 to 1950 masl. Using random numbers, 15 plots were selected in forests that had been grazed each summer at least since 1930, and another 15 plots were randomly selected in forests that had not been grazed since 1930. The information as to whether an area had been grazed or remained ungrazed was derived from Wildi and Ewald (1987) as well as Bebi (1999), and from direct observations in the field.

The spatial distribution of the trees in each plot was sketched. In each plot, the species, heights, and diameters of all trees were recorded; from the trees larger than 8 cm in diameter, tree cores were taken at a height of 0 to 30 cm. To estimate the age of young trees that had a diameter of up to 8 cm, bud-scale scars and the yearly shoot whirls were counted.

FIGURE 18.1 Very open wood pasture at the beginning of the 20th century. Today, a dense Norway spruce forest grows on this slope.

Additionally, the heights of all trees were assessed. All tree cores were processed following standard dendroecological procedures (Stokes and Smiley, 1968). To compare the stand structure of grazed and ungrazed forests in terms of the number of trees, species composition, and tree height, Wilcoxon rank sum tests were performed in S-PLUS (MathSoft Inc., 1999). To compare the age structure of grazed and ungrazed plots, analyses of variance (ANOVA) and multiple comparisons between the plots were performed in S-PLUS (MathSoft Inc., 1999).

IMPACT OF CATTLE ON YOUNG TREES

On seven wood pastures, consisting of 44% woodland, 25% half-open area (with young trees and shrubs), and 31% open area (pasture without trees), the impact of cattle on young trees was studied. The wood pastures were grazed with heifers or cows, with stocking densities ranging from 0.4 to 2.8 livestock units per hectare. Grazing durations ranged from 12 to 114 d (Table 18.1).

To quantify tree damage during defined grazing periods, the tree condition of 165 naturally regenerated young trees (*Picea abies* [L.] Karst.), European larches (*Larix decidua* Miller), and rowans (*Sorbus aucuparia* [L.]) was assessed. For the selection of sample trees, each of the seven test areas was subdivided into 50 m × 50 m squares, and at the intersections, the nearest young tree within a 10-m radius was selected and marked as a sample tree. Tree condition was assessed by

shoot length measurements and by recording tree lesions. In detail, the total height of the sample trees, the length of the apical shoot, and the length of three lateral shoots (±0.5 cm) at different heights were measured. Lateral-shoot browsing was assessed using an intensity scale consisting of four categories: 0 to 20%, 21 to 40%, 41 to 60%, and over 60% of the lateral shoots browsed. Additionally, it was registered whether or not the apical shoot was browsed and whether the sample trees had other damage, such as broken lateral shoots, trampling marks, or fraying scars. As the test areas had been grazed before by domestic and wild ungulates (*Cervus elaphus*, *Capreolus capreolus*, and *Rupicapra rupicapra*), all existing tree lesions were recorded immediately before grazing commenced. After the cattle-grazing period, tree condition was assessed again, and changes in tree condition were interpreted as damage mainly caused by cattle. However, the fences around the test areas probably did not entirely prevent wild ungulates from entering, thus part of the tree damage occurring during the test period of cattle grazing might have been caused by wild intruders. To characterize the exclusive influence of wild ungulates on young trees during the summer, assessment of tree condition was repeated after a period equal in length to that of the preceding cattle-grazing period. The following spring, the condition of the sample trees in test areas 1, 2, 3, and 6 was assessed again, to record the damage caused by wild ungulates during the winter.

TABLE 18.1
Description of the test areas (A1 to A7)

	A1	A2	A3	A4	A5	A6	A7
Altitude (masl)	1560–1580	1580–1600	1600–1700	1720–1800	1800–1850	1850–1950	1950–2000
Aspect	N	N	S	S	S	S	S
Slope, average (%)	40	50	65	60	60	60	60
Size (ha)	1.0	19.3	6.5	3.0	3.4	9.3	5.4
Start of pasturing	June 8	June 20	May 26	July 4	August 3	July 13	August 25
Grazing days	78	114	41	10	12	64	32
Stocking rate (livestock units ha[-1])[a]	1.2	0.5	1.0	2.8	1.6	0.5	0.4

[a]LU; i.e. 600 kg body weight according to BLW/BUWAL (1994).

Plant Species Diversity and Herbage Selection by Cattle

Before grazing commenced, on six test areas, lists of plant species were compiled on 20 cm × 20 cm plots, which were systematically selected using a grid of 50 m × 50 m. Each plot was characterized by a site category: S.1 = dense forest; S.2 = open forest; S.3 = tree regeneration or dwarf-shrub area; S.4 = nutrient-poor grassland; S.5 = nutrient-rich grassland; S.6 = wet grassland; and S.7 = avalanche track.

Directly after the grazing period, the plant species in the plots were classified as either "grazed" or "not grazed." Plant species were then subdivided into four functional botanical groups: grasses (Poaceae, Cyperaceae, and Juncaceae), legumes, forbs, and shrubs. Based on these data, the frequency of functional botanical groups in the herbage available and in the herbage consumed by the cattle was assessed, and the selection index, as the percentage of functional group consumed relative to the percentage of functional group available, was calculated. Additionally, the selection of plant species found on 11% of the plots and the intensity of grazing on different site categories were determined.

RESULTS

Structure and Species Composition of Grazed and Ungrazed Forests

The grazed forest was less dense than the ungrazed forest and had a higher percentage of European larch (Table 18.2).

The gap size (areas without trees >3 m) within the plots of grazed forest (mean size = 39 m^2) and ungrazed forest (mean size = 24 m^2) differed significantly. The young trees grew faster in the grazed forest. There were less dominant trees (>25 m) in the grazed forests, and these trees were much older than in the ungrazed forests. The most frequently found structure type in the grazed forest was multi-layered–open, whereas in the ungrazed forests, the most frequently found structure type was uniform–dense.

Impact of Cattle on Young Trees

On average, 9% of the young trees were browsed (either apical or lateral shoots) within the cattle grazing period. The sum of browsed trees was strongly correlated with the actual stocking rate (Figure 18.2).

Apical-shoot browsing was found on 4% of the sample trees. Tree damage during the cattle-grazing period due to reasons other than browsing (fraying, trampling, and breaking) was nearly equal to that of damage from browsing, but no tree was both browsed and otherwise damaged. During the summer of 2000, when the effects of wild ungulates were measured on test areas 1 and 2, there was no apical-shoot browsing. However, lateral-shoot browsing in that period nearly equaled lateral-shoot browsing during the cattle-grazing period. During the winter of 2000 to 2001, the percentage of trees browsed by wild ungulates on test areas 1, 2, 3, and 6 was nearly triple of that found in the preceding cattle-grazing period. During the cold season, trees on south-facing slopes were browsed by wild ungulates twice as frequently as trees growing on north-facing slopes.

TABLE 18.2
Percentage of *Larix decidua* and *Picea abies* in grazed and ungrazed forests

	All Trees		Trees <3 m		Trees ≥3 m	
	L. decidua	*P. abies*	*L. decidua*	*P. abies*	*L. decidua*	*P. abies*
Grazed forest	12.7[a]	86.8	8.3[a]	91.4	16.8[a]	82.6[a]
Ungrazed forest	2.7[b]	95.5	3.0[b]	93.7	2.5[b]	96.9[b]

[a,b]Means of grazed and ungrazed plots with different lowercase superscripts differ at $p < .05$.

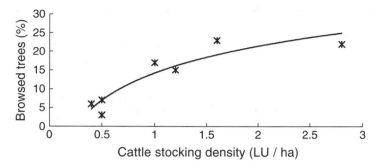

FIGURE 18.2 Logarithmic regression of cattle-stocking density in livestock units per hectare (x) and percentage of browsed trees (y); $y = 10.463 \ln(x) + 14.213$; $R^2 = 0.869$.

PLANT SPECIES DIVERSITY AND HERBAGE SELECTION BY CATTLE

In the 136 vegetation relevé plots of the test areas, 118 plant species were found (29% grass species, 8% legumes, 49% forbs, and 14% shrubs). Only 25 out of these 118 species occurred on 11% of the plots (10 grasses, 3 legumes, 10 forbs, and 2 shrubs). These are listed in Table 18.3.

Nardus stricta was the most frequently found grass species, followed by *Luzula sylvatica* and *Agrostis capillaris*. The most frequently found legume was *Trifolium pratense*. *Homogyne alpina* was the most frequently found forb species, followed by *Hieracium murorum* and *Potentilla aurea*. The shrubs *Vaccinium myrtillus* and *Vaccinium vitis-ideae* were found on 56% and 40% of the plots, respectively.

The grass species *A. capillaris, Poa* sp., and *L. sylvatica* were grazed most intensively. The legumes *Trifolium repens* and *T. pratense* were grazed on 53% and 38% of the plots, respectively. Although the forb species *Arnica montana* and the shrub *Vaccinium myrtillus* have an assumed nutritional value of 0, they were grazed on 37% and 36% of the plots, respectively.

In the tree regeneration or dwarf-shrub area, the highest number of plant species was found (84 plant species; Table 18.4), followed by the nutrient-poor grassland (75 plant species). The open forest had a herbaceous layer with more plant species than the dense forest. The plant species diversity in the avalanche tracks, where 46 different plant species were found, was still

higher than in the nutrient-rich and wet grasslands. The nutrient-poor, nutrient-rich, and wet grasslands were grazed most intensively by the cattle. However, the plant species growing in both the open forest and the tree regeneration or dwarf-shrub area were also grazed on almost one quarter of the plots. In the dense forest and the avalanche tracks, only 16 and 18% of the recorded species, respectively, were found to have been grazed by cattle.

Over all test areas, 49% of the records of grazed species were grass species (Table 18.5), and 45% of the recorded grass species found in the plots were grazed (data not given in table). The percentage of grass species in the herbage consumed was higher than in the herbage available, resulting in a selection index of 1.7 on an average and a selection index of 1.5 in all the test areas (Table 18.5). In contrast, the selection indices of all other functional groups (legumes, forbs, and shrubs) were 0.8, on average, in most of the single-test areas.

DISCUSSION

The main objective of this study was to analyze the influence of subalpine forest grazing on plant-species diversity, forest structure, and tree regeneration to discuss the value of traditional animal husbandry as a management strategy. As the comparison between grazed and ungrazed forests in our study showed, grazed forests were less dense with larger gaps than ungrazed forests. Instead of the usual uniform–dense forest type, the grazed forests showed a multilayered open structure. Usually, subalpine forests tend to form stands of uniform height, although the

TABLE 18.3
Assumed nutritional value and frequency of occurrence and grazing of herbaceous and shrubby plant species found on at least 15 plots (≥11% of all plots)

Plant Species	Nutritional Value[a]	Frequency on the Plots in All Areas	Plots with This Species Grazed (Percentage of Records)
Agrostis capillaris	2	23	71
Poa sp.	3	11	67
Luzula sylvatica	1	24	66
Luzula sp.	1	16	55
Trifolium repens	3	11	53
Phleum alpinum	2	18	42
Trifolium pratense	3	12	38
Arnica montana	0	14	37
Festuca rubra	2	21	36
Vaccinium myrtillus	0	56	36
Nardus stricta	1	34	35
Anthoxanthum odoratum	1	21	34
Alchemilla vulgaris	2	15	30
Calamagrostis sp.	—[b]	14	26
Ranunculus sp.	—	15	24
Melampyrum sp.	—	17	22
Avenella flexuosa	—	18	20
Lotus corniculatus	—	11	20
Potentilla aurea	—	26	17
Hieracium murorum	—	32	16
Crocus albiflorus	—	11	13
Viola biflora	—	11	7
Vaccinium vitis-idaea	—	40	6
Oxalis acetosella	—	26	3
Homogyne alpina	—	43	2

[a] Feeding value according to Dietl (1990): high = 3, medium = 2, low = 1, worthless/noxious = 0.
[b] Feeding value not stated by Dietl (1990).

ages of trees may differ greatly (Ott et al., 1997). The utilization of subalpine forests as wood pastures counteracts the natural tendency of the subalpine spruce-dominated forests to form uniform stands of the same height, thus favoring the development of a multilayered structure, and consequently facilitating the regeneration of the forest (Ott et al., 1997). There were larger gaps in the grazed forests than in the ungrazed forests in our study. The enhanced light due to the formation of gaps and the multilayered open structure is known to lead to a more pronounced ground vegetation with a high number of species (Ellenberg, 1996). The size of most of the gaps did not exceed the limits for sufficient avalanche protection, and the number of trees with a height relevant for avalanche protection still exceeded the minimum number given by Meyer-Grass (1987). In the grazed forests of our study, there were fewer dominating trees taller than 25 m, but these trees were significantly older than in the ungrazed forests, reaching up to 400 years in age. The reason for this phenomenon can probably be found in the past differences in logging activities. In the grazed forests, a certain

TABLE 18.4

Total number of herbaceous and shrubby plant species found on the different site categories and percentage of plants grazed on the plots within the different site categories

Site Category	Number of Species	Percentage of Records Grazed[a]
Dense forest	44	16
Open forest	63	24
Tree regeneration or dwarf shrub	84	24
Nutrient-poor grassland	75	40
Nutrient-rich grassland	29	59
Wet grassland	32	30
Avalanche track	46	18

[a]Percentage of times that a recorded species was found grazed on the different plots.

TABLE 18.5

Percentage of functional botanical groups (as available/as consumed) per test area and selection index

Functional Groups	Number of Species	Percentage of functional groups in all plots (as available/as consumed) selection index[a]						
		All Areas	Area 1	Area 2	Area 4	Area 5	Area 6	Area 7
Grass species	26	29/49; 1.7	30/51; 1.7	31/63; 2.0	20/31; 1.6	26/38; 1.5	27/47; 1.7	35/53; 1.5
Legume species	5	8/6; 0.8	8/4; 0.5	9/4; 0.4	9/15; 1.7	6/4; 0.7	8/4; 0.5	6/7; 1.2
Forbs species	78	49/34; 0.7	51/43; 0.8	46/8; 0.2	55/49; 0.9	46/54; 1.2	56/39; 0.7	43/33; 0.8
Shrub species	9	14/11; 0.8	10/2; 0.2	14/25; 1.8	16/5; 0.3	22/4; 0.2	10/10; 1.0	17/7; 0.4

[a]Selection index = functional group consumed (%)/functional group available (%).

Note: On Area 3 grazing on herbaceous and shrubby plants was not measured.

number of mature trees per area were usually left as shelter trees for the grazing animals, and in some regions, these trees were even removed (Blumer, 1983). One reason for the lower density of grazed forest stands might be browsing and trampling by cattle. In our own previous experiments, very high cattle-stocking rates on small ranges and grazing durations exceeding the herbage supply were found to cause remarkable browsing damage (Mayer et al., 2002). However, the results of the present field study suggest that the usual stocking rates on traditional wood pastures of approximately 1 to 2 livestock units per hectare do not hinder tree regeneration remarkably. In addition, Liss (1988) found that cattle grazing does not severely impair the regeneration of subalpine forests if stocking rates are low. It has to be taken into account that subalpine forests are also used by wild animals such as the hare, red deer (*Cervus elaphus*), and roe deer (*Capreolus capreolus*). These animals are known to frequently browse the shoots of young trees, especially in winter (Bergström and Guillet, 2002), as demonstrated by the fact that in Switzerland, 15% of the young trees were found to be damaged by wild animals (Brassel and Brändli, 1999). However, according to the present results, extensive grazing by domestic and wild ungulates is not likely to alter the forest structure significantly. Thus, the main reason for the lower density of the grazed forests might be the result of tree-cutting activities. To improve the growth of the herbage plant species on wood pastures, farmers used to cut both mature and regenerating trees (Janett, 1943). Unintentionally, the reduction in competition due to the cutting of trees also improved growth conditions for the remaining young trees. The higher percentage of larches in grazed forests might be attributable to both the enhanced light supply because of the cutting of trees and the fact that farmers usually spared the larches when cutting young trees (Janett, 1943).

Concerning the ground vegetation within the wood pastures, the high number of species (118) and the high proportion of forbs stand out. As expected, the open forest was found to have a more diverse herbaceous layer than the dense forest. The highest number of plant species was found in the tree regeneration or dwarf-shrub area, followed by the nutrient-poor grassland. Generally, an important characteristic of wood pastures is the variety of different vegetation types, ranging from dense forests to open grasslands. This heterogeneity, caused by traditional animal husbandry, is known to lead to an increased biodiversity of the subalpine areas (Sickel et al., 2004).

For the evaluation of forest grazing as a management strategy to maintain biodiversity in subalpine regions, it is also important to know more about the grazing habits and preferences of cattle. Based on the former practice of keeping the forest open for grazing, in our study, the herbaceous plant species in the open forest and in the tree regeneration or dwarf-shrub areas were grazed more intensively than in the dense forest. Thus, the open areas (either nutrient-poor, nutrient-rich, or wet grasslands) were grazed most intensively by the cattle. This is in line with the findings of Sickel et al. (2004) that cattle prefer to graze in the areas identified as most valuable and species-rich.

Grass species were grazed most intensively by the cattle in our study. Also, Estermann et al.

(2001) as well as Pordomingo and Rucci (2000) observed that cows preferred grass species. This suggests a decline of grasses in grazed forests over the years, which allows an increased growth of forbs, as was already found in Liss (1988), and hence an increase in biodiversity. Of all domestic animals, cattle are said to have the lowest ability of selective feeding (e.g. Rösch, 1992). The findings of our study confirm this statement. The relatively high intake of *Nardus stricta*, *Vaccinium myrtillus*, or *Arnica montana*, which all have a low nutritional value, suggests that the reachability of food is an important factor in the feeding behavior of cattle. Thus, the comparatively low percentage of forbs consumed can also only partly be explained by active selection. Cattle may have difficulties in finding and ingesting these plants, because many forbs found on the subalpine wood pastures are very small (e.g. *Melampyrum* sp.) compared to the grass species, or have rosette-like basal leaves (e.g. *Hieracium* sp.). Nevertheless, it is interesting to see that in Area 4, the test area with the highest stocking rate, the percentage of forbs consumed is highest over all the ranges. This suggests that with an increased stocking rate, the cattle have to exploit food resources of difficult reachability.

Considering that the impact of cattle on the forest flora and fauna depends on the intensity and duration of grazing, the animal breed, vegetation, and type of soil (Malkamäki and Haeggström, 1997), our results suggest that forest grazing at an extensive level, together with periodic selective logging operations of the forest owners, is a valuable tool for managing the subalpine forests to maintain or even increase their biodiversity. This is in line with Dullinger et al. (2003), whose results indicate a significant long-term decline of plant-species diversity following the abandonment of traditional low-intensity cattle grazing in the Austrian Alps. The long-ranging maintenance of biodiversity in landscapes is only possible if the landscapes show some kind of dynamics, at least on a small scale (Krüsi et al., 1995). Forest grazing, as described in the preceding text, can be one way of generating such dynamics.

CONCLUSIONS

The management of subalpine conifer forests as wood pastures, as still practiced in parts of the Swiss Alps, favors the development of a heterogeneous forest structure with higher herbaceous and woody-plant species diversity and can be recommended as a valuable management strategy to maintain or even increase biodiversity. The main reasons for this are the low stocking density of 1 to 2 livestock units per hectare, with the cattle staying on pasture during some weeks in the growing season, and the interest of the forest owners in improving pasture quality by conducting more selective logging operations in these areas (although usually on Swiss mountain slopes, timber can hardly be harvested cost efficiently anymore due to low timber prices). In the small gaps derived from selective logging operations, forest regeneration is enhanced, because temperature is the limiting factor in this area, and young trees need direct light to grow well. Forest grazing, as described in the preceding text, was found to be compliant with other very important functions of the subalpine forests, such as avalanche protection.

SUMMARY

In this study, plant species diversity, forest structure, and tree regeneration on subalpine wood pastures were investigated. The study was carried out on heterogeneous wood pastures traditionally stocked with cattle and located in the Dischma Valley (46°46 N, 9°53E) between 1560 and 2000 masl at Davos, Canton of Grisons, Switzerland. The forests in the study area were dominated by *Picea abies* (L.) Karst (Norway spruce) and *Larix decidua* Miller (European larch). The herbaceous vegetation was a very diverse (118 species). To analyze structure, species composition, and regeneration of grazed forests, dendroecological methods were used, and tree condition before and after grazing was assessed. Additionally, the selection of herbaceous plant species was studied by classifying the species into grazed and not grazed on systematically selected plots. Apical-shoot browsing was found on only 4% of the young trees. The grazed forest had a higher diversity of tree species, was less dense, and had larger gaps than the ungrazed forests. On the wood pastures, the open forest had a more diverse herbaceous layer than the dense forest. Grass species were grazed most intensively. The management of subalpine forests as wood pastures favors the development of a heterogeneous forest structure with a higher herbaceous and woody-plant species diversity. The exposure of wood pastures to extensive grazing for several weeks in the growing season was found to be compliant with other very important functions in the management of subalpine forests, such as avalanche protection, and can therefore be recommended as a sustainable management strategy in subalpine regions.

References

Bebi, P. (1999). Erfassung von Strukturen im Gebirgswald als Beurteilungsgrundlage ausgewählter Waldwirkungen. Dissertation thesis, Swiss Federal Institute of Technology (ETH), Zürich, 125 pp.

Bergström, R. and Guillet, C. (2002). Summer browsing by large herbivores in short-rotation willow plantations. *Biomass and Bioenergy,* 23(1): 27–32.

Blumer, E. (1983). *Die Betreuung der Wälder im Glarnerland.* Tschudi Druck und Verlag AG, Glarus, Switzerland, 149 pp.

BLW/BUWAL (Swiss Federal Office for Agriculture Swiss Federal Office for Environment, Forest and Landscape), 1994. Wegleitnng für den Gewässerschutz in der Landwirtschaft, Lehrmittelzentrale Berne, Switzerland, 100 pp.

Bosshard, W. (Ed.). (1986). Der Naturraum und dessen Nutzung im alpinen Tourismusgebiet von Davos. Reports of the Swiss Federal Institute for Forest, Snow and Landscape Research, 289. Birmensdorf, Switzerland, 336 pp.

Brassel, P. and Brändli, U.B. (Eds.). (1999). *Schweizerisches Landesforstinventar.* Ergebnisse der Zweitaufnahme 1993–1995. Haupt, Berne, Stuttgart, Vienna, 442 pp.

Brugger, S. (2003). Vegetation und Lawinen. Diploma thesis, Department of Geography, University of Zurich, Switzerland, 65 pp.

Delucchi, M. (1993). Waldweide aus forstlicher Sicht. *Bündner Wald,* 46: 12–15.

Dietl, W. (1990). Alpweiden naturgemäss nutzen. *Landfreund*, 11, AGFF, Swiss Federal Research Station of Agronomy and Agroecology, Zurich-Rechenholz, Switzerland. 6 pp.

Dullinger, S., Dirnbock, T. et al. (2003). A resampling approach for evaluating effects of pasture abandonment on subalpine plant species diversity. *Journal of Vegetation Science*, 14(2): 243–252.

Ellenberg, H. (1996). *Vegetation Mitteleuropas mit den Alpen*. 5. Auflage. Eugen Ulmer Verlag.

Estermann, B.L., Wettstein, H.R., Sutter, F., and Kreuzer, M. (2001). Nutrient and energy conversion of grass-fed dairy and suckler beef cattle kept indoors or on high altitude pasture. *Animal Research*, 50: 477–493.

Günter, T.F. (1984). Nutzungsgeschichte von Davos. Eine empirische Untersuchung über Nutzungsänderungen unter dem Einfluss des Tourismus seit 1900. Dissertation thesis, University of Zurich, Switzerland, 169 pp.

Günter, T.F. (1986a). Nutzungsgeschichte. In Der Naturraum und dessen Nutzung im alpinen Tourismusgebiet von Davos. Ergebnisse des Man and Biosphere-Projektes Davos. *Berichte der Eidgenössischen Forschungsanstalt für Wald, Schnee und Landschaft WSL*, 289.

Günter, T.F. (1986b). Das Testgebiet Davos. In Der Naturraum und dessen Nutzung im alpinen Tourismusgebiet von Davos. Ergebnisse des Man and Biosphere-Projektes Davos. *Berichte der Eidgenössischen Forschungsanstalt für Wald, Schnee und Landschaft WSL*, 289: 21–29.

Janett, A. (1943). Über die Regelungen von Wald und Weide. *Schweiz Z Forstwes*, 94(4): 105–117.

Krause, M. and Peyer, K. (1986). Böden. In Der Naturraum und dessen Nutzung im alpinen Tourismusgebiet von Davos. Ergebnisse des Man and Biosphere-Projektes Davos. *Berichte der Eidgenössischen Forschungsanstalt für Wald, Schnee und Landschaft WSL*, 289.

Krüsi, B.O. et al. (1995). Huftiere, Vegetationsdynamik und botanische Vielfalt im Nationalpark. Ergebnisse von Langzeitbeobachtungen. *Cratschla*, 3(2): 14–25.

Laely, A. (1984). Der Wald in der Geschichte der Landschaft Davos. Davoser Heimatkunde: Beiträge zur Geschichte der Landschaft Davos. *Davos, Genossenschaft Davoser Revue*, pp. 62–126.

Landolt, E., Krüsi, B.O. et al. (1986). Vegetationskartierung und Untersuchungen zum landwirtschaftlichen Ertrag im MaB6-Gebiet Davos. *Veröffentlichungen des Geobotanischen Institutes der ETHZ*, Stiftung Rübel, Heft 88a.

Liss, B.M. (1988). Der Einfluss von Weidevieh und Wild auf die natürliche und künstliche Verjüngung im Bergmischwald der ostbayerischen Alpen. *Forstwissenschaftliches Centralblatt*, 107(1): 14–25.

Malkamäki, E. and Haeggström, C.A. (1997). Short term impact of Finnish landrace cattle on the vegetation and soil of a wood pasture in SW Finland. *Acta Botanica Fennica*, 159: 1–25.

Mayer, A.C., Stöckli, V., Konold, W., Estermann, B.L., and Kreuzer, M. (2002). Effects of grazing cattle on subalpine forests. In Bottarin, R. and Tappeiner, U. (Eds.), *Interdisciplinary Mountain Research*. Blackwell Verlag GmbH, Berlin, Wien, pp. 208–218.

Meyer-Grass, M. (1987). *Waldlawinen: Gefährdete Bestände, Massnahmen, Pflege des Gebirgswaldes*. Leitfaden für die Begründung und forstliche Nutzung von Gebirgswäldern, Bern, 379 pp.

Ott, E., Frehner, M., Frey, H.U., and Lüscher, P. (1997). *Gebirgsnadelwälder. Ein praxisorientierter Leitfaden für eine standortgerechte Waldbehandlung*. Verlag Paul Haupt, Bern, 288 pp.

Pordomingo, A.J. and Rucci, T. (2000). Red deer and cattle diet composition in La Pampa, Argentina. *Journal of Range Management*, 53: 649–654.

Rösch, K. (1992). Einfluss der Beweidung auf die Vegetation des Bergwaldes. *Forschungsbericht 26, National Park Berchtesgaden*, Germany, 100 pp.

Schönenberger, W. and Frey, W. (1988). Untersuchungen zur Ökologie und Technik der Hochlagenaufforstung; Forschungsergebnisse aus dem Lawinenanrissgebiet Stillberg. *Swiss Journal of Forestry*, 139: 735–819.

Sickel, H., Ihse, M. et al. (2004). How to monitor semi-natural key habitats in relation to grazing preferences of cattle in mountain summer farming areas: an aerial photo and GPS method study. *Landscape and Urban Planning*, 67(1–4): 67–77.

Stokes, A. and Smiley, T.L. (1968). *An Introduction to Tree Ring Dating*. University of Chicago Press, Chicago.

Ten Klooster, L. (2000). Waldweide als Teil der alpinen Kulturlandschaft. Diploma thesis, Universität für Bodenkultur Wien, Wien/Wageningen, 126 pp.

Wildi, O. and Ewald, K. (Eds.). (1987). Der Naturraum und dessen Nutzung im alpinen Tourismusgebiet von Davos. Ergebnisse des MaB-Projektes Davos. *Berichte der Eidgenössischen Anstalt für das Forstliche Versuchswesen*, Birmensdorf: 336.

Part IV

Effects of Grazing on Mountain Forests

19 Patterns of Forest Recovery in Grazing Fields in the Subtropical Mountains of Northwest Argentina

Julietta Carilla, H. Ricardo Grau, and Agustina Malizia

INTRODUCTION

In many areas of the Andes, anthropogenic degradation due to grazing, fire, and forest exploitation led to the replacement of native forest by grasslands (Kappelle and Brown, 2001). However, in some areas, this tendency began to revert due to different socioeconomic processes, including rural emigration, economic changes toward a lower dependence on natural resources, and management decisions that excluded some productive areas for conservation purposes (Aide and Grau, 2004). These areas of secondary forest succession provide opportunities for ecological restoration by allowing the recovery of biodiversity associated with forests. In addition, recovering forests provide ecological services such as the production of timber and the sequestration of atmospheric carbon (Silver et al., 2000). To evaluate the conservationist and economic values of these secondary forests, it is necessary to understand the floristic tendencies during secondary succession and the recovery rates of biodiversity, composition, and biomass parameters.

Patterns of secondary forest succession are influenced by the preabandonment conditions (previous land use, vegetation structure and microenvironmental characteristics), the availability of propagules in the early stages of succession, and the interactions inter- or intraspecific between secondary forest trees (Pickett et al., 1987). For example, in many temperate forests, those forests monodominated by pioneer species have slow growth rates due to intensive intraspecific competition (self-thinning phase) until large trees die, releasing resources and providing opportunities for new recruitment and faster growth of the surviving trees (Oliver and Larson, 1996).

The upper-montane forest of northwestern Argentina is characterized by grasslands, shrublands, mature forests, and successional forests that became established on grasslands and shrublands in which grazing pressure has decreased. The most abundant secondary forest types are the monodominant forests of *Alnus acuminata* and *Podocarpus parlatorei* (Arturi et al., 1998; Brown et al., 2001). In this study, we analyzed 10 years of structural and compositional changes in different successional forest stages that range from young to old mature forests, where secondary forests have established on old grasslands and shrublands. Our objectives were: (1) to describe floristic trends and relationships between different successional forest stages; (2) to quantify and analyze the rates of change in structural and demographic parameters, such as mortality, recruitment, composition, and basal area of the main tree species; and (3) to discuss the management implications of the observed patterns and processes, in particular, in relation to the demography of the most abundant species. We hypothesized that the secondary forests observed correspond to successional stages in which pioneer species will be replaced by non-pioneers or climax tree species, tending to reach mature phases.

METHODS

STUDY AREA

The studied sites were located between 1600 and 1800 m elevation in the upper-montane forest of the Sierra de San Javier (ca. 26°47 S and 65°22 W), a protected area since 1974, belonging to the Universidad Nacional de Tucumán, Argentina (Figure 19.1). The vegetation of the area corresponds to the Argentinean *yungas* (Cabrera and Willink, 1980) and is characterized by a mosaic of forests, grasslands, and shrublands (Moyano and Movia, 1989; Arturi et al., 1998). These forests are representative of floristic and physiognomic forest types that extend latitudinally for 1500 km, from 15° S, approximately, in the Cochabamba department, Bolivia (Navarro et al., 1996), to 28.5°S in Catamarca Province, Argentina (Brown et al., 2001), along the eastern slopes of the Andes. Characteristic tree species are *Alnus acuminata*, *Crinodendron tucumanum*, and *Podocarpus parlatorei* in early to mid successional stages, and *Ilex argentina*, *Prunus tucumanensis*, *Juglans australis*, *Cedrela lilloi*, and species from the Myrtaceae family in mature forests (for botanical families and authorities see Table 19.1). Shrublands are dominated by *Baccharis articulata* Pers., *B. tucumanensis* Hook et Arn. (Asteraceae), *Lepechinia graveolens* (Regel.) Epl. (Laniaceae), and *Chusquea lorentziana* Griseb. (Bambuceae). Grasslands are dominated by *Festuca hieronymii* Haeckel, *Deyeuxia polygama* (Griseb.) Parodi An., and *Stipa eriostachia* H.B.K. (Poaceae) (Giusti et al., 1997).

DATA COLLECTION

During 1991, permanents plots were established (Table 19.1) in ten forests differing in successional age and characterized by different dominant species: two *Alnus acuminata*–dominated forests (aaj, the youngest, and aa12, the oldest, two *Crinodendron tucumanum*-dominated forests (ct and ctv, young and old, respectively), four forests dominated by *Podocarpus parlatorei* (pp9, pp8, pp1, and pp5, ordered in

increasing age), and two mature forests dominated by species of the Myrtaceae family (m11 and m7). Plots were set using contiguous 20 m × 20 m quadrats (the number of quadrats varied between plots from 2 to 12 (Table 19.1). The total area surveyed was 2.64 ha. Trees were identified at the species level, following Morales et al. (1995) and Zuloaga and Morrone (1999a, 1999b), labeled with numbered tags, and mapped in an x–y coordinate system. We measured the diameter at breast height (dbh) of all trees >10 cm in diameter and estimated tree height visually. Permanent plots wcrc remeasured after 5 and 10 years of establishment (December 1996 and December 2001). For each forest in 1991 and 2001, we estimated total tree density (individuals/ha), basal area (m²/ha), mortality (%), recruitment (new individuals >10 cm/ha), and species richness (mean number of species/quadrat). Given that species richness is area-dependent, and because our plots differed in area, we used the mean number of species per 20 m × 20 m quadrat as an index of species richness. Finally, for each tree, we registered the "most likely successor," defined as the tallest juvenile tree <10 cm dbh growing under the projection of each measured tree (Horn, 1975).

To estimate the age of the *Alnus* forests, we sampled the largest *A. acuminata* individuals of each plot with increment borers and dated them using dendrochronology methods (Grau et al., 2003). For all other forests, we estimated their age based on diameter–growth relationships. In the Myrtaceae- and *Podocarpus*-dominated forests, we calculated the relation between the mean diameter and annual growth rate of the largest *P. parlatorei* individuals, whereas in *Crinodendron*-dominated forests, age was estimated using the same relationship mentioned earlier, but with *C. tucumanum* individuals.

DATA ANALYSIS

To explore the floristic relationships and successional trends of the different forests in 1991 and 2001, we performed an ordination of the forests' composition data using nonmetric multidimensional scaling (NMDS) (Kruskall and Wish, 1978), based on a matrix of Bray–Curtis distanccs (Legendre and Legendre, 1998). The

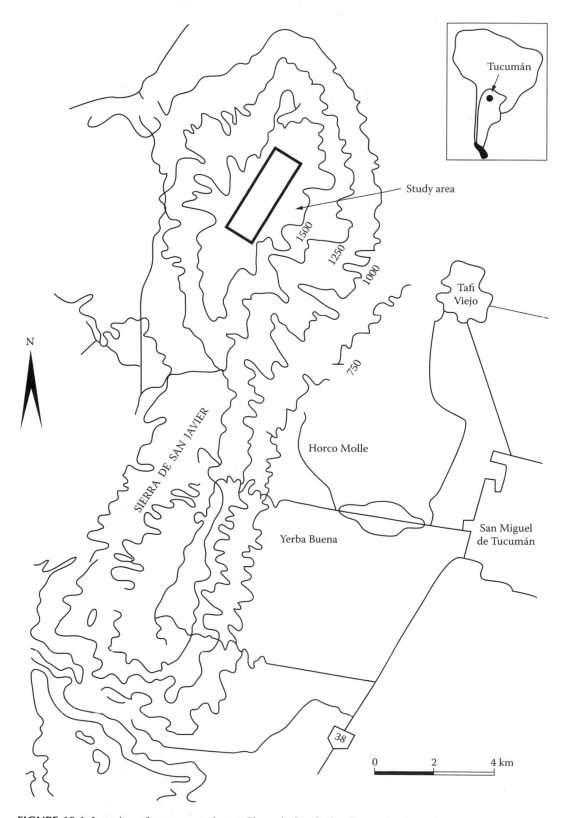

FIGURE 19.1 Location of permanent plots at Sierra de San Javier, Tucumán, Argentina.

TABLE 19.1
Ages and area (m²) of forest plots and their main structural characteristics

Forest	References	Surface	# 20 m × 20 m Quadrats	Age	BA1991	BA2001	Delta BA	Density 1991	Density 2001	Delta Density	Total Mortality	Total Recruitment	New Species	Richness Average 1991	Richness Average 2001
aa12	Old *Alnus acuminata*	2400	6	80	28.2	25.0	-3.2	350.0	445.8	95.8	27.1	212.5	1	4.7	4.8
aaj	Young *Alnus acuminata*	2400	6	40	14.1	18.8	4.6	445.8	450.0	4.2	11.3	50.0	1	1.8	2.3
ct	Young *Crinodendron tucumanum*	2400	6	55	20.2	23.2	3.0	275.0	329.2	54.2	16.7	108.3	2	3.2	3.5
ctv	Old *Crinodendron tucumanum*	2400	6	187	21.5	22.5	1.0	220.8	354.2	133.3	18.9	166.7	1	2.9	3.6
mi1	Old growth Myrtaceae	4000	10	>500	36.9	34.2	-2.7	462.5	430.0	-32.5	14.5	30.0	1	6.4	6.5
mi7	Old growth Myrtaceae	4800	12	>500	35.6	37.3	1.7	425.0	393.8	-31.3	29.5	97.9	0	5.7	5.2
pp1	Old *Podocarpus parlatorei*	2400	6	430	30.0	28.3	-1.7	275.0	300.0	25.0	24.7	91.7	0	4.2	4.2
pp5	Old *Podocarpus parlatorei*	2400	6	437	43.5	41.4	-2.0	558.3	512.5	-45.8	17.4	54.2	5	2.7	3.5
pp8	Middle-aged *Podocarpus parlatorei*	2400	6	331	54.9	56.0	1.1	416.7	429.2	12.5	16.2	95.8	3	4.3	4.5
pp9	Young *Podocarpus parlatorei*	800	2	220	31.9	33.1	1.1	550.0	537.5	-12.5	22.7	112.5	2	2	4

ᵃBasal area (m²/ha), density (individuals/ha), mortality (%), total recruitment (individuals/ha), new species (individuals of <10-cm dbh), and richness average (individuals of >10-cm dbh/quadrat).

matrix of data included tree species abundances in all forest plots, in both years. The advantage of NMDS over other ordination methods is that it does not assume any data distributions and is robust to different distribution along the underlying gradients (Kenkel and Orlóci, 1986). To explore possible future trends in forest composition, we performed an additional NMDS ordination including the "future" composition, based on the most likely successor species (i.e. the expected future composition, assuming that the most likely successor will replace current canopy trees in the "next" generation, Horn, 1975). The most likely successor was defined as the tallest juvenile growing underneath the crown of each tree. The final stress for a two-dimensional configuration was 9.884 and 16.391 for each NMDS, respectively, which did not differ significantly from three-dimensional configuration stress. Stress values lower than 20 indicate a relatively good fit between the graph configuration and Bray–Curtis similarity matrix (Legendre and Legendre, 1998) and, therefore, we used the two-axes configuration.

To determine the tree species that were most important in separating forests in the ordination space, we used nonparametric Kendall correlation coefficients (Sokal and Rohlf, 1995) between tree species abundances and NMDS axis scores. For this, we only used canopy species based on the adults' mean height (>12 m height).

To analyze changes in species richness between forests and between both dates (1991 and 2001), we used a two-way ANOVA analysis.

RESULTS

We recorded a total of 1080 tree individuals of >10 cm dbh, belonging to 20 tree species and 17 botanical families. Of these, 13 were canopy species, and 7 were understory species (Table 19.2). According to the forest's age estimation, plots ranked between 40 years old (in young *Alnus* forests) to more than 500 years (in Myrtaceae mature forest) and represented a wide rank of successional ages (Table 19.1).

In the NMDS ordination based on the 1991 and 2001 forest composition, we identified four groups along the NMDS axis: (1) *Alnus* forests (aaj and aa12, negative side of axis 1); (2) *Crin-*

odendron forests (ct and ctv, positive side of axis 2); (3) *Podocarpus* forests (pp1, pp5, pp8, pp9, center and positive side of axis 1); and (4) Myrtaceae or mature forests (mi7 and mi11, negative side of axis 1) (Figure 19.2). The successional trajectories (changes in the ordination space between 1991 and 2001) showed a clear trend of convergence toward the center of the ordination diagram. Kendall correlations between both axis scores and species abundances showed 11 significant correlation coefficients: *Alnus acuminata* was negatively correlated, and *Podocarpus parlatorei* and *Cedrela lilloi* were positively correlated with axis 1. *Crinodendron tucumanum* and *A. acuminata* were positively correlated with axis 2, whereas *Blepharocalix saliscifolius, Dunalia lorentzii, Ilex argentina, Myrcianthes mato, M. pseudomato,* and *Prunus tucumanensis* were negatively correlated (Table 19.2).

Forest ordination including most likely successors also showed a clear trend to convergence of all forests into the negative portion of axis 1 and axis 2 (Figure 19.3). Kendall's correlations between both axes and 1991 to 2001 to future abundances did not show significant correlations. Considering the most likely successor species, the number of new species were highest in *Podocarpus* forests pp5 (five new species) and pp8 (three new species) (Table 19.1). They include *B. salicifolius, D. lorentzii, M. mato, M. pseudomato, P. tucumanensis,* and *I. argentina,* all species characteristic of mature forests. The number of new species in *Alnus* and *Crinodendron* forests was very low, between 1 and 2.

Species richness differed significantly among forests at each date and between 1991 and 2001 (two-way ANOVA: Forest: $F(9,114) = 19.6, p < .001$; year $F(1,114) = 3.98, p = .04$; Forest by year NS), showing the maximum richness estimated in Myrtaceae mature forests (6.5 spp./quadrat in 2001) and the minimum in the young *Alnus* forest (2.3 spp./quadrat in 2001) (Table 19.1). Forest richness was significantly higher in 2001 than in 1991.

In the old *Alnus* forest (aa12), mortality (27%) and recruitment (212 individuals/ha) were comparatively high. In young *Alnus* and *Crinodendron* forests, total recruitment was moderately high (50 and 109 individuals/ha,

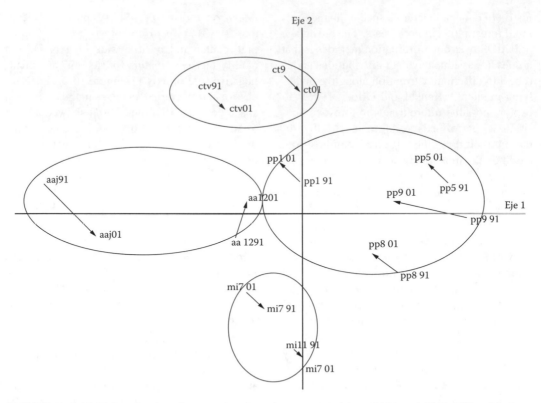

FIGURE 19.2 NMDS ordination diagrams based on forest composition. (1991 and 2001. Ellipses indicate arbitrarily defined homogeneous groups. Both axes explain 80% of the total variation (57 and 23% for axis 1 and axis 2, respectively).

respectively). Some species, particularly the treelet *Solanum grossum*, showed the highest values of recruitment (79 and 146 individuals/ha, respectively) (Appendix 1). In *Podocarpus* and Myrtaceae forests, recruitment and mortality showed intermediates values (Table 19.1).

Alnus and *Crinodendron* forests presented the lowest values in basal area and the most marked changes between 1991 and 2001. The oldest plots of both forest types (aa12 and ctv) presented the maximum density increments. Podocarpus forests showed the highest values of basal area (more than 50 m²/ha), undergoing the smallest change during the 10 years of the study. Myrtaceae forests showed intermediate values and changes for the basal area and density parameters (Table 19.1).

DISCUSSION

Our analysis suggests the existence of three successional pathways in the upper-montane forests of the Sierra de San Javier. Two forest types dominated by *Alnus acuminata* and *Crinodendron tucumanum*, respectively, were comparatively similar in their successional dynamics, whereas the forests dominated by *Podocarpus parlatorei* showed different characteristics. Supporting our hypothesis, there is an apparent trend toward a compositional convergence in the future (Figure 19.3), but the rates of change were very variable.

Alnus and *Crinodendron* forests showed the lowest values of basal area (between 19 and 25 m²/ha), but given the young age of these forests, they represent relatively high rates of accumulation of biomass. These results are similar to those reported by Morales and Brown (1996) who observed a basal area of 26.9 m²/ha for a similar secondary upper-montane forest located in the Bermejo River Basin of Argentina (22° S). In young *Alnus* (aaj) and *Crinodendron* (ct) forests, the great increase of basal area in the last 10 years was due to the high growth

TABLE 19.2
Tree species recorded within all forests, botanical families, and tree types

Species	Family	Tree Type	Axis 1	Axis 2
Alnus acuminata H.B.K.	Betulaceae	C	0.50[a]	0.37[a]
Blepharocalyx salicifolius (H.B.K.) O. Berg	Myrtaceae	C	0.21	0.66[a]
Cedrela lilloi C. DC.	Meliaceae	C	0.43[a]	0.15
Crinodendron tucumanum Lillo	Eleocarpaceae	C	0.15	0.43[a]
Dunalia lorentzii (Damner) Sleumer	Solanaceae	C	0.18	0.40[b]
Ilex argentina Lillo	Aquifoliaceae	C	0.13	0.42[b]
Juglans australis Griseb.	Juglandaceae	C	0.24	0.24
Myrcianthes callicoma McVaugh	Myrtaceae	C	0.27	0.24
Myrcianthes mato (Griseb.) McVaugh	Myrtaceae	C	0.14	0.71[a]
Myrcianthes pseudo-mato (D. Legrand) McVaugh	Myrtaceae	C	0.15	0.44[a]
Podocarpus parlatorei Pilg.	Podocarpaceae	C	0.81[a]	0.05
Prunus tucumanensis Lillo	Rosaceae	C	0.04	0.49[a]
Sambucus peruviana H.B.K.	Caprifoliaceae	C	0.26	0.17
Allophylus edulis (St. Hill) Radlk.	Sapindaceae	U	—	—
Azara salicifolia Griseb.	Flacourtiaceae	U	—	—
Duranta serratifolia (Griseb.) Kuntze	Verbenaceae	U	—	—
Kaunia lasiophthalma (Griseb.) R. King. and H. Robinson	Compositae	U	—	—
Prunus persica (L.) Batsch (exotic)	Rosaceae	U	—	—
Solanum grossum C.V. Morton	Solanaceae	U	—	—
Vassobia breviflora (Sendnt.) Hunz.	Solanaceae	U	—	—

Note: C = canopy, U = understory; Kendall correlation coefficients between tree species abundances for forests and NMDS axis scores are reported. Botanical nomenclature follows Morales et al. (1995) and Zuloaga and Morrone (1999a, 1999b).

[a] $p < .01$

[b] $p < .05$

rate of three abundant species: *A. acuminata*, *P. parlatorei*, and *S. grossum*, having less importance was the recruitment of new individuals.

In contrast to the high growth rates in the young *Alnus* and *Crinodendron* forests, old *Crinodendron* (ctv) showed low growth, and old *Alnus* (aa12) showed a reduction in basal area due to the mortality of large trees (mainly *A. acuminata* individuals). A common pattern in the two types of forests is an abundant recruitment of *S. grossum*, which could indicate a forest species substitution by understory species, and a decrease of the dominant canopy species abundance. This pattern suggests that *Alnus* forests sequestrate biomass rapidly during the first years of succession but, in part, this biomass is not retained, due to the short life span of this species and because it is not rapidly replaced by other canopy tree species. Such forest dynamics may slow down succession toward mature forest composition, which is reflected in a low recruitment of mature forest species.

Contrarily, the *Podocarpus* forests accumulated biomass slowly; the high basal area showed little change through time, suggesting that these forests were undergoing intense

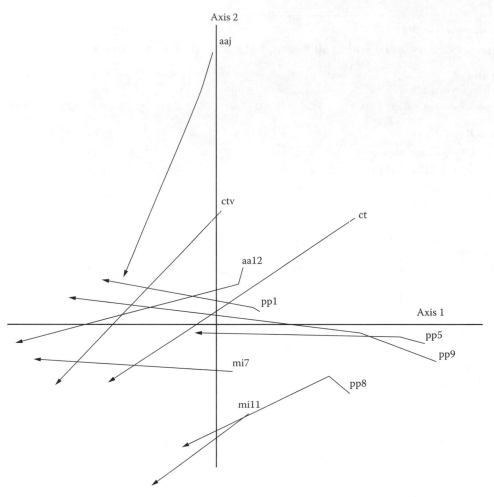

FIGURE 19.3 NMDS ordination diagrams based on forest composition. 1991, 2001, and future composition based on most likely successor species. Arrows represent successional trajectories (i.e. movement in the ordination diagram of each forest plots through time). Both axes explain 70% of the total variation (40 and 30%, for axis 1 and axis 2, respectively). Forest codes are: aa12 and aaj for *Alnus* forests; ct and ctv for *Crinodendron* forests; pp1, pp5, pp8, and pp9 for *Podocarpus* forests; and mi11 and mi7 for Myrtaceae or mature forests.

intraspecific competition, thus leading to very slow growth of dominant individuals. *Podocarpus* forests (pp8 and pp5) were being replaced by mature forest species (such as species of the Myrtaceae family, *I. argentina* and *P. tucumanensis*), although slowly. Similar patterns have been found by Ramadori (1998) in upper-montane secondary forests of the Bermejo River Basin (22°S), where monodominant *P. parlatorei* stands originated in abandoned grasslands and were later replaced by mature forest species. According to Ramadori's results, the recovery rate after fire for abandoned grassland is slower than after agriculture. Our results also

indicate that forests such as *Podocarpus* in late stages of succession could reach basal area values (average, 37 m²/ha) similar to those of mature forests (average, 36 m²/ha), with extreme values of 50 m²/ha, the highest recorded to date for northwest Argentina's subtropical forests.

Alnus forests showed a rapid structure recovery, but compositional recovery toward mature forest is limited for the low regeneration rate of mature forest species. These results are consistent with other studies in Argentinean montane forests, which showed that compositional recovery may take longer than structural recovery (Grau et al., 1997; Easdale, 1999). In addition, *Alnus* forests' basal area decreased after a few decades because of the short longevity of dominant species that are not rapidly replaced by other canopy trees. For these forests, management considerations should be to plant mature forest species that could potentially use the resources liberated by the old *Alnus* trees as they die. The question of why species of mature forests did not recruit under *Alnus* forests in our plots remains unanswered. Potential explanations include the effect of distance to seed sources and edaphic factors such as allelopathic effects (Murcia, 1997).

Podocarpus forests seem to have a great capacity for biomass sequestration reflected in high basal area values. However, the intense intraspecific competition produced a very slow rate succession in old secondary stands. A possible management practice could be selective exploitation (thinning) to liberate suppressed individuals from mature forest species, which are generally abundant in the understory. Moreover, species with economic value such as *P. parlatorei*, *C. lilloi*, and *J. australis*, considered late-pioneer species, which establish early in succession, need a gap for becoming part of the mature forest canopy (Morales and Brown, 1996).

In our study, we assumed that time is the most important factor conditioning forest composition, and that environmental variables and land use history did not differ significantly among plots. These assumptions need further testing. Despite these limitations, our study is the first to describe long-term successional and demographic trends in subtropical Argentinean upper-montane forests. Our results emphasize the importance of long-term studies to understand the dynamics of high-elevation forests and to manage them for their important ecological services.

SUMMARY

Northwest Argentina's upper-montane forests occur in a mosaic of different physiognomies, which, in part, reflect different stages of postgrazing forest succession. We analyzed 10 years of changes in structure and composition of secondary forest permanent plots dominated by *Podocarpus parlatorei*, *Alnus acuminata*, and *Crinodendron tucumanum*, at 1600 to 1800 m elevation, in Sierra San Javier, Tucumán, Argentina. Plots were measured in 1991 and in 2001 and were compared with mature forests dominated by species of the Myrtaceae family. Myrtaceae forests showed the highest values of species richness, whereas early successional forests dominated by *A. acuminata* showed the lowest. Successional trends in species composition indicated convergence toward mature forests, but secondary forests differed in terms of demographic rates and patterns of succession. *A. acuminata* forests stored biomass faster, reaching 25 m²/ha of basal area in a few decades. However, due to the short life span of *A. acuminata* and the low recruitment rate of mature forest species, biomass started to decrease in a few decades, and composition tended to be dominated by understory trees, mainly *Solanum grossum*. *Podocarpus parlatorei* forests reached very high basal area values (more than 50 m²/ha) and showed recruitment of mature forests species. However, possibly due to the intense intraspecific competition of the dominant trees, these forests showed very small changes in structure and were characterized by slow growth rates. Forests dominated by *C. tucumanum* were similar to *A. acuminata* forests in terms of successional patterns, whereas mature forests showed intermediate characteristics between *A. acuminata* and *P. parlatorei* forests.

ACKNOWLEDGMENTS

Jose Gallo helped in the field. Christian Körner and two anonymous reviewers provided helpful comments on the manuscript. Financial support was provided by grants from the Consejo de Investigaciones de la Universidad Nacional de Tucumán (CIUNT) and the Agencia Argentina Científica y Tecnológica (FONCYT).

References

Aide, T.M. and Grau, H.R. (2004). Globalization, rural–urban migration, conservation policy, and the future of Latin American ecosystems. *Science* 305: 1915–1916.

Arturi, M.F., Grau, H.R., Aceñolaza, P.G., and Brown, A.D. (1998). Estructura y sucesión en bosques montanos del Noroeste de Argentina. *Revista de Biología Tropical*, 46: 525–532.

Brown, A.D., Grau, H.R., Malizia, L.R., and Grau, A. (2001). Argentina. In Kappelle, M. and Brown, A.D. (Eds.), *Bosques nublados del Neotrópico*. INBIO, San José, Costa Rica, pp. 622–659.

Cabrera, A.L. and Willink, A. (1980). *Biogeografía de América Latina*. Organización de los Estados Americanos, Washington, D.C.

Easdale, T.A. (1999). Relación de disturbios y factores ambientales con la diversidad, composición y estructura de comunidades leñosas en el Valle de Los Toldos, Yungas Argentinas. Laboratorio, de investigaciones ecológicas de las Yungas, Universidad Nacional de Tucumán, Tucumán, Argentina.

Giusti, L., Slanis, A., and Aceñolaza, P.G. (1997). Fitosociología de los bosques de aliso (Alnus acuminata H.B.K. ssp. acuminata) de Tucumán, Argentina. *Lilloa*, 38: 93.

Grau, H.R., Arturi, M.F., Brown, A.D., and Aceñolaza, P.G. (1997). Floristic and structural patterns along a chronosequence of secondary forest succession in Argentinean subtropical montane forests. *Forest Ecology and Management*, 95: 161–171.

Grau, H.R., Easdale, T.A., and Paolini, L. (2003). Subtropical Dendroecology. Dating disturbances and forest dynamics in subtropical mountains of NW Argentina. *Forest Ecology and Management*, 177: 131–143.

Horn, H. (1975). Forest succession. *Scientific American*, 232: 90–98.

Kappelle, M. and Brown, D.A. (2001). Introducción a los bosques nublados del Neotrópico: una síntesis regional. In Kappelle, M. and Brown, A.D. (Eds.). *Bosques nublados del Neotrópico*. INBIO, San José, Costa Rica.

Kenkel, N.C. and Orlóci, L. (1986). Applying metric and non-metric multidimensional scaling to ecological studies: some new results. *Ecology*, 67: 919–928.

Kruskal, J.B. and Wish, M. (1978). Multidimensional scaling. *Sage University Papers Series on Quantitative Applications in the Social Sciences*, 07-011. Sage Publications, Beverly Hills, CA.

Legendre, P. and Legendre, L. (1998). *Numerical Ecology* (2nd ed.). Elsevier Science, Amsterdam.

Morales, J.M. and Brown, A.D. (1996). Bosques montanos con diferente intensidad de explotación. *Bosques y Desarrollo*, 17: 51–52.

Morales, J.M., Sirombra, M., and Brown, A.D. (1995). Riqueza de árboles en las Yungas argentinas. In Brown, A.D. and Grau, H.R. (Eds), *Investigación, Conservación y Desarrollo en Selvas Subtropicales de Montaña*. Laboratorio de investigaciones ecológicas de las Yungas, Universidad Nacional de Tucumán, Tucumán, Argentina, pp. 163–174.

Moyano, M.Y. and Movia, C.P. (1989). Relevamiento fisonómico estructural de las sierras de San Javier y El Periquillo (Tucumán, Argentina) I: Área de las Yungas. *Lilloa*, 37: 123–135.

Murcia, C. (1997). Evaluation of Andean alder as a catalyst for the recovery of tropical cloud forests in Colombia. *Forest Ecology and Management*, 99: 163–170.

Navarro, G., Arrázola, S., Antesana, C., Saravia, E., and Atahuachi, M. (1996). Series de vegetación de los valles internos de los Andes de Cochabamba (Bolivia). *Revista Boliviana de Ecología y Conservación Ambiental*, 1: 3–20.

Oliver, C.D. and Larson, B.C. (1996). *Forest Stand Dynamics*. Updated edition. John Wiley & Sons, New York.

Pickett, S.T.A., Collins, S.L., and Armesto, J.J. (1987). Models, mechanisms and pathways of succession. *Botanical Review*, 53: 335–371.

Ramadori, E.D. (1998). Sucesión secundaria en bosques montanos del Noroeste Argentino. Doctoral thesis, Facultad de Ciencias Naturales, Universidad Nacional de La Plata, Argentina.

Silver, W.L., Ostertag, R., and Lugo, A.E. (2000). The potential for carbon sequestration of abandoned tropical agriculture and pasture lands. *Restoration Ecology*, 8: 396–407.

Sokal, R.R. and Rohlf, F.J. (1995). *Biometry — The Principles and Practice of Statistics in Biological Sciences* (3rd ed.). Freeman and Company, New York.

Zuloaga, F.O. and Morrone, O. (1999a). *Catálogo de las plantas vasculares de la Republica Argentina I*. Missouri Botanical Garden Press, MO.

Zuloaga, F.O. and Morrone, O. (1999b). *Catálogo de las plantas vasculares de la Republica Argentina II*. Missouri Botanical Garden Press, Missouri, USA.

APPENDIX 1
Tree species abundance of forest types in 1991 and 2001

Species	aa12		aaj		ct		ctv		mi11		mi7		pp1		pp5		pp8		pp9	
	1991	2001	1991	2001	1991	2001	1991	2001	1991	2001	1991	2001	1991	2001	1991	2001	1991	2001	1991	2001
Alnus acuminata	14	8	100	95	5	4	8	6	1						1				1	1
Allophylus edulis										2										
Azara salicifolia																1		3		
Blepharocalyx salicifolius									11	11	4	3					4			
Cedrela lilloi									2	2			1	1	3	4		1		
Crinodendron tucumanum	3	3			25	24	21	21	1	1	20	17	9	11	5	7				
Dunalia lorentzii	10	10		1					29	25	20	18	4	2						
Duranta serratifolia									20	14										
Kaunia lasiophthalma	1	3		1							1									
Ilex argentina									8	7										
Juglans australis											7	6	3	4	1	1	7	7		
Myrcianthes calicoma																	2	2		
Myrcianthes mato									60	58	53	44					14	13		
Myrcianthes pseudo-mato									6	6										
Podocarpus parlatorei	11	11	2	2	14	18		1	20	19	11	9	17	12	121	99	62	53		
Prunus persica						1														
Prunus tucumanensis								2	8	8	6	4								
Solanum grossum	26	58	2	2	22	30	13	43	1	3	54	60	20	28	1	7	5	6	5	5
Sambucus peruviana	19	14	3	7	3	2	11	12	17	15	27	27	12	14	2	4	6	18	3	3
Vassovia breviflora									1	1										

Note: Forest codes are: aa12 and aaj for *Alnus* forests; ct and ctv for *Crinodendron* forests; pp1, pp5, pp8, and pp9 for *Podocarpus* forests; and mi11 and mi7 for Myrtaceae or mature forests.

20 Climatic and Anthropogenic Influences on the Dynamics of *Prosopis ferox* Forests in the Quebrada de Humahuaca, Jujuy, Argentina

Mariano Morales and Ricardo Villalba

INTRODUCTION

Prosopis ferox forests constitute an important floristic community in the intermontane arid valleys of northwestern rural Argentina (NOA). In the Quebrada de Humahuaca, *P. ferox* forests have been widely used as local sources of fuel and wood for rural construction. The fruits and leaves of *P. ferox* represent a major source of fodder for caprine and ovine cattle. The impacts of anthropogenic activity in the region are therefore reflected in the morphology and structure of the remnant forest.

With the Spanish conquest, important changes occurred in the resource-use systems in the Quebrada de Humahuaca. Among these changes, the replacement of native plants and indigenous animals by European species produced a distinct and increasing change in the landscape. The overexploitation by cattle farming increased the process of desertification, favored by the neglect of agricultural soil preservation techniques and the predation of forested communities (Lorandi 1997). The shortage of trees in most of the region led to the overuse of native species, which were suitable for use as fuel to cook food. In this manner, the *P. ferox* woods were reduced to small patches (León 1997).

The impoverishment of the environment, along with the process of proletarianization occurring in the Puna at the beginning of the 20th century, brought about the migration of the rural population to urban centers and the subsequent abandonment of fields (Reboratti 1994). A major process of regional migration started in 1930; the rural population left to work in mills and in other agriculture-related activities in the subtropics, and in railway and road construction. In this manner, the dependency on a monetary income gradually increased, weakening the access to the diverse alternative production methods that had so far sustained the traditional economy of subsistence in the region (Campi and Lagos 1994). The strong rural migration that took place in the Quebrada de Humahuaca caused the abandonment of land intended for pasture and crops.

Furthermore, important climatic changes have been registered in the NOA region since the middle of the 19th century. A steady increase in precipitation in the region has been registered since around 1950 (Minetti and Vargas 1997). This increase in precipitation, which does not seem to have historic precedence, has also been documented in dendrochronological data. The variations in the thickness of growth rings in two species of the region, *Juglans australis* (*nogal criollo*) and *Cedrella lilloi* (*cedro tucumano*), indicate that regional rainfall had reached a historic minimum in the 1860s and 1870s, oscillated around the mean from the beginning of the 20th century until the end of the 1940s, and has increased steadily over the last few decades (Villalba et al. 1998).

During previous explorations in the region, we observed an apparent increase in the cover area of *P. ferox* in the surroundings of the town of Humahuaca. To identify the environmental changes related to this dynamic process of forest expansion, we evaluated the past and present roles of climatic and anthropogenic factors on the structure and dynamics of the *P. ferox* forests in the Quebrada de Humahuaca during the 20th century. Previous studies of species of the genus *Prosopis* indicate that the trees themselves create environmental conditions that facilitate the development of other associated species (Aggawarl et al. 1976; Simpson and Solbrig 1977; Archer et al. 1988). It has been shown that the woody species of *Prosopis* in the arid zones of western Argentina modify the microclimatic conditions under their crowns, thereby generating different environmental conditions in the nearby open areas are, then, adequate habitats for companion species (Rossi and Villagra, 2003). Our study, although concentrating on the dynamics of a single species, has, therefore, important implications with respect to biodiversity.

MATERIALS AND METHODS

STUDY AREA

In Argentina, *P. ferox* grows between 2600 and 3800 masl in arid environments (annual mean precipitation ~300 mm) from the northern sector of the Calchaquí Valley, Salta, the Quebrada de Humahuaca, Jujuy, and the Río Grande de San Juan Valley on the border between Jujuy and Bolivia (Legname 1982). In Bolivia, it can be found in the arid inter-Andean valleys between 2600 and 3800 masl in the departments of Potosí, Tarija, and Chuquisaca (Saldías-Paz, 1993; Lopez, 2000). Floristically, the region belongs to the phytogeographic pre-Puna province, with numerous elements of the Monte province (Cabrera 1976). Among the common companion species of *P. ferox* are *Trichocereus pasacana (cardón)*, *Opuntia sulphurea*, *Opuntia soehrendsii*, *Opuntia tilcaren-*

sis, *Parodia* spp., *Baccharis boliviensis*, *Baccharis salicifolia*, *Gochnatia glutinosa*, *Maihueniopsis* spp., *Schinus latifolius*, *Senna crassiramea*, *Lycium venturii*, *Proustia cuneifolia*, and *Aphylloclados spartioides* (Beck et al. 2003). *P. ferox* forms an open forest with individuals grouped in patches, which are usually associated with Cactacea species and several Asteraceae shrubs.

The temperature in the study area markedly declines with altitude, whereas precipitation depends largely on the topographic location in relation to the bearing of the mountain chains, which intercept the humid air masses. Climatic data are taken from sites in Humahuaca (23°10 S, 65°20 W) and La Quiaca (22°06 S, 65°36 W), which are situated at 2940 and 3460 m elevation, respectively, and can be considered representative of the climatic conditions of the study sites. Whereas annual temperatures in Humahuaca and La Quiaca lie between 10°C and 9.5°C, the total precipitation on average is 175 and 322 mm, respectively. The water deficit is more distinct between April and December; soil water is partially replenished during summer months (January–February). For La Quiaca, the annual temperature difference between the hottest month (December) and the coldest month (June) is 8.7°C; this is less than the mean daily temperature range for any month of the year, which ranges from 14.3°C in January to 23.5°C in July.

COLLECTION AND PROCESSING OF SAMPLES

The different habitats of *P. ferox* were determined based on geomorphological features of the landscape. We used aerial photographs for the delimination of the vegetation units and subsequent verification in the field. The communities of *P. ferox* were grouped according to their location on plateaus, alluvial cones, hillsides, or in riparian environments. The density of individuals determined the size of the plots, but every plot included at least 40 individuals. Sampling was carried out following conventional dendrochronological methods (Stokes and Smiley, 1968). Growth rings were correctly dated by year of formation.

DETERMINATION OF POPULATION STRUCTURE

To establish the age structure of the plots, the age of each individual was determined by counting growth rings, starting at the outermost ring next to the bark and ending with the innermost ring around the pith. For those individuals in whom the rings could not be unequivocally recognized with a dissecting microscope, the growth ring count was carried out using histological cuts. The same procedures were used to determine the age of individuals displaying groups of very thin growth rings. For these samples, errors in their ages range from 2 to 5 years.

COLLECTION AND ANALYSIS OF CLIMATIC DATA

To analyze annual fluctuations in precipitation along the Quebrada de Humahuaca over the past 100 years, instrumental data were collected from meteorological stations in the NOA; the data were taken from Bianchi and Yañez (1992), the publications of the National Meteorological Service, and other international databases from institutions such as the Oak Ridge National Laboratory (ORNL) and the International Research Institute (IRI). From these data, information from weather stations located between 21.5° and 29° S and between 62.5° and 69° W was compiled. From the total of 82 initially selected precipitation records, only 32 fulfilled the criteria of quality, reliability, and minimum time span (56 years) required for our study. The dominant patterns of variability in the precipitation of NOA were determined using principal component analysis (Cooley and Lohnes, 1971) of the 32 selected records.

COLLECTION AND ANALYSIS OF ANTHROPOGENIC DATA

Regional demographic changes (rural migration process) and land use changes (changes in stocking rates) were reconstructed using historical documents. The information on demographic changes and stocking rates was compiled from the National Population Censuses beginning in 1869 (first record) and the National Agropecuarian Censuses beginning in

1908. This information is available at the National Institute for Statistics and Censuses (INDEC), Buenos Aires, Argentina.

RESULTS

POPULATION STRUCTURES

Common patterns of tree establishment were recorded in 12 sampling plots of *Prosopis ferox* located in different environments aross the Quebrada de Humahuaca. The most distinct common feature is the period of increased establishment from around the mid-1970s until the year 1990. This pattern was observed clearly in plots 5, 9, 10, 13, 14, 15, 16, and 42 (Figure 20.1). In plot 12, there was a greater rate of establishment during the 1960s and at the beginning of the 1970s (Figure 20.1). In plot 11, establishment took place from the end of the 1960s until the mid-1980s (Figure 20.1). In plot 41, new recruits had established between 1933 and 1970, with a decrease in establishment in the following decades (Figure 20.1). Apart from plot 13, establishment in the 1990s was either zero or greatly reduced.

An idea of the regional temporal evolution of establishment can be obtained from the sum of population structures of all the plots. In this way, the regional signal emerges clearer, as events affecting individual plots are minimized in the regional mean (Villalba and Veblen, 1997). The total sum of age structures over all plots reflects important temporal changes in the establishment process. Individuals that established between the beginning of the 1930s and mid-1950s were scarce. The number of established individuals increased gradually between the mid-1950s and mid-1970s. From then on, there was a marked increase in establishment, which ended in the beginning of the 1990s, when it decreased considerably (Figure 20.1m).

VARIATION IN PRECIPITATION

The dominant precipitation pattern in the NOA region shows a positive trend during the period from 1930 to 1998 (Figure 20.2). A significant increase in rainfall, beginning in 1973, was observed in all records. During this humid period, which in general lasted until 1992, the

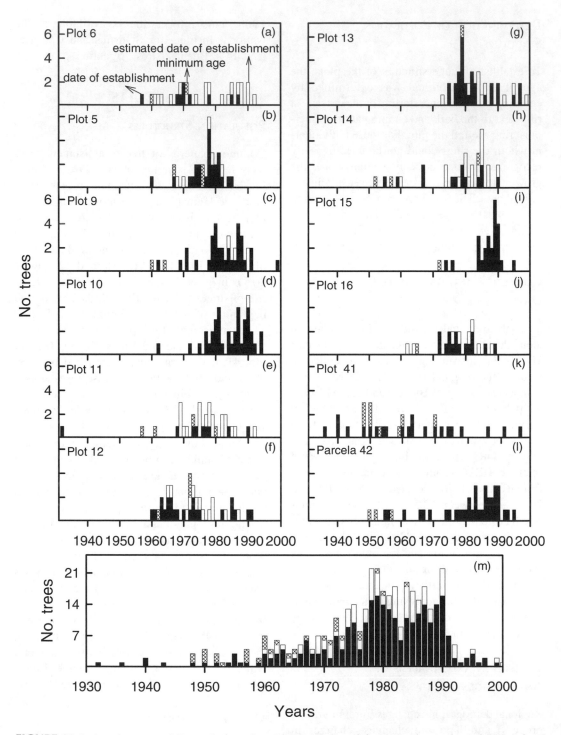

FIGURE 20.1 Age structure of *Prosopis ferox* in each plot (a–l) and for the entire study region (m) from 1930 onward.

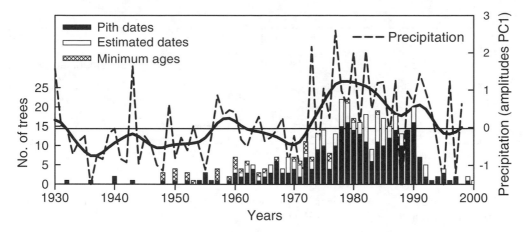

FIGURE 20.2 Comparison between the pattern of establishment of *Prosopis ferox* and the first principal component of precipitation from 32 meteorological stations in northwestern Argentina.

mean annual precipitation values were much higher than in previous decades and in later years, with the exception of 1996 (Figure 20.2).

DEMOGRAPHIC AND LAND USE CHANGES

During the colonial period until the end of the 19th century, the majority of the population of Jujuy was concentrated in the Quebrada and the Puna, which were strategic regions in linking the economy of Argentina with its principal market, Potosí (Larrouy 1927). This pattern was modified toward the end of the 19th century as, among other reasons, other regions increased in importance, in particular the central and subtropical valleys, where the new agroindustries formed the principal regional economic activity of the emerging national market.

The steady increase of the economy in subtropical regions, largely allocated with sugar mill activities, speeded up the process of population migration from the Puna to the lowlands. This process, which started at the end of the 19th century, persisted until the mid-20th century. During the last few decades, a strong urbanization process led to a concentration of the population in the principal cities. Figure 20.3a clearly shows this exponential increase in population in the central and subtropical valley regions since the end of the 19th century. In contrast, the population in the Puna and the Quebrada regions grew very slowly (Figure 20.3a). Nevertheless, a process of urbanization concentrating the population in the towns

located along the Quebrada de Humahuaca was recorded. In the Humahuaca Department, the 1947 census indicated that the majority of the population was rural (82%), and the urban population was very small.

From the middle of the 20th century onward, the migration of the rural population to small towns was observed. In the 1960 census, the rural proportion of the population had decreased to 43% (Figure 20.3b). Unfortunately, there exist no recent data about the relationship between rural and urban populations, but we believe that the trends have remained stable, which has led to a further decline in rural population levels over the last decades.

A very important reduction in regional cattle stocking rates has been observed in association with these migration processes. In the Humahuaca and Tilcara departments, the stocking rate decreased steadily from the beginning of the 20th century until the 1970s. From then on, there was a very distinct decline in cattle numbers, particularly in ovine and caprine cattle (Figure 20.4a and Figure 20.4b).

DISCUSSION

The dendrochronological studies of *Prosopis ferox* based on the population structures of 12 plots located in different environments of the Quebrada de Humahuaca show four periods with distinct characteristics: a stage of scant establishment up until the 1960s, a period of

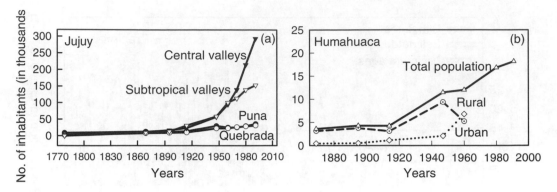

FIGURE 20.3 Demographic variation in (a) the four principal regions of the Jujuy Province, and (b) the rural and urban population in the Department of Humahuaca.

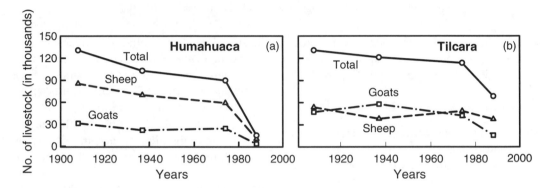

FIGURE 20.4 Temporal variation in cattle stocking rates in the departments of (a) Humahuaca and (b) Tilcara.

low establishment between 1960 and 1970, a marked increase between 1974 and 1990, and another period of reduced establishment during the last 10 years. This pattern is consistent with the observed recovery of the forests and their expansion into open areas.

Some of the observed changes in this regional pattern of establishment coincide with the recorded changes in precipitation. Although there is a positive trend in rainfall for the interval 1930 to 1998, a distinct leap was recorded in the year 1973, which coincides with a marked increase in the number of established individuals of *P. ferox* in the region. This interval of higher precipitation represents the most humid period in the 20th century. The decreased establishment observed during the 1990s coincides with lower precipitation during the same period.

Furthermore, the demographic changes recorded at a regional level, the subsequent

decrease in cattle stocking rates, and the resulting decline in grazing pressure are also factors associated with the observed changes in the pattern of establishment of *P. ferox*. The Population and Agrarian Censuses provide evidence of an important change in the relationship between the rural and the urban population in the middle of the 20th century, and a marked reduction in cattle stocking rates between 1974 and 1988. This reduction in cattle stocking facilitated the sudden increase in the establishment of *P. ferox* observed between the mid-1970s and 1990.

Based on these observations, reduced establishment before 1960 can be attributed to the pressure on the vegetation caused by high densities of ovine and caprine livestocks. In addition to the impact of browsing, there occurred a relatively dry period, which hindered the establishment of new recruits. As cattle stocking decreased and precipitation increased, the number of established individuals increased. From the mid-1970s until the beginning of the 1990s, the number of individuals recruited increased markedly, which was a consequence of reduced grazing pressure and the sudden increase in precipitation. The lower establishment rates recorded in the last 10 years can be attributed mainly to the decrease in rainfall, as grazing pressure in the 1990s remained stable or even decreased.

The observed pattern of establishment may indicate that both climatic variation and land use changes regulate the dynamics of *P. ferox* forests in the Quebrada de Humahuaca. The reduced rates of establishment during the 1990s, despite cattle stocking similar to or lower than in the 1980s, suggest that variations in rainfall play an important role in the establishment of new individuals.

The observed recovery and expansion of *P. ferox* forests may create new habitats for other species beneath the tree crowns, thereby increasing the spatial heterogeneity of the ecosystems and the local biodiversity. In agreement with Aide and Grau (2004), our study indicates that a reduction of human-induced disturbances on the landscape facilitates conservation efforts, as the decline in human pressure and the impact of associated activities often allows ecosystem recovery. Nevertheless, the recovery of *P. ferox* forests and the increase in local biodiversity cannot be solely attributed to the decrease in human impact. Particular climatic conditions, such as the humid events recorded during the 1970s and 1980s, are necessary for the establishment and survival of new recruits. The interaction between social and natural factors largely determines the future development of ecosystems and their species richness. For this reason, it is important that we quantify the relative importance of these environmental forcing factors to establish management guidelines for the intermontane and subtropical valleys of the Andes.

SUMMARY

The recent increase of *Prosopis ferox*, both in population density and distribution range in the Quebrada de Huamahuaca, Jujuy, Argentina, appears to be related to major changes in land use and regional precipitation. *P. ferox* grows between 2600 and 3800 m elevation in the upper-elevation intermontane valleys in northwestern Argentina and southern Bolivia. Across its range of distribution, *P. ferox* has been largely used as a local fuel source, as construction material, and as fodder for livestock. To determine the factors affecting the recent changes in the population structure of *P. ferox*, we compared the age structure of the woodlands with human demographic and land use changes, and with regional variations in precipitation. Based on age structures from 12 stands, four periods of recruitment were identified. Reduced recruitment was recorded until 1960 followed by a gradual increase between 1960 and 1970. An abrupt increase was observed around 1974, which persisted to 1990. Finally, low recruitment was recorded during the past 10 years. These regional patterns of recruitment are consistent with the recent expansion of the woodlands. Variations in recruitment reflect regional variations in precipitation since the middle of the 20th century. Furthermore, demographic and agrarian censuses show significant changes in the relationship between rural and urban populations, and a substantial reduction of livestock density between 1974 and 1988. Reduced grazing by livestock during this interval might have also favored the recruitment of *P. ferox* trees. Similar or lower grazing pressure during the 1990s suggests a greater influence of precipitation than livestock on the reduced tree establishment during this decade.

ACKNOWLEDGMENTS

These studies were financed by the Agencia Nacional de Promoción Científica y Tecnológica (PICTR 2002-123) and the Instituto Interamericano para el Estudio del Cambio Global (IAI). The authors wish to thank Alberto Ripalta, Susana Monge, Sergio Londero, and Tromen Villalba for their collaboration in the field.

REFERENCES

Aggawarl, R.K., Gupta, J.P., Saxena, S.K., and Muthana, K.D. (1976). Studies on soil physicochemical and ecological changes under twelve year old desert tree species of Western Rajasthan. *Indian For*, 102: 863–872.

Aide, T.M. and Grau, H.R. (2004). Globalization, migration and Latin American ecosystems. *Science*, 305: 1915–1916.

Archer, S., Bassham, C.R., and Maggio, R. (1988). Autogenic succession in a subtropical savanna: conversion of grassland to thorn woodland. *Ecological Monographs*, 58(2): 111–127.

Beck, S., Paniagua, Z., and Yevara Garate, M. (2003). Flora y vegetación: las áreas de Rodero, Tilcara y Volcán. In Reboratti, C. (Ed.), *La Quebrada*. La Colmena, Buenos Aires, Argentina, pp. 47–70.

Bianchi, A.R. and Yañez, C. (1992) Las precipitaciones del noroeste argentino. Instituto Nacional de Tecnología Agropecuaria, Salta, Argentina, 388 pp.

Cabrera, A. (1976). *Regiones fitogeográficas Argentinas*. Enciclopedia Argentina de agronomía y jardineria. Tomo 2, fasc. 1, ACME, Buenos Aires, Argentina.

Campi, D. and Lagos, M. (1994). Auge azucarero y mercado de trabajo en el noroeste Argentino. 1850–1930, *Andes. Antropología e Historia*, 6 :179–208.

Cooley, W.W. and Lohnes, P.R. (1971). *Multivariate Data Analysis*. John Wiley & Sons, New York.

Larrouy, P.A. (1927). *Documentos del Archivo de Indias para la historia del Tucumán*. Tomo segundo, siglo XVIII. (En Santuario de Nuestra Señora del Valle, Vol. 4. Tolosa).

Legname, P.R. (1982). *Arboles indígenas del Noroeste Argentino*. Ministerio de cultura y educación. Fundación Miguel Lillo.

León, C. (1997). *Evolución económica y recursos naturales en el NOA*. De Hombres y Tierras: 49–59. Proyecto GTZ, Salta.

Lopez, R.P. (2000). La Prepuna boliviana. *Ecología en Bolivia*, 34: 45–70.

Lorandi, A.M. (1997). *El contacto hispano-indígena y sus consecuencias ambientales*. De Hombres y Tierras: 26–38. Proyecto GTZ, Salta.

Minetti, J.L. and Vargas, W.M. (1997). Trends and jumps in the annual precipitation in South America, south of the 15 S. *Atmósfera*, 11: 205–221.

Reboratti, C. (1994). *La naturaleza y el hombre en la Puna*. Desarrollo agroforestal en comunidades rurales del noroeste Argentino. Serie Nuestros Ecosistemas. Proyecto GTZ. Salta.

Rossi, B. and Villagra, P. (2003). Effects of Prosopis flexuosa on soil properities and the spatial pattern of understorey species in arid Argentina. *Journal of Vegetation Science*, 14: 543–550.

Saldias-Paz, M. (1993). Mimosoideae. In Killeen, T., Garcia, E., and Beck, S.G. (Eds.), *Guía de Arboles de Bolivia*. Herbario Nacional de Bolivia, Missouri Botanical Garden, pp. 420–456.

Simpson, B.B. and Solbrig, O.T. (1977). Introduction. In Simpson, B.B. (Ed.), *Mesquite. Its Biology in Two Desert Scrub Ecosystems*. US/IBP Synthesis Series 4. Dowden, Hutchinson and Ross, Stroudsburg, PA, pp. 1–26.

Stokes, M.A. and Smiley, T.L. (1968). *An Introduction to Tree-Ring Dating*. University of Chicago Press, Chicago, IL, p. 73.

Villalba, R. and Veblen, T. (1997). Regional patterns of tree population age structure in northern Patagonia: climatic and disturbance influences. *Journal of Ecology*, 85: 113–124.

Villalba, R., Grau, H.R., Boninsegna, J.A., Jacoby, G.C., and Ripalta, A. (1998). Tree-ring evidence for long-term precipitation changes in subtropical South America. *International Journal of Climatology*, 18: 1463–1478.

Part V

Land Use Effects on Mountain Biodiversity: Socioeconomic Aspects

21 Conservation of Biodiversity in the Maloti–Drakensberg Mountain Range

Terry M. Everson and Craig D. Morris

INTRODUCTION

The Maloti–Drakensberg mountains (29°30 S; 29°00 E) form a 300-km border between the landlocked mountain kingdom of Lesotho and South Africa (Figure 21.1). The range extends over an alpine and montane area of 5000 km². The alpine area is the only true alpine region in southern Africa (Linder 1990). The region is recognized as one of southern Africa's eight "hot spots" of botanical diversity in terms of species richness and endemism (Cowling and Hilton-Taylor 1944). The area is characterized by high levels of habitat diversity (wetlands, alpine tundra, grassland, heathlands, woodlands, and forests) created by a range of environmental variables such as altitudinal zones, aspects, and slope (Carbutt and Edwards 2004). The alpine and subalpine areas of the Maloti–Drakensberg support a network of unique high-altitude wetland systems that are a major source of water for Lesotho and South Africa (Stewart 2000).

In Lesotho, very little of the high plateau is formally protected. Sehlabathebe National Park (6,500 ha) is the only formally protected area. In South Africa, there is a high degree of protection and conservation of biodiversity through the establishment of national parks. Here, livestock are excluded, and grazing by wildlife is minimal in comparison to the adjacent communal rangeland areas where stocking densities are high (1 to 2 animal units [AU] per ha). The Drakensberg Park (242,813 ha) is the largest protected area on the great escarpment of the southern African subcontinent. It comprises a northern and much larger southern section that supports 2,153 plant species, of which 247 are endemic (Hilliard and Burtt 1987). It is also home to the greatest gallery of San (bushman) rock art in the world. These unique qualities led to the declaration of the park as one of the few dual (natural and cultural) world heritage sites (Derwent, Porter, and Sandwith 2001). On both sides of the international boundary, and bisecting the Drakensberg Park, are communal tenure populations that are dependent on the mountains for all or part of their livelihood. The biodiversity and cultural and other features of this region are increasingly under threat from development, unsustainable rangeland management, and invasion by alien species. Management of these catchments is a complex problem involving not only environmental issues but also issues such as social dynamics and land tenure. Past failures of development initiatives to solve the problems of environmental degradation in South Africa have been attributed to the lack of consultation and involvement of the rural population. The challenge to conservation efforts is (1) to conserve this unique mountain region while (2) ensuring that the development needs of the local populations are met. Two case studies are presented here in which participatory projects were carried out in two tribal areas in the Drakensberg to address problems of environmental degradation and biodiversity conservation.

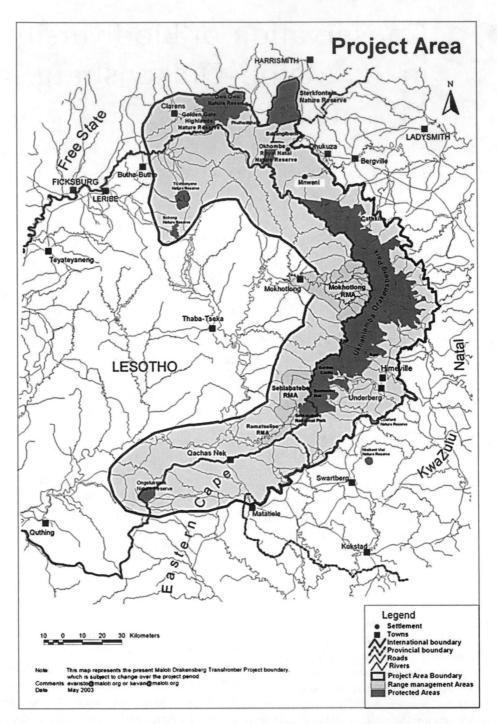

FIGURE 21.1 Map showing location of the two study areas (Okhombe and Mnweni).

CASE STUDY 1: MNWENI MOUNTAIN CATCHMENT

One of the most important high-yielding water catchments in South Africa is the Amangwane mountain region of Mnweni. Much of the water of the Mnweni River is transferred to Gauteng, the industrial center of South Africa. Situated in the Drakensberg, the land has steep slopes that have a high potential for soil erosion. The Mnweni community relies on the natural resources in the area for its daily living. The mountain slopes and plateaus are designated as communal grazing land, thatch grass (*Hyparrhenia hirta*) for roofing provides an income, and the available trees are used for firewood and poles. The area is very high in plant species diversity, has spectacular mountain vistas, and, therefore, has the potential for income through tourism. A major concern of the communities in Mnweni is the high soil erosion and degradation of the natural resources.

The Rand Water Mnweni Trust was established in 1999 to contribute to improving the lives of people living in the Mnweni catchments by facilitating activities compatible with conservation and sustainable development. Income from the Mnweni Trust is used to promote employment for conservation-related projects. The Donga Reclamation and Conservation Project was initiated with funding from the Mnweni Trust to address the problem of land degradation. This is one of the first sustainable job creation projects in the area. Twenty people from each of the three communities in Mnweni are employed for 6 months on a rotational basis. The main objectives of the project are to:

1. Enhance the capacity of the people of Mnweni to manage their natural resources through the development of institutions, training in rehabilitation and monitoring, and environmental education.

2. Plan and implement interventions to reduce rates of land degradation through the following conservation measures:

   (i) Physical structures (e.g. contours, stone lines, stone packs, etc.)

   (ii) Vegetative measures (e.g. planting of vetiver grass (*Vetiveria zizanioides*), indigenous grasses and trees, brush packing, etc.)

   (iii) Establishment of alternative fodder species

3. Assess the species-level diversity of plants in grazed and protected areas.

During the 2-year duration of the project, approximately 40 *dongas* (erosion gullies) have been rehabilitated. The most successful erosion control method was the planting of a stoloniferous, exotic grass species, *Pennisetum clandestinum* (Kikuyu). The most significant feature of this rehabilitation technique was the return of indigenous grass species and increase in biodiversity. It was apparent that Kikuyu grass was a good pioneer species that enabled later successional species to establish. At one of the severely eroded sites, 11 new grass species were recorded in the donga (18 m^2) that was previously bare.

A biodiversity study showed that mean species richness in 100-m^2 plots in the communally grazed area was significantly lower (35) than the adjacent conserved areas (49). This has been attributed to heavy grazing by livestock and unseasonal burning practices. To increase the biodiversity and ecotourism potential of the area, one of the villages in Mnweni has taken the initiative and has set aside an area of approximately 50 ha above 1800 m where livestock have been excluded and a burning program is being implemented.

Disturbance in the form of fire can increase diversity by changing the species composition. The fire climax grasses in the montane area are well adapted to survive regular burning (Everson and Tainton 1984). However, incorrect fire management results in the invasion of unpalatable species such as *Elionurus muticus* and *Koelerig capensis*. In the conserved areas, a biennial spring burn resulted in high overall diversity (Shannon diversity index = 2.33) (Figure 21.2). In contrast, a recovery period between burns of only 1 year or more than 3 years had a detrimental effect on species diversity (1.75 to 1.98). Summer burning promoted the diversity of nongrass species (2.44) but reduced the palatable grass species (Morris, Dicks, Everson, and Everson 1999). These

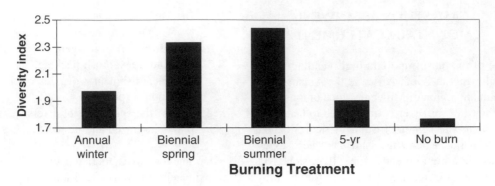

FIGURE 21.2 Mean Shannon diversity index in different burning treatments in conserved areas.

results indicate that a varied burning regime can increase the diversity of these rangelands.

CASE STUDY 2: COMMUNAL GRAZING IN OKHOMBE

One of the major threats to biodiversity is overgrazing of the rangeland. In Lesotho, the government has turned to community-based approaches for range conservation and improvement. The most successful of these has been the development of Range Management Areas and Grazing Associations (Ivy and Turner 1996). Through this program, the Government of Lesotho aims to involve the community in implementing range management plans that will reverse the decline in rangeland conditions.

In South Africa, there are no formal grazing management programs in the communal areas of the Drakensberg. Past failures to solve the problems of environmental degradation have been attributed to the lack of consultation and involvement of the rural population (Everson and Hatch 1999). To address this problem, a participatory project was initiated with the Okhombe community in the Amazizi tribal area of the Drakensberg.

A veld (grassland) condition technique was used to assess the productive capacity of the grassland. This is a standard agricultural approach to grassland assessment (Tainton 1999) and is based on agronomic features (ability to support livestock production) and on ecological features (successional status of species, long-term stability of the plant community, and ability to protect soil from crosion). The veld

condition is calculated by determining the species composition at 200 points randomly placed at each site. The grassland within a site is rated against a site in the same ecological zone that is in excellent condition (benchmark site) and has a veld condition index of 100% (Table 21.1). The species are divided into their respective ecological status groups (Tainton 1999), and the percentage of each is multiplied by the grazing value (the quantity and quality of plant material produced for grazing). The grazing value is assessed according to production, palatability, nutritional value, growth vigor, and digestibility (Van Oudtshoorn 1999). Decreaser species are desirable, palatable species that are abundant in good grassland but decrease in number when the grassland is overgrazed. Increaser species are generally unpalatable species that are common in poor-condition grasslands. They are further divided into three categories: Increaser I (indicative of underutilization), Increaser II (indicative of overgrazing), and Increaser III (indicative of selective grazing), depending on their response to utilization (Table 21.1). Increaser II species are further subdivided into a, b, and c categories according to increasing severity of over-utilization.

The veld (grassland) condition of the 36 study sites ranged from poor (15.9%) to reasonable (64.8%), with an average of 46.3%. The degraded sites were characterized by a large percentage of Increaser II species, indicating long-term overgrazing. The veld condition of these sites could be improved with good management practices such as resting. However, some of the sites were severely degraded with a high percentage of Increaser III species

TABLE 21.1
Effects of Grazing on Species Composition and Veld Condition

Group	Species	Grazing Value	Benchmark (%)	Benchmark Score	Site (%)	Site Score
Increaser I	*Alloteropsis semialata*	3	2	6	0	0
	Eulalia villosa	3	1	3	0	0
	Trachypogon spicatus	3	2	6	0	0
	Tristachya leucothrix	9	20	180	3	27
	Digitaria tricholaenoides	6	0	0	0.5	3
	Hyparrhenia aucta	6	0	0	0	0
	Andropogon eucomus	1	0	0	0	0
	Subtotal		25	195	3.5	30
Decreaser	*Brachiaria serrata*	3	1	3	0	0
	Diheteropogon amplectens	8	1	8	0	0
	Monocymbium ceresiiforme	6	2	12	0	0
	Themeda triandra	10	45	450	1.5	15
	Melinis nerviglumis	2	0	0	0	0
	Panicum natalense	2	0	0	0.5	1
	Subtotal		49	473	2	16
Increaser IIa	*Eragrostis capensis*	2	1	2	7	14
	Harpochloa falx	3	3	9	0	0
	Heteropogon contortus	6	4	24	19	114
	Setaria nigrirostris	5	0	0	0	0
	Subtotal		8	35	26	128
Increaser IIb	*Eragrostis curvula*	5	1	5	3	15
	Eragrostis plana	3	1	3	3	9
	Eragrostis racemosa	2	1	2	22.5	45
	Hyparrhenia hirta	3	1	3	0.5	1.5
	Sporobolus africanus	3	0	0	0	0
	Digitaria monodactyla	0	0	0	1	0
	Setaria sphacelata v *torta*	2	0	0	0	0
	Subtotal		4	13	30	70.5
Increaser IIc	*Microchloa caffra*	1	1	1	0	0
	Forbs and sedges	0	6	0	12	0
	Paspalum dilatatum	7	0	0	9	63
	Paspalum notatum	3	0	0	0	0
	Aristida congesta	0	0	0	0	0
	Cynodon dactylon	3	0	0	0	0
	Melinis repens	1	0	0	0	0
	Subtotal		7	1	21	63
Increaser III	*Diheteropogon filifolius*	0	2	0	0	0
	Elionurus muticus	0	5	0	0	0
	Aristida junciformis	0	0	0	9	0
	Rendlia altera	0	0	0	8.5	0
	Subtotal		7	0	17.5	0
	Total		100	717	100	307.5
	Score (%)			**100**		**42.89**

Note: Comparison between benchmark site (representative grassland in optimum condition for livestock production) and poor-condition sample site. Grazing value ranges from poor (0) to excellent (10).

(e.g. *Aristida junciformis*), which indicates selective grazing. *Aristida junciformis* is a hard, wiry-leaved, unpalatable grass that has virtually no grazing value. If selective overgrazing continues, *A. junciformis* will be virtually impossible to eradicate with normal management practices. The recommended grazing capacity of the area is 0.5 AU ha^{-1}. The current stocking densities of most of the sites were much higher than the recommended number of animal units (1 to 2 AU ha^{-1}). These stocking rates were, however, much lower than those in the upland pastures of Ethiopia where stocking rates in communally grazed areas were 4.2 AU ha^{-1} (Mohamed-Saleem and Woldu 2002). Here, heavy grazing changed the species composition but did not reduce species richness.

Under the grazing system that prevails in Okhombe, there have been major changes in species composition. The most distinctive feature of the rangeland at Okhombe is the loss of highly palatable grass species such as *Themeda triandra*. These species are preferred by livestock and have been eliminated with overgrazing. The dominance of poor grazing grasses such as *Eragrostis plana* and *Sporobolus africanus* is an indicator of overgrazing. However, the grass cover formed by these species is generally more dense, making it more resistant to physical degradation. This is mainly due to the dense nature of the *Eragrostis curvula* tufts and to the stoloniferous nature of the invader grass *Paspalum dilatatum*.

To address land degradation through overgrazing, a national Landcare project was implemented in Okhombe. The Landcare initiative promotes community involvement in natural resource management. Project activities include the monitoring of soil erosion, runoff, and basal cover by a group of volunteer community members. In addition, the community has developed a rotational grazing plan to improve the condition of the rangeland and reduce the severe soil erosion that is threatening loss of species. Although the project has been successful in establishing a livestock committee that has developed a three-camp rotational grazing plan for three villages, theft of fencing has been a major problem in the implementation of the plan. One solution to this may be the "social" fencing system used in Lesotho whereby herd-ers are paid in either money or livestock to manage the movement of livestock between camps. The exclusion of cattle from the cropping lands, nevertheless, has increased the abundance of thatch grass resulting in increased income for the community.

CONCLUSION

Species richness in communally grazed areas was significantly lower than in adjacent conserved areas. Community initiatives that have been successful in increasing biodiversity include reclamation of bare areas with Kikuyu grass. Biennial spring burning promoted species diversity in contrast to annual burning and protection from fire. Summer burning promoted the diversity of nongrass species.

Community involvement in grazing management of the Maloti–Drakensberg region is essential if maintenance of biodiversity and livestock grazing are to be compatible. Grazing management initiatives in communal areas of South Africa include the setting up of rotational grazing camps that are managed by herders who are paid by the community. The introduction of resting and burning regimes will prevent overgrazing and promote biodiversity conservation. It is apparent that transfrontier cooperation between South Africa and Lesotho can contribute to the management of biodiversity in the region and, at the same time, address the socioeconomic needs of the local people.

SUMMARY

The Maloti–Drakensberg mountain range, situated between Lesotho and South Africa, is globally important for biodiversity conservation. It is high in endemism (247 plant species) and is functionally an important area for water production. However, extensive areas are threatened by excessive livestock grazing, increased cultivation on steep mountain slopes, unseasonal burning, and invasion of alien plant species. In the communal rangelands of Mnweni, a biodiversity study showed that species richness in 100 square meter plots was significantly lower (35) than in adjacent conserved areas (49). In the conserved areas, a biennial spring burn resulted in high overall diversity (Shannon diversity index = 2.33) in contrast to annual burning (1.98) and protection from fire (1.75). Summer burning promoted the diversity of non-grass species (2.44). Long-term conservation of the region's biodiversity depends on community involvement in establishing grazing-management strategies, environmental management, and protected area management. In the Okhombe ward, overgrazing has resulted in loss of palatable grass species such as *Themeda triandra* and the dominance of *Eragrostis plana* and *Sporobolus africanus*. The initiation of job creation for conservation and monitoring programs is a significant step in addressing the socioeconomic needs of the local people, at the same time addressing threats to biodiversity such as unsustainable use of natural resources and soil erosion.

References

Cowling, R.M. and Hilton-Taylor, C. (1944). Patterns of plant diversity and endemism in southern Africa: an overview. In Huntley, B.J. (Ed.), *Botanical Diversity in Southern Africa*. National Botanical Institute, Pretoria, South Africa, pp. 31–52.

DerWent, S., Porter, R., and Sandwith, T. (2001). Maloti-Drakensberg: Transfrontier Conservation and Development Programme. Ezenvelo Kwazulu Natal Wildlife, Pietermaritzburg, South Africa.

Everson, C.S. and Tainton, N.M. (1984). The effect of thirty years of burning on the Highland Sourveld of Natal. *J Grass Soc South Afr* 1, 3: 15–20.

Everson, T.M. and Hatch, G.P. (1999). Managing veld (rangeland) in the communal areas of southern Africa. In Tainton, N.M. (Ed.). *Veld Management in South Africa*. University of Natal Press, Pietermaritzburg, South Africa, pp. 381–388.

Hilliard, O.M. and Burtt, B.L. (1987). *The Botany of the Southern Natal Drakensberg*. National Botanical Institute, Annals of Kirstenbosch Botanic Gardens, Vol. 15., p. 253.

Ivy, D. and Turner, S. (1996). *Successful Natural Resource Management in Southern Africa — Range Management Areas and Grazing Associations Experience at Sehlabathebe, Lesotho*. Gamberg Macmillan, Windhoek, Namibia.

Linder, H.P. (1990). On the relationship between the vegetation and floras of the Afromontane and the Cape regions of Africa. *Mitt Inst Allg Bot Hamburg*, 23: 777–790.

Mohamed-Saleem, M.A. and Woldu, Z. (2002). Land use and biodiversity in the upland pastures of Ethiopia. In Körner, C. and Spehn, E.M. (Eds.), *Mountain Biodiversity — A Global Assessment*. Parthenon, London.

Morris, C.D., Dicks, H.M., Everson, T.M., and Everson, C.S. (1999). Brotherton Burning Trial: Effects of Treatments on Species Composition and Diversity. Unpublished ARC report, Pietermaritzburg, South Africa.

Stewart, G. (2000). The Maloti-Drakensberg mountains — conservation challenges in a region of international significance. *Journal of the Mountain Club of South Africa*, pp. 147–159.

Tainton, N.M. (1999). *Veld Management in South Africa*. University of Natal Press, Pietermaritzburg, South Africa.

Van Oudtshoorn, F. (1999). *Guide to Grasses of Southern Africa*. Briza Publications, Pretoria, South Africa.

22 Effects of Anthropogenic Disturbances on Biodiversity: A Major Issue of Protected-Area Management in Nepal

Khadga Basnet

INTRODUCTION

Nepal has been giving a high priority to biodiversity conservation for more than three decades by creating and managing protected areas (protected areas), which cover more than 19% of the total area of the country. There are nine national parks, three wildlife reserves, one hunting reserve, and three conservation areas and buffer zones. Most of the protected areas lie along the international borders with China and India (Figure 22.1), and represent mainly two Global 200 ecoregions, as identified by the World Wildlife Fund. These include alpine shrubs and meadows in the mountains and Tarai-Duar savannas and grasslands in the lowland *Tarai* (see Box 22.1). For more than three decades, a large number of management problems, including biological, sociocultural, economical, political, and developmental issues (Basnet 2003a, 2003b), have been encountered in managing these protected areas. Livestock grazing and pasture management proved to be one of the major issues of all the protected areas of Nepal (Richard et al. 2000). The main objective of this chapter is to explore the impacts of anthropogenic disturbances (e.g. livestock grazing and resource harvesting) on the biodiversity of alpine pastures in mountain parks, particularly focusing on:

1. Pasture management and pattern, frequency, and intensity of livestock grazing across the protected areas of Nepal

2. Effects of livestock grazing on wildlife and plant communities

3. Effects of resource harvesting on local species

These important questions from the biodiversity conservation point of view are investigated through literature review, field research, and case studies from the mountain protected areas (e.g. Shey Phoksundo National Park) of Nepal (Basnet 2000; Richard et al. 2000). There are nine mountain protected areas that cover about 21,040 km², including >2,954 km² (Biodiversity Profile Project, 1995a,b; Basnet 2003b) in the alpine shrub and meadow ecoregion. Tarai parks, which cover more than 3,428 km², come under Tarai-Duar savannas and grasslands ecoregion (Box 22.1). In the tarai parks, livestock grazing is strictly prohibited, but in the mountain parks, local residents can use pastures and manage them for their livestock grazing. Nepalese rangelands (grasslands, pastures, and shrub), which are not confined exclusively to the protected areas, usually have high biodiversity as they range from tropical savannas to alpine meadows and even to cold, arid steppes in the north of the Himalayas. Grazing land covers about 1.7 million hectares (12% of the total area of Nepal) and almost 70% of the total grazing land lies above 3,000 m (Table 22.1). Because of difficult terrain and steep slopes, almost 63% of the rangeland forage is not accessible to livestock (His Majesty's Government of Nepal, 1993; Pariyar 1998).

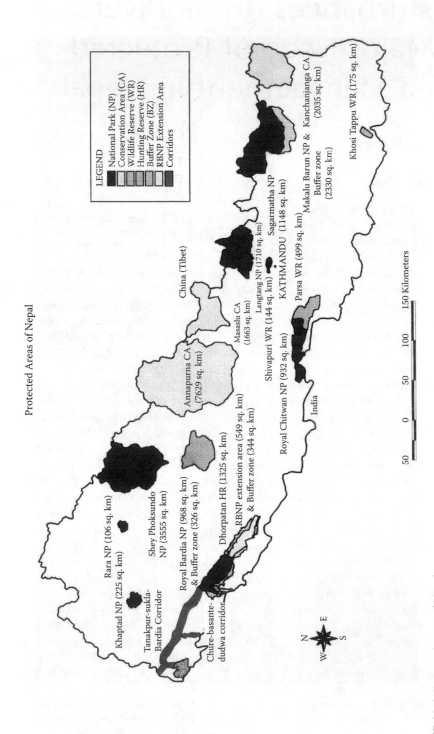

FIGURE 22.1 Protected areas of Nepal.

During the second half of the 1990s, WWF identified more than 240 ecoregions as the world's most unique and biologically representative places, meriting focused conservation. They are called Global 200 ecoregions. Two prominent and distinct ecoregions of Nepal include:

Alpine shrub and meadow: Himalayan alpine belt grasslands or meadows are classified according to their locations, type of vegetation, and diversity. They include: (1) Trans-Himalayan, (2) Northern Himalayan, (3) Western Himalayan, (4) Eastern Himalayan alpine shrub and meadow, and (5) High-altitude cold steppe (Wikramanayake et al. 1998). All the mountain parks belong to the Eastern Himalayan alpine shrub and meadow ecoregion, which is one of the two prominent Global 200 ecoregions that stretches along the northern part of Nepal. It is characterized by: (1) the highest species richness in the Eastern Himalayas, (2) the highest area of endemism, (3) the highest concentration of plants and animals, (4) the highest numbers of globally threatened species, and (5) a large number of natural and anthropogenic threats including landslides, overgrazing, encroachment, illegal harvesting of NTFPs (nontimber forest products), and illegal hunting.

Tarai-Duar savannas and grasslands: The tropical alluvial grasslands, distributed in the lowlands of Nepal and Assam, are the last remnants of a once-extensive ecosystem in southern Nepal and northern India (Lehmkul, 1994; Peet et al. 1997). These ecosystems (known as *Charkose Jhadi* in Nepal) are the tallest grasslands (up to 6-m tall) in the world, and are now confined mainly to protected areas (Bell and Oliver 1992; Peet et al. 1997). Some of the distinguished characteristics of the ecoregion include:

1. A large number of rare, endangered and threatened, and endemic wildlife species together with some habitat specialists.

2. The highest densities of tigers, rhinoceros, and ungulate biomass in Asia (Wikramanayake et al. 1998).

3. Records of more than 100 species of mammals (including endemic species) and about 500 species of birds (including 18 endemic species).

4. High floral diversity with more than 248 grass species and 9 grassland assemblages (Lehmkul 1994, Peet et al. 1997).

5. Migration of megafauna (e.g. elephants) and breeding grounds for endangered bird species (e.g. Sarus crane).

6. Large number of human settlements and disturbances.

These unique habitats, at present, are facing several direct (e.g. deforestation) and indirect (e.g. population growth) threats and ecological degradation throughout their range.

PASTURE MANAGEMENT AND LIVESTOCK GRAZING

PASTURE MANAGEMENT

In Nepal, the highland resource management is closely related to the local religious institutions such as the *gompas* (monasteries) and their administrators, the lamas (priests), and also the villages. Two parallel systems of resource management, the gompa system and the *talukdar* system managed by talukdars (collective name for government revenue collectors such as *jimuwal* for irrigated fields and *mukhiya* for upland fields), are common in high-mountain areas of Nepal. Both of these systems involved community participation in the management process, which followed unwritten rules and regulations formulated by lamas in the past. Parajuli (1998) explored how these two systems operate in one of the highland villages of Shey Phoksundo National Park (SPNP) (Box 22.2). In the gompa system of resource management, all natural resources were managed by a religious hierarchic institution locally known as *dratsang* (a religious committee with different members in a hierarchic system). It is mainly responsible for monitoring and regulating forests, wildlife,

TABLE 22.1
Distribution of rangelands in Nepal according to Land Resources Mapping Project (LRMP), 1986

Ecological Regions	Altitudinal Range (m)	Total Land Area of Nepal		Grazing Land Area of Nepal	
		km²	Percent	km²	Percent
Tarai (tropical zone)	<500	21,220	14.4	496.6	2.92
Churia range (subtropical zone)	501–1,000	18,790	12.7	205.5	1.2
Middle mountains (subtropical and temperate zones)	1,001–3,000	43,503	29.5	2,927.8	17.2
High mountains (subalpine and alpine zones)	3,001–5,000	29,002	19.7	5,071.3	29.8
High Himalayan (nival zone)	Above 5,000	34,970	23.7	8,315.4	48.9
Total		147,484	100.00	17,016.6	100.00

pastures, and livestock grazing, including rotation of herds in seasonal pastures and commercial harvesting of pasture products. Lamas also fix the dates for harvesting medicinal plants, fodder, and grass. Thus, an annual calendar, with specific dates for upward, downward, inward, and outward movements of livestock in different pastures is maintained. Violators of the rules are fined in cash or kind, and the revenue generated is used in religious ceremonies or community development. The talukdar system was introduced in 1911 to collect revenue from various land users. Besides liaising between government and local people, the talukdar had the responsibility of maintaining local security, settling disputes, and controlling and managing agriculture land, forests, and pastures by providing general guidelines based on the traditional rules and regulations made by the lamas. Thus, talukdars actually implemented the lama's system. This system ended in 1996 when the government assigned village development committees (the lowest political unit) to collect local revenues.

LIVESTOCK GRAZING

Animal husbandry is an integral part of the subsistence agriculture in both mountain and tarai. It is the main occupation of more than 68% of the total households of SPNP (King Mahendra Trust for Nature Conservation [KMTNC

2004a]). Literature review and case studies show that livestock grazing is common in both tarai and mountain parks of Nepal (Table 22.2), but the grazing pressure is higher in the Tarai parks for several reasons:

1. Tarai parks are mostly surrounded by human settlements, agriculture land, and degraded forests (BPP 1995c).
2. Grazing areas outside the protected areas are limited, and the frequency of livestock grazing is high.
3. Human population has been growing in the buffer-zone areas, where the resources are limited.
4. There are a large number of unproductive livestock (Dhakal 1995; Shrestha 1998).

In contrast to the tarai parks, local residents are allowed to graze their livestock in the mountain parks, where human population is relatively low with sparsely scattered settlements, and alpine meadows are relatively large. Besides, grazing pattern in the mountain parks is still guided by traditional transhumance systems (e.g. rotational grazing). For example, livestock move upward in summer and reach the maximum altitude (3000 to greater than 5000 m) during July and August and then slowly return to the lower region (< 3000 m). During December and January, livestock grazing is at the lowest altitude of the area.

TABLE 22.2
Protected areas (including their buffer zones) of Nepal showing their altitudinal ranges, areas, and anthropogenic disturbances

Mountain Parks and Buffer Zones	Altitude Range (m)	Protected Area (Km²) Core	Protected Area (Km²) Buffer	Grazing	Competition	MAPs/Grass Harvest	Diversity Loss	Sources
Khaptad National Park	1400–3300	225	NA	X	X	X	X	BPP 1995b, Richard et al. 2000
Langtang National Park	792–7245	1710	420	X	X	X	X	BPP 1995a, Richard et al. 2000
Makalu Barun National Park	435–8463	1500	830	X	X	X	X	BPP 1995a
Rara National Park	2300–4048	106	NA	X	X	X	X	BPP 1995a, Richard et al. 2000
Sagarmatha National Park	2845–8848	1148	275	X	X	X	X	BPP 1995a
Shey Phoksundo National Park	2000–2732	3555	449	X	X	X	X	BPP 1995a, Richard et al. 2000
Shivapuri National Park	1336–2732	144	NA	X	X	X	X	BPP 1995b, KMTNC 2004b
Dhorpatan Hunting Reserve	2850–5500	1325	NA	X	X	X	X	BPP 1995a, Richard et al. 2000
Annapurna Conservation Area	1151–8091	7629	0	X	X	X	X	BPP 1995a, Richard et al. 2000
Kanchenjunga Conservation Area	1200–8586	2035	0	X	X	X	X	BPP 1995a, Richard et al. 2000
Manaslu Conservation Area	2000–8156	1663	0	X	X	U	U	BPP 1995a
Total area of the mountain protected areas		21040	1974					
Tarai Parks and Buffer Zone								
Royal Bardia National Park	152–1441	968	328	XX	XX	XX	X	BPP 1995c, Richard et al. 2000
Royal Bardia National Park Extension Area[a]	153–1247	549[a]	344[a]	XXX	XXX	XX	X	Basnet et al. 1998
Royal Chitwan National Park	150–815	932	750	XX	XX	XX	X	BPP 1995c
Koshi Tappu Wildlife Reserve	75–81	175	173	XXX	XXX	XX	X	BPP 1995c, Richard et al. 2000
Parsa Wildlife Reserve	100–950	499	NA	XX	XX	XX	X	BPP 1995c, Richard et al. 2000
Royal Shuklaphanta Wildlife Reserve	92–270	305	243.5	XX	XX	XX	X	BPP 1995c, Richard et al. 2000
Total area of the Tarai protected areas		3428	1838.5					
Total area of all the protected areas		24468	3812.5					
Total percent of the protected areas		19.2%						

Note: X, XX, and XXX indicate moderate, high, and very high intensities of disturbances, respectively; NA = not available; U = unknown.

[a] Not gazetted.

As a result, although the frequency of grazing is very high, the intensity of grazing is not that serious in comparison to the Tarai parks.

Box 22.2—Shey Phoksundo National Park

The park, gazetted in 1984, is located in Dolpa and Mugu districts of the midwestern development region of Nepal (Figure 22.1). Covering about 3555 km², it is the largest national park, that represents trans Himalayan flora, fauna, and ecosystems in Nepal. With extensive alpine grassland (Figure 22.2) within an elevation ranging from 2000 to 6883 m, SPNP also represents the alpine shrub and meadow ecoregion designated by Global 200 (Wikramanayake et al. 1998). SPNP is home to more than 30 species of mammals (including 3 protected species in Nepal), about 200 avifauna, and 6 reptilian species. More than 35 bird species found in SPNP are internationally important (BPP 1995a, Basnet 1998). Indicator species of the upper region of the park include blue sheep (*Pseudois nayaur* Hodson), snow leopard (*Panthera uncia* Shreber), wolf (*Canis lupus* Linnaeus), and the Himalayan *thar* (*Hemitragus jemlahicus* H. Smith). The lower region of the park includes the common leopard (*Panthera pardus* Linnaeus) and the musk deer (*Moschus chrysogaster* Hodson). The Great Tibetan sheep (*Ovis ammon hodsoni* Blyth), wild yak (*Bos grunniens* Linnaeus), Tibetan antelope (*Pantholopos hodgsoni* Abel), Tibetan wild ass (*Equus kiang* Moorcroft), and the Tibetan gazelle (*Gazella gazella* Pallas) are some of the unique wild fauna that may occur intermittently around the Nepal–Tibetan border of the park (Figure 22.1). More than 407 medicinal and aromatic plants have been reported from the park (Ghimire et al. 2001). Based on species diversity, the park can be divided into three zones: (1) lower zone (below 2800 m), (2) middle zone (2800 to 4500 m), and (3) upper zone (above 4500 m) (see also Basnet 1998). The middle zone is the highest in species richness and habitat diversity. The buffer zone (outside the park) of the park includes more than 449 km² of nine village development committees. There are about 3000 people in the park, 13,000 people in total, and more than 5466 households in the park and buffer zone of Dolpa District alone (Basnet 1998). Subsistence agriculture, animal husbandry, and trade are the main income sources of local people.

IMPACTS OF LIVESTOCK GRAZING IN SPNP

Livestock grazing generates a large number of direct and indirect impacts on park management and wildlife species. Some of the direct effects often cited (e.g. KMTNC 2004a, 2004b, 2004c) include:

1. Competition for forage between livestock and wildlife.
2. Degradation of wildlife habitats and biodiversity loss.
3. Poaching wildlife (e.g. snow leopard, Tibetan wolf) that prey upon livestock.
4. Livestock trampling and killing a large number of wildlife species (e.g. small mammals).

Indirect impacts include:

1. Harvesting medicinal and aromatic plants (MAPs) from the pastures.
2. Transferring diseases and parasites to wildlife species.
3. Soil erosion and compaction. In this section, competition between livestock and wildlife, habitat degradation and biodiversity loss, and harvesting MAPs from the pastures have been presented as case studies.

WILDLIFE–LIVESTOCK COMPETITION

The issue of wildlife–livestock competition was examined in blue sheep (*Pseudois nayaur*) and livestock in SPNP (Basnet 1998, 2003a). The

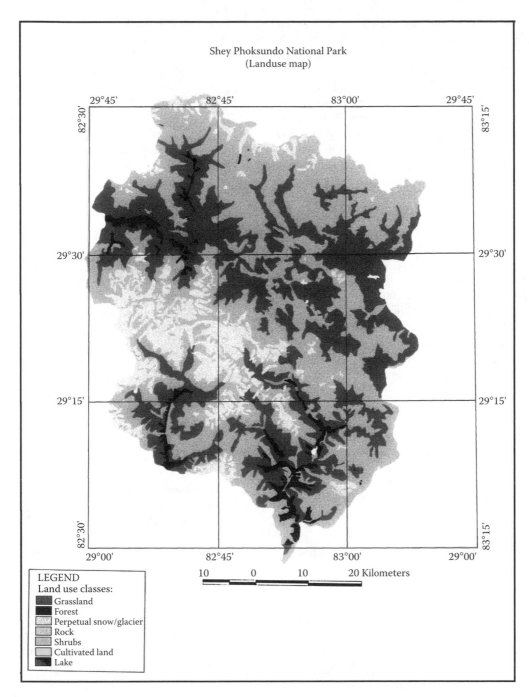

FIGURE 22.2 Land use map of Shey Phoksundo National Park, Dolpa.

study focused on three different but related aspects: (1) productivity and vegetation cover of alpine pastures, (2) status of blue sheep and livestock, and (3) seasonal movement of blue sheep and livestock along the altitudinal gradient and their overlap. Pasture productivity and ground cover were assessed using 12 20 m × 20 m plots and 36 20-m transects in upper Dolpa, (Karang at 4100 m, Pauwa at 5000 m, Shey at 4150 m, and Pericopuwa at 3900 m) and lower Dolpa (Ankhe at 2800 m and Suligad at 2600 m) during a peak period of livestock grazing (Basnet 1998, 2003a). From each plot, a subplot of 20 cm × 20 cm was randomly selected and all the vegetation was clipped and weighed to estimate biomass per unit area of the pastures. Information on the livestock holdings of local people, abundance of both wildlife and livestock and their movement along the altitudinal gradient, spatial and temporal (seasonal) overlapping of wildlife and livestock, and food composition in the park and buffer zone was gathered using the rapid rural appraisal and additional questionnaire survey.

The study showed that:

1. The pastures were productive with standing biomass of 2 to 13 t per hectare, 47 to 66% ground cover by vegetation, and high species diversity with >407 species in all the alpine pastures of Dolpa (Ghimire et al. 2001).

2. Pasture productivity was significantly higher ($p < .05$) in lower Dolpa than in upper Dolpa.

3. The pastures were not overgrazed or degraded except in a few small areas in Shey, Pericopuwa, Pungmo, and Jagadula, where large herds of livestock were kept continuously for several weeks during summer.

Regarding the status of blue sheep and livestock:

1. Blue sheep are widely distributed on the grassy slopes with cliffs above the timberline (3900 m), sometimes descending as low as Ankhe

(2800 m) during winter (Wegge 1979; Yonzon 1990).

2. More than 400 individuals were sighted mostly in the trans-Himalayan pastures such as Naure (4540 m); Shey (4150 m); Yak-yong (4600 m); Angjir, Kagmara, Pungmo (3700 m); Namdo (4082 m); Key (3830 m); and Vijer and Pauwa (4500 m); some of which (e.g. Shey, Pericopuwa, and Kagmara) have been known for stable populations of blue sheep for years (Schaller 1974; Yonzon 1990; Prieme and Oksnebjerg 1992; Richard 1994; Basnet 1998).

3. Livestock density was relatively low with 5466 households, each holding an average of 2.2 cows, 8.9 sheep or goats, 0.2 buffalo, and 1 yak.

4. The livestock number has been decreasing recently due to changing socioeconomic conditions (Dhakal 1998) and increasing frequency of wildlife predation by the Tibetan wolf.

Regarding the seasonal movement and overlap of wildlife and livestock:

1. Blue sheep move to a lower region (as low as 2800 m) of their range during winter and forage on shrubs and forbs that emerged after livestock grazing in summer, whereas the livestock move to even lower regions (below 2800 m), where the winter cold is less severe and the food is relatively abundant.

2. There is some overlapping between blue sheep and livestock during summer and early fall when the grass is relatively abundant.

When there is a scarcity of food, blue sheep spatially separate themselves, escaping to very steep slopes which the majority of livestock cannot reach (Schaller 1973, 1974). Studies on wildlife–livestock interactions in other protected areas of Nepal (e.g. Kanchenjunga Conservation Area) and India (e.g. Pin Valley National Park) also showed similar results

(Bhatnagar et al. 2000; Timilsina and Basnet 2000; Watanabe and Otaki 2002). The study concluded that livestock–blue sheep competition in SPNP was questionable because of the park's stable populations of blue sheep and abundant resources for the last three decades (Schaller 1974; Yonzon 1990; Richard 1994; Basnet 1998) and proposed long-term experimental research to answer the question of competition or facilitation.

BIODIVERSITY LOSS AND HABITAT DEGRADATION

Plant diversity in relation to grazing pressure in three alpine pastures of SPNP was studied by a team of students and instructors of San Francisco State University (Carpenter and Klein 1995). The objective of the study was to gather information about the floristic composition of pastures located at different distances from a major livestock thoroughfare to compare patterns of dominance and species diversity to heavy, intermediate, and low levels of grazing pressure. They selected three pastures:

(1) Roman Campsite, where a large number of livestock graze and pass through, (2) Murwa Meadow, with medium traffic of livestock and grazing, and (3) Norbu Knoll, which is less accessible to livestock (Table 22.3). Species diversity of plants, ground cover, and plant heights were measured using point samples along 20-m transects. Data analysis showed that

1. Plant cover decreased with increased grazing intensity from 94.2% at Norbu Knoll to 77.4% at Roman Campsite.

2. Species of Graminaceae, Primulaceae (*Androsace globifera*), Asteraceae (*Anaphalis triplinervis*), and Plantaginaceae (*Plantago* sp.) families made up 69.4% and 55.5% of total plant cover at the heavy- and medium-grazed sites, Roman Campsite, and Murwa Meadow, respectively, but at Norbu Knoll with less grazing pressure, only two species

TABLE 22.3
Rangeland conditions and frequency of most common taxa in three grazing sites in the Murwa River Valley of Shey Phoksundo National Park

Conditions	Roman Campsite	Murwa Meadow	Norbu Knoll
Grazing intensity	High	Medium	Low
Sample size (N = number of hits)	1092	699	796
Number of taxa	27	33	61
Plant cover	77.5%	85.8%	94.3%
Common taxa	7.4%	9.1%	3.3%
Rare taxa	63.0%	63.6%	68.9%
Diversity (alpha)	5.32	7.49	15.79
Plant height	1.3 ± 1.5	6.8 ± 13.8	9.1 ± 15.4
Dung cover	1.0%	1.3%	0.4%
Frequency of taxa			
Graminaceae	40.5%	22.0%	25.3%
Androsace globifera	—	19.2%	—
Monocot (five star?)	—	14.4%	—
Anaphalis triplinervis	15.0%	—	—
Plantago species	13.8%	—	—
Rhododendron lepidotum	—	—	12.4%
Total cover by dominant species	69.4%	55.5%	37.6%

Source: From Carpenter, C. and Klein, J. (1995).

(*Graminaceae* species and *Rhododendron lepidotum*) comprised 37.6% of plant cover.

3. Species richness was inversely proportional to presumed grazing intensity. For example, 61 taxa were recorded at low intensity of grazing (Norbu Knoll), 33 at medium intensity of grazing (Murwa Meadow), and 27 at high intensity of grazing (Roman Campsite).

4. A log series index (alpha) showed that Norbu Knoll had the highest species diversity (alpha = 15.79) compared to 7.49 in Murwa Meadow and 5.32 in Roman Campsite.

5. Mean plant height varied significantly ($p < .05$) from each site (Table 22.3).

These findings support the intermediate disturbance hypothesis: Low and high levels of grazing intensity result in a decrease in standing crop and species richness of the rangeland community.

HARVESTING PRESSURE

People and Plants Initiative of the World Wildlife Fund (WWF, UK) and UNESCO have been conducting long-term research on conservation of plant resources and community development in SPNP (Ghimire et al. 2000, 2001). Many of the more than 407 MAPs in SPNP (Ghimire et al. 2001) are facing threats from commercial harvesting. One of the main objectives of the ongoing research project is to explore the impacts of harvesting MAPs on the pasturelands. Highly threatened MAP species (*Nardostachys grandiflora, Neopicrorhiza scrophulariiflora, Dactylorhiza hatagirea, Delphinium himalayai,* and *Jurinea dolomiaea*) of high economical and ecological values (Box 22.3) were selected as study units for monitoring. Different pastures growing these species were selected at an altitude of 3900 to 4300 m inside the park and at 3763 to 4270 m in the buffer zone and 20 5 m × 5 m plots, each subdivided into 25 1 m × 1 m subplots were established and marked permanently. Seven plots inside the park were chosen for experimental harvesting

of *N. grandiflora,* and *N. scrophulariiflora.* Using Latin square and randomized block designs, these plots were assigned for five different levels of harvesting with five replicates: (1) no harvesting, (2) 25% harvesting, (3) 50% harvesting, (4) 75% harvesting, and (5) 100% harvesting of *N. scrophulariiflora.* In case of *N. grandiflora,* the five levels of harvesting were: (1) no harvesting, (2) 10% harvesting, (3) 25% harvesting, (4) 50% harvesting, and (5) 75% harvesting. Within each subplot, plants were tagged and their growth, mortality, reproductive capacity, and phenology were recorded at regular intervals throughout a year. The remaining plots, both in the park and buffer zone, were located along a gradient of different human pressure to assess structure and population dynamics of all the selected species (Box 22.3) in natural conditions. This experiment showed that

1. There was a general trend of decreasing ramet (vegetative offshoot) recruitment and survivorship with the increasing intensity of harvesting.

2. A moderate harvesting intensity (25% harvesting) had the least impact on ramet density of *N. scrophulariiflora* (in case of *N. grandiflora,* 10% harvesting had the least impact).

3. Beyond these minimum levels of harvesting, there was a reduction in the rate of ramet recruitment.

4. *N. globiflora* was more vulnerable to harvesting than *N. scrophulariiflora.*

A similar trend was observed in the population dynamics of these species in the buffer-zone plots. The outcomes of the experimental and observational assessment in the park and buffer zone suggested that commercial harvesting has been a major problem in protecting species and managing alpine pastures (Ghimire et al. 2001).

Currently, 23 different medicinal and aromatic products of more than 21 plant species are traded from Dolpa District. This has caused heavy pressure on local pasturelands and MAPs. The district forest office records showed that commercial harvesting for export has been increasing every year since 1992. For example, 5 t of raw dry products commercially extracted from SPNP and its buffer zone in 1992–1993, increased to 9 to 12 t in 1995–1996, and 12 t in 1996–1997 (Shrestha et al. 1998). In 1997–1998, this amount reached more than 37.8 t (Shrestha et al. 1998; Ghimire et al. 2001). Because of the change in government policy from 2001, this amount has been multiplied several times. Besides, illegal harvesting by traders and local herders may also account for another significant portion of the total MAPs extracted from SPNP pastures.

Box 22.3—Major MAP Species Used in the Study

Some MAP species, which are highly threatened due to their high economic values (e.g. phytochemical and ethnobotanical) and unsustainable harvesting, were selected for monitoring. They included:

1. *Nardostachys grandiflora* DC. It is a perennial herb with woody rhizome growing 10 to 60 cm in alpine and subalpine regions (3200 to 5000 masl). Morphological variation is observed in different habitats. It is a clonal plant that spreads by multiplication of ramets (vegetative offshoot) and also by seeds. The plant has a restricted distribution above 4000 m in SPNP.
2. *Neopicrorhiza scrophulariiflora* Pennel. It is a perennial herb with elongated rhizome occurring in alpine and subalpine regions (3500 to 4800 masl). It is also a clonal plant that spreads by multiplication (multiplication is more rapid than in *N. grandiflora*) of ramets and also by seeds. The plant has a restricted distribution above 4000 m in SPNP. Ramets composing clonal populations of both *N. grandiflora* and *N. scrophulariiflora* undergo recruitment and mortality throughout the growing season. Both of them were subjected to experimental treatments

(i.e. different intensities of harvesting) during 1998 and 2000 to understand the impacts of harvesting.
3. *Dactylorhiza hatagirea* D. Don, *Delphinium himalayai* Munz and *Jurinea dolomiaea* Boiss were monitored to study their structure, demography, and population dynamics without experimental treatments.

CONCLUSION

Livestock grazing and pasture management is a complex process that generates biological, socioeconomical, and political issues. Three case studies from SPNP, a representative of mountain parks of Nepal showed that:

1. Wildlife–livestock competition was not as serious as has been perceived.
2. Livestock grazing intensity was inversely related to the number of plant species.
3. Commercial harvesting (including collection by herders) of MAPs beyond minimum levels (e.g. 10% in the case of *N. grandiflora* and 25% of *N. scrophulariiflora*) seemed to be the main factor in degradation of pastures and biodiversity loss.

Although these results are compatible with the findings from other Himalayan protected areas (Bhatnagar et al. 2000; Timilsina and Basnet 2000; Watanabe and Otaki 2002; Basnet et al. 2003), they cannot be generalized because livestock grazing and other conditions (e.g. ecological, sociocultural, and location) vary across the mountain protected areas of Nepal (Table 22.2). This is a justification for long-term integrated research and a holistic approach of biodiversity conservation that involves local participation, techniques, and resources. Emerging conservation approaches (e.g. buffer-zone development, landscape-level conservation, and transboundary conservation) with such promises have been reflected in the recent management plans of mountain protected areas of Nepal (KMTNC 2004a, b, c). Success depends on their proper implementation.

SUMMARY

Nepal has been giving a high priority to biodiversity conservation for more than three decades, by creating and managing 16 protected areas (national parks, wildlife reserves, hunting reserves, conservation areas, and their buffer zones), which cover more than 19% of the total area of the country. During this period, a large number of management issues have been encountered in managing these protected areas, human disturbance being significant in all. Most protected areas lie along the international boundaries with China and India and represent mainly two ecoregions: alpine shrub and meadow in the mountains and Tarai-Duar savannas and grasslands in the lowland tarai. The main objective of this chapter was to explore the impacts of anthropogenic disturbances (e.g. livestock grazing and resource harvesting) on biodiversity of alpine pastures through literature review, field surveys, and case studies. Specifically, it focused on: (1) pasture management and pattern, frequency, and intensity of livestock grazing across the protected areas, (2) effects of livestock grazing on wildlife and plant communities, and (3) effects of resource harvesting on local species. All the case studies come from Shey Phoksundo National Park, which, at an altitude of 2000 to 6883 m, shelters more than 26 mammalian species, 200 bird species, 6 herpetofauna species, and 407 medicinal and aromatic plants. Qualitative and quantitative assessments and case studies show that:

1. The local system of pasture management has been disappearing.

2. Livestock grazing is common in both tarai and mountain parks but the grazing pressure is significantly higher in the tarai parks.

3. Blue sheep and livestock competition is not obvious because the availabilities of space and food are not limited during their overlapping.

4. Heavily grazed pastures have significantly lower plant cover (77.5%) and species diversity (alpha = 5.32) than moderately grazed pastures, which had 94.3% of plant cover and 15.8 species diversity.

5. Commercial harvesting beyond minimum levels (10% in *N. grandiflora* and 25% in *N. scrophulariiflora*) showed significantly lower density, recruitment, and survivorship of the plants.

A community-based participatory approach of protected-area management is a possible option to minimize these human disturbances. Nepal has already initiated this approach adopting buffer-zone development programs, landscape-level conservation, and transboundary conservation, which have reflected in recent management plans of the protected areas.

ACKNOWLEDGMENTS

WWF Nepal Program supported the project. Department of National Parks and Wildlife Conservation and SPNP provided permission and field logistics for the study, and SPNP staff extended their support in the field. Chris Carpenter, Suresh Ghimire, Camille Richard, Eva Spehn, and two anonymous reviewers provided comments on the previous version of the manuscript. I am grateful to all of them.

References

Basnet, K. (1998). Biodiversity Inventory of Shey Phoksundo National Park: Wildlife Component. Publication Series No. 34. WWF Nepal Program, Kathmandu, Nepal.

Basnet, K. (2000). Representation of grassland ecosystems in the Himalayan ecoregions. In Richard, C., Sah, J.P., Basnet, K., Karki, J.B., and Raut, Y. (Eds.). *Grassland Ecology and Management in Protected Areas of Nepal*, Vol. I. International Centre for Integrated Mountain Development, Kathmandu, Nepal, pp. 7–14.

Basnet, K. (2003a). Wildlife-livestock competition: a major issue in park management? In *Proceedings of International Seminar on Mountains (March 6-8, 2002)*, Royal Nepal Academy of Science and Technology, Kathmandu, Nepal, pp. 59–68.

Basnet, K. (2003b). Transboundary biodiversity conservation initiative: An example from Nepal. *Journal of Sustainable Forestry*, 17: 205–226.

Basnet, K., Shrestha, K.M., Sigdel, E.R., and Ghimire, P. (1998). Royal Bardia National Park-Extension Area. A Biodiversity Inventory. WWF Nepal Program, Kathmandu, Nepal.

Basnet, K., Dhakal, D.P., and Thapa, S.B. (2003). Sagarmatha National Park Management Issues. Department of National Parks and Wildlife Conservation and UNDP, Nepal.

Bell, D.J. and Oliver, W.L.R. (1992). Northern Indian tall grasslands: management and species conservation with special reference to fire. In Singh, K.P. and Singh, J.S. (Eds.), *Tropical Ecosystems: Ecology and Management*. Wiley Eastern Limited, New Delhi, India, pp. 109–123.

Bhatnagar, Y., Raut, G.S., Jhonsingh, A.J.T., and Stuwe, M. (2000). Ecological separation between ibex and resident livestock in a trans-Himalayan protected area. In Richard, C., Basnet, K., Sah, J.P., and Raut, Y. (Eds.). *Grassland Ecology and Management in Protected Areas of Nepal*, Vol. III. International Centre for Integrated Mountain Development, Kathmandu, Nepal, pp. 70–84.

Biodiversity Profiles Project (BPP) (1995a). Biodiversity Profile of the High Himalaya/High Mountains Physiographic Zone. Department of National Parks and Wildlife Conservation, Kathmandu, Nepal.

BPP (1995b). Biodiversity Profile of the Midhills Physiographic Zone. Department of National Parks and Wildlife Conservation, Kathmandu, Nepal.

BPP (1995c). Biodiversity Profile of the Terai/Siwalik Physiographic Zone. Department of National Parks and Wildlife Conservation, Kathmandu, Nepal.

Carpenter, C. and Klein, J. (1995). Plant Species Diversity in Relation to Grazing Pressure in Three Alpine pastures, Shey Phoksundo National Park, Dolpa district Nepal. Publication Series No. 20, WWF Nepal Program, Kathmandu, Nepal.

Dhakal, N.H. (1995). Socio-Economic Survey of Royal Bardia National Park and Buffer Zone. WWF Nepal Program, Kathmandu, Nepal.

Dhakal, N.H. (1998). Baseline survey report of Northern Mountains Area Conservation Project in Shey Phoksundo National Park Area. WWF Nepal Program, Kathmandu, Nepal.

Ghimire, S.K., Lama, Y.C., Gurung, T.N., and Thomas, Y.A. (2000). Conservation of Plant Resources, Community Development and Training in Applied Ethnobotany at Shey Phoksundo National Park and Its Buffer Zone, Dolpa. Publication Series No. 40. WWF Nepal Program, Kathmandu, Nepal.

Ghimire, S.K., Lama, Y.C., Tripathi, G.R., Schmidt, S., and Thomas, Y.A. (2001). Conservation of Plant Resources, Community Development and Training in Applied Ethnobotany at Shey Phoksundo National Park and Its Buffer Zone. Publication Series No. 41. WWF Nepal Program, Kathmandu, Nepal.

His Majesty's Goverment of Nepal (HMG) (1993). Livestock Master Plan. His Majesty's Government of Nepal, National Planning Commission, Asian Development Bank, and Agriculture Projects Service Centre, Kathmandu, Nepal.

KMTNC (2004a). Shey Phoksundo National Park Management Plan. King Mahendra Trust for Nature Conservation, Jawalakhel, Lalitpur, Nepal.

KMTNC (2004b). Shivapuri National Park Management Plan. King Mahendra Trust for Nature Conservation, Jawalakhel, Lalitpur, Nepal.

KMTNC (2004c). Rara National Park Management Plan. King Mahendra Trust for Nature Conservation, Jawalakhel, Lalitpur, Nepal.

Lehmkul, J.F. (1994). A classification of subtropical riverine grassland and forest in Chitwan National Park, Nepal. *Vegetatio*, 111: 29–43.

LRMP (1986). Land Resource Mapping Project. His Majesty's Government of Nepal and Government of Canada.

Parajuli, D.B. (1998). Indigenous System of Pasture Resource Management in Kunasa Area within Shey Phoksundo National Park. WWF Nepal Program, Kathmandu, Nepal.

Pariyar, D. (1998). Rangeland Resource Biodiversity and Some Options for Their Improvements. National Biodiversity Action Plan, Kathmandu, Nepal.

Peet, N.B., Watkinson, A.R., Bell, D.J., and Brown, K. (1997). The Management of Tall Grasslands for the Conservation of Biodiversity and Sustainable Utilization. Department of National Parks and Wildlife Conservation and University of East Anglia.

Prieme, A. and Oksnebjerg, B. (1992). Field Study in Shey Phoksundo National Park: Expedition Snow Leopard 1992. Department of National Parks and Wildlife Conservation, Kathmandu, Nepal.

Richard, C. (1994). Natural Resources Use in Protected Areas of the High Himalaya: Case Studies from Nepal. IOF Project Technical Paper.

Richard, C., Basnet, K., Sah, J.P., and Raut, Y. (2000). *Grassland Ecology and Management in Protected Areas of Nepal*. Vol. II and III, ICIMOD, Kathmandu, Nepal.

Schaller, G.B. (1973). On the behavior of Blue Sheep (*Pseudois nayaur*). *Journal of Bombay Natural History Society*, 69: 523–537.

Schaller, G.B. (1974). A Wildlife Survey of the Shey Gompa Area. *New York Zoological Society*, New York.

Shrestha, J.N. (1998). Royal Bardia National Park: Socio-Economic Assessment and Gender Analysis. WWF Nepal Program.

Shrestha, K.K., Ghimire, S.K., Gurung, T.N., Lama, Y.C., and Aumeeruddy, Y. (1998). Conservation of Plant Resources, Community Development and Training in Applied Ethnobotany at Shey Phoksundo National Park and Its Buffer Zone. Publication Series No. 33, WWF Nepal Program, Kathmandu, Nepal.

Timilsina, L.P. and Basnet, K. (2000). A case study of blue sheep in Kanchenjunga. In *Proceedings of the International Symposium on the Himalayan Environment*. Tribhuvan University and Hokkaido University, Kathmandu, Nepal, pp. 188–194.

Watanabe, T. and Otaki, Y. (2002). Study for conservation of Blue sheep (*Pseudois nayaur*) in the Kanchenjunga conservation area, Eastern Nepal: Interaction between Blue sheep and other animals. In *Proceedings of International Seminar on Mountains (March 6–8, 2002)*, Royal Nepal Academy of Science and Technology, Kathmandu, Nepal, pp. 69–78.

Wegge, P. (1979). Aspects of the population ecology of blue sheep in Nepal. *Journal of Asian Ecology*, 1: 10–20.

Wikramanayake, E.D., Dinerstein, E., Loucks, C., Wettengel, W., and Allnut, T. (1998). *A Biodiversity Assessment and Gap Analysis of the Himalayas*. WWF-US and UNDP.

Yonzon, P. (1990). The 1990 wildlife survey of Shey Phoksundo National Park, Dolpa, West Nepal. Department of National Parks and Wildlife Conservation, Kathmandu, Nepal.

23 Agricultural Development and Biodiversity Conservation in the Páramo Environments of the Andes of Mérida, Venezuela

Maximina Monasterio, Julia K. Smith, and Marcelo Molinillo

INTRODUCTION

The páramo of the northern Andean highlands is one of the most important biogeographic regions of the Andes. Extending insularly around the equator (11° N to 8° S), it protects the headwaters of the high catchments that drain toward the Pacific, i.e. the Orinoco and Amazon rivers, and plays a fundamental role in the stability of the highlands.

Moreover, the páramo is characterized by a high biological landscape and cultural diversity, a biota unique in its adaptations, its hydrological environmental services, and its large potential for tourism (Monasterio and Celecia, 1991). In fact, the páramo has been assigned high global priority for conservation (Biodiversity Support Program, 1995). However, in the tropical Andean countries of Venezuela, Colombia, and Ecuador, there is a growing need for new agricultural land due to the high demand in the national markets for crops cultivated exclusively in cold mountain environments, as well as due to the growing population. The ongoing advancement of the agricultural frontier in each of the three countries has led to alarming annual losses of natural páramo areas (Hess, 1990; Verweij, 1995; Drost et al., 1999).

In tropical mountain environments, in which the high biodiversity is endangered by dynamic processes of transformation and degradation, meeting the production needs of the local population while protecting fragile ecosystems represents a challenge for the design of effective conservation strategies. This chapter addresses the relationship between agricultural dynamics in the páramo region and the conservation of areas providing fundamental hydrologic services for agriculture. The conceptual and methodological approach used is based on the analysis of three spatial scales, all of them integrated through water as an environmental service of the high páramo. The scale integration approach departs from the premise that the natural páramo ecosystems in which agroecosystems are embedded are components of a complex production system as important as the crop production systems themselves. The ecological functions of the páramo determine the stability of these environments transformed by agricultural processes (Monasterio and Molinillo, 2002). In this context, the high páramo belt is interpreted as an area for the conservation of biodiversity, water capture, and hydrological equilibrium of outflow rates.

This chapter aims at analyzing the relationships between agricultural dynamics and the conservation of páramo ecosystems, especially the wetlands of marshes and swards, areas of origin, and accumulation of the water used for páramo agriculture in the central core of the Cordillera de Mérida (Venezuela) above 3000 m (Figure 23.1). The present chapter has

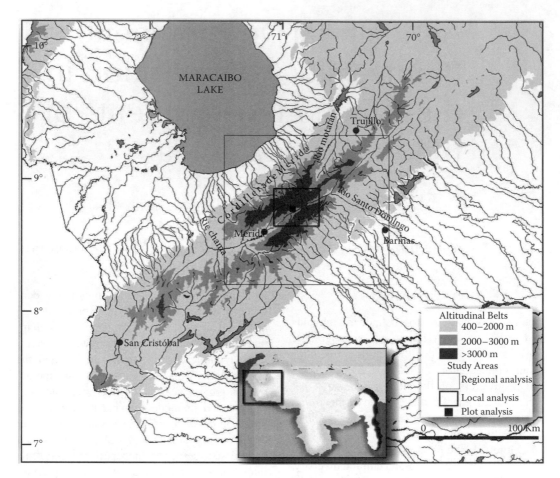

FIGURE 23.1 Study area in the Cordillera de Mérida (western Venezuela). The dense hydrographic network arising in this region flows toward the Orinoco River and the Maracaibo Lake catchments. The páramo belt, where the main watercourses originate, spatially corresponds with land above 3000 m asl. The large rectangle is the perimeter of the regional study, the small rectangle circumscribes the central páramo core, which was analyzed at a local scale, and the small black square indicates the study area at a plot scale.

been developed within the framework of the project called "Ecological and Social Sustainability of Agricultural Production in the Cordillera de Mérida: Flow of Environmental Services of the High Andean Páramos for Potato Agriculture." Project sponsored by Universidad de Los Andes, Venezuela. It analyzes the importance of environmental services for production in the agricultural zone.

METHODS

Information gathering and analysis are carried out at three spatial scales: regional, local, and plot scale. In Figure 23.2, a diagram of the scales considered and their integration through

water as a service of the páramo is presented. On a regional scale, the spatial distributions of grass páramo and wetland páramo have been surveyed using a Landsat 7 image (006-054) established in January 2001. Based on this information, the life zones with a potential to supply other zones with water were identified.

The main páramo agricultural area in the upper basin of the Chama River was identified and analyzed. Areas with fallow and intensive agriculture as well as the zones within the páramo where the water sources used by each community for agriculture are located were mapped (marshes and swards). The analysis was based on processing the orbital spectral image Spot 4 651-332 (© CNES 2001, Spot

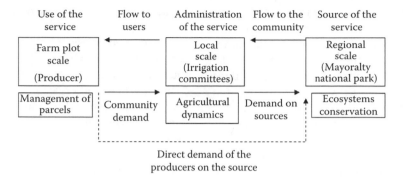

FIGURE 23.2 Simplified diagram of the links between scales that constitute the transversal theme of water as an environmental service of the páramo. The integration of scales allows a dynamic approach to the relationships between priority issues for conservation and development for the different actors and spaces. From Monasterio, M. and Molinillo, M. (2002).

Image S.A.). To classify the agricultural zone, the NDVI (Normalized Difference Vegetation Index) was used first and then pattern filter analyses. Based on pixel heterogeneity, three levels of land use intensity were derived: (1) fields of intensive agriculture (high-pixel heterogeneity PHG), (2) fallow farming (medium PHG), and (3) abandoned fields (low PHG).

On a plot scale, the vegetation of 20 marsh and sward sites was recorded in El Banco watershed. These were sampled using random quadrats. In each 1-m^{-2} quadrate, the number of species was determined, and vegetation, dry matter, and dung cover (grazing indicator) were measured; soil moisture was also determined. In each sward area, between 10 and 20 quadrats were established, and the following variables determined: altitude, slope, stoniness, forage supply, and distance to water sources. In addition, farmers were interviewed to analyze their grazing strategies, agricultural practices, and private and communal water management.

RESULTS AND DISCUSSION

PÁRAMO ECOLOGICAL BELTS AND WATER SUPPLY

The páramo of the Cordillera de Mérida extends above 3000 m asl. The altitudinal gradient is related to climatic and land use gradients that constitute true ecological belts at different altitudes. Within this altitudinal zoning, the following levels can be identified (from low to high altitude): the Andean belt, the high-Andean belt, and the periglacial belt.

The analysis on a regional scale shows that the high-Andean and periglacial belts are indispensable sources of water to maintain páramo agriculture.

The upper limit for agriculture is confined to the high-Andean belt (3000 to 4000 m). The natural ecosystems above the present agricultural frontier have a colder periglacial climate and a sparser vegetation cover with giant Andean rosettes of the *Espeletia* genus. At these altitudes, land use is limited to seasonal low-intensity grazing in the bottom of glacial valleys covered by marshes and swards, which have a large potential for tourism.

In the periglacial belt (4000 to 4800 m), the cycle of nightly freezing and daily thaw prevents all forms of agricultural activity because of the recurrent frosts. Two plant formations colonize this environment: the desert páramo and the periglacial desert (Monasterio, 1980a). The flora, with a high number of endemic plants, has developed strategies and spectacular life-forms (giant rosettes of *Espeletia* and acaulescent cushions of different genera) that help to reduce solifluction caused by the daily freeze–thaw cycles and stabilize the mobile soils in these areas. The formation of microterraces by these species decreases soil displacements produced by cyroreptation processes. In this context, topographic and climatic conditions and a biota adapted to extreme hydric and thermic stress constitute key elements of a unique and fragile environment, which cannot be directly utilized because of its harsh climate, low productivity, and high erosive potential.

These páramo environments are the headwaters of three important river basins: the Chama, Santo Domingo, and Motatán, three large hydrographic networks originating from superficial drainage in the highest páramos and from the approximately 400 glacial lagoons above 3800 masl. Water resources coming from the páramo are essential for intensive, year-round cultivation of potatoes, vegetables, and more recently, garlic.

AGRICULTURE, GRAZING AND WATER RESOURCES

At a local scale, the analysis dealt with the agricultural zone and the páramo ecosystems of the high Chama River watershed. The agricultural zone between the Sierra Nevada and Sierra La Culata is the main area benefiting from water generated by this watershed. This is an area of about 8400 ha extending along the Chama River valley and the transversal inter-Andean valleys that feed into it. On the one hand, the valley bottoms tend to be occupied by intensive agriculture with irrigation and highly productive crops (roots, vegetables, flowers, etc.), replacing the natural páramo ecosystem (especially in relatively flat areas of high soil fertility). On the other hand, fallow agriculture, generally without irrigation, is practiced in the steeper valley slopes at the upper limit of the agricultural belt (above 3400 m). Fallow agriculture is characterized by alternating crop and fallow cycles. Because of this management practice, uncultivated areas go through a successional process, which tends toward the regeneration of the natural ecosystem (Sarmiento et al., 1993). Hence, the agricultural frontier appears as a gradual transition zone between successional mosaics and the natural ecosystem.

On a local scale, there are two landscape elements, glacial lagoons and peatlands, which represent the most important water sources and water retention basins. Most lagoons are of glacial origin and relatively small. They are scattered mainly above 3800 masl, where glaciers were more active during the Pleistocene era. Together, they occupy more than 300 ha and constitute the main storage compartment for surface waters. Closely linked to lagoons are the marshes covered by tussock grass and sward vegetation, which occupy more than 3000 ha between 3600 and 4200 masl. They act as true sponges, dampening runoff because of their deep soils, high soil organic matter levels, and irregular microrelief. These units are the most important temporal water reservoirs (approximately 70% of total storage capacity), which discharge slowly during the dry season. Because of a permanent water supply and the presence of tender forage, marshes and swards are the main areas where cattle grazing is concentrated (Molinillo and Monasterio, 1997a). Farmers are currently discussing strategies to protect vegetation and soils in these marshes and swards to reconcile these practices with cattle raising. Hence, analyzing the spatial location and conservation status of these wetlands in relation to grazing impacts is essential for a sustainable management of irrigation water.

AGRICULTURAL INTENSIFICATION AND SUCCESSIONAL DYNAMICS IN THE HIGH-ANDEAN PÁRAMO

The traditional land use system of the páramo is based on a shifting cultivation, with short periods of crop production alternating with longer fallow periods. After one or two cultivation cycles, plots are abandoned, and successional processes produce a trend toward a regeneration of natural páramo vegetation (Sarmiento et al., 2003). The use of the fallow not only favored soil fertility restoration (Sarmiento et al., 1993), but also had a positive effect on water dynamics, maintaining soil humidity, increasing organic matter contents, and decreasing runoff and soil loss (Sarmiento, 2000). This type of management has been the basis for the maintenance of a sustainable agriculture in these communities.

The decrease in importance or abandonment of the fallow strategy has been an important trend in the evolution of land use strategies in the last few decades. Especially since the beginning of irrigation in the late 1960s and up to the present, large areas with traditional land use of fallow agriculture were abandoned or transformed into intensive agriculture (Velásquez, 2001). The other important trend in this process of evolution, agricultural intensification, has been of such magnitude that it shows in the positive correlation between cultivated area and potato production between the late 1950s and late 1990s (Figure 23.3). The leap in productivity from the 1970s was fundamentally determined by an increase in irrigated area and the conversion of fallow agriculture into intensive agriculture.

Plot management under intensive agriculture has translated into an increase in water demands, related to an increase in the number of crops per year and surface runoff. In this last decade, the introduction of garlic cultivation in the inter-Andean valleys has further increased water demands. Traditional crops were substituted with garlic. The advance of garlic and substitution of traditional crops happened in those plots with access to an established irrigation system. These changes are not only related to an increase in water use, but also to an increase in the private capture of water resources against communal use and the gradual drop in recent years in water levels of glacial lagoons. They have also produced a decrease in the carrying capacity for grazing in the agricultural belt and an increase in the permanence time of cattle in the high-Andean belt.

These direct influences of agricultural dynamics on high-Andean grazing are part of an agropastoral strategy developed with the introduction of livestock husbandry as an integral component of wheat production technology instigated at the time of the Spanish conquest (Monasterio, 1980b). Since then, transhumance (animal movements), especially of cattle, and their permanence in the different altitudinal and vegetation belts has been determined largely by the agricultural calendar (Molinillo and Monasterio, 1997a). Animals were brought down from their grazing areas in the high-Andean belt not only because of the need for traction in plowing but also to complement their diet with cultivated forage and wheat stubble during the dry season. In the last decades, potato intensification, the introduction of garlic, and the decrease in importance and ultimately disappearance of wheat cultivation in the páramo have meant the loss of natural pasture and stubble fields for animals in the agricultural belt (Figure 23.4). As a direct

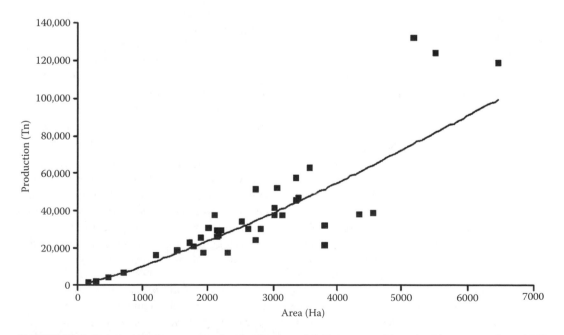

FIGURE 23.3 Relationship between production levels and surface area cultivated with potatoes from 1957 to 1997 in the Mérida state (the one with the highest páramo surface area and potato production in Venezuela). $R^2 = 0.9041$. Data source: MAC (Venezuela Ministry of Agriculture), 1958 to 1997.

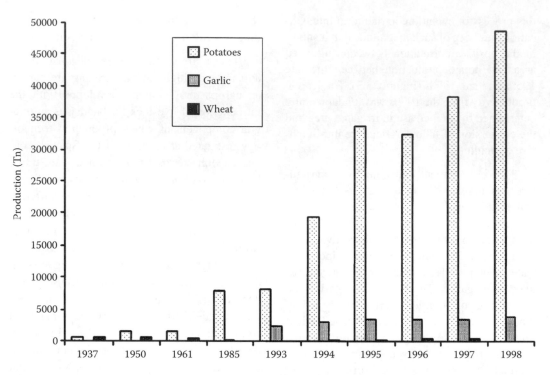

FIGURE 23.4 Production levels for potatoes, garlic, and wheat in the Rangel county (one of the counties with the highest páramo surface area and potato production in Mérida state) from 1937 to 1998. Data source: Velásquez, N. (2001), and Venezuelan Ministry of Agriculture, 1995 to 1998.

consequence, although not necessarily because of an increase in animal numbers, grazing has intensified in the swards and marshes of the high-Andean belt. The decrease in regeneration time of the large sward patches, which are grazed in rotational fashion, has accelerated the successional mechanisms through which swards are degraded (Figure 23.5).

Even though there has been a clear impact from cattle rearing, species richness in the swards of El Banco watershed (Chama River affluent) does not show a clear correlation with grazing intensity (Figure 23.6). This could be largely because of the fact that a large proportion of the increase in native weed species (*Acaulimalva acaule, Aciachne pulvinata, Geranium* sp., etc.) and exotics (*Rumex acetosella, Taraxacum officinalis*, etc.) during succession is counterbalanced by a fall in the number of native forage species (of the genera *Calamagrostis, Carex, Muehlenbergia, Agrostis*, etc.). These results seem more evident when the cover of good forage species as compared

to weeds is analyzed (Figure 23.7). Similar results were recorded for swards in the Colombian páramo (Pels and Verweij, 1992; Verweij and Budde, 1992; Verweij, 1995), and they agree with the model of Milchunas et al. (1988) on the effects of grazing in environments without a history of grazing: the large impact on vegetation physiognomy and the markedly increased potential of weed colonization.

In present environments with swards in the valley bottom of El Banco watershed, tussock grass species (species of the genera *Calamagrostis, Agrostis*, and *Festuca*) have been relegated to areas inaccessible to cattle because of water logging (e.g. *Calamagrostis toluencis*). They also colonize experimentally fenced off areas (Molinillo, 1992) where cattle were excluded (e.g. *Sporobolus tenuissinus*). A moderate to low stocking density favored short grass species (e.g. *Calamagrostis coarctata, Carex albolutescens, C. humboldtiana, Muehlenbergia ligularis,* and *Eleocharis acicularis*), which today dominate swards and marshes but which

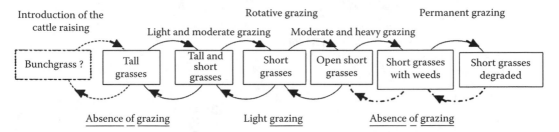

FIGURE 23.5 Simplified diagram of grass species succession under grazing in dry páramos. Successional cycles are controlled by the relationship between agriculture and cattle grazing. (From M. Molinillo and M. Monasterio, 2002.)

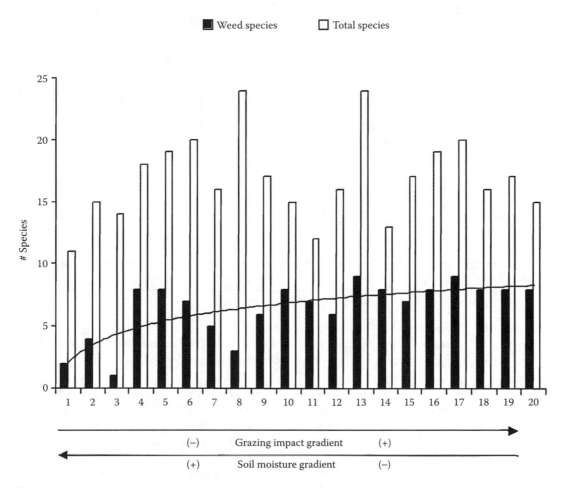

FIGURE 23.6 Number of weeds and total species richness in 20 units of marshes and swards as grazing impact increases and soil moisture levels decrease in glacial valley bottoms between 3800 and 4100 m in the El Banco watershed (Sierra La Culata, Cordillera de Mérida). The logarithmic relationship ($R^2 = 0.5206$) is for the number of weed species (both native and introduced).

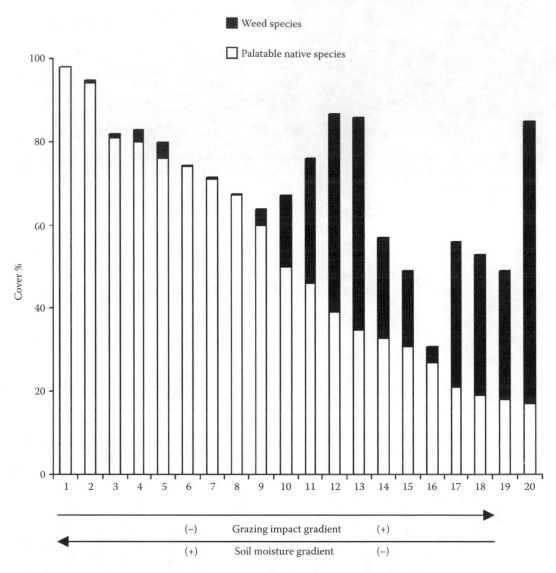

FIGURE 23.7 Changes in cover of palatable native and weed species (both native and introduced) in 20 units of marshes and swards as grazing impact increases and soil moisture levels decrease in glacial valley bottoms between 3800 and 4100 m in the El Banco watershed (Sierra La Culata, Cordillera de Mérida).

decrease in importance under heavier loads or larger animal permanence times. The appearance of species considered as native colonizers (e.g. *Agrostis breviculmis, Aciachne pulvinata, Acaulimalva acaule, Geranium* sp., *Lachemilla pinnata,* etc.) and exotics (*Rumex acetosella* and *Taraxacum officinalis*) has always been the case under these heavy grazing conditions.

The equilibrium in species composition and diversity in páramo wetland environments, controlled to a large extent by grazing frequency and intensity, have been strongly influenced by

agricultural trends during the last decades in the lower belts. In the El Banco watershed, potato agriculture intensification and the introduction of garlic have been fast processes that, together with the decrease and loss of wheat cultivation, led to a decrease in the carrying capacity of the agricultural belt. The consequent increase in animal permanence times in the high-Andean belt, has increased grazing impact on precisely the areas that have the capacity to store more water in these páramos. Because species composition and richness depend on grazing intensity (animal numbers and permanence times on each wetland) and on soil moisture levels, the decrease in the number of native forage species and the increase in weeds can be a good indicator of the conservation status and water storage capacity of each sward unit.

SCALE INTEGRATION: RURAL DEVELOPMENT AND WETLAND CONSERVATION IN THE HIGH-ANDEAN BELT

The analysis at different spatial scales shows the tight relationship between the páramo as water sources and crop production in the agricultural belt. In this tight link, other important environmental services for agriculture are also involved, such as erosion control through soil retention, phytomass accumulation as green manure in fallow agriculture, special microclimatic conditions within slopes that decrease the incidence of crop pests, and tender forage from the valley bottom pastures for extensive cattle grazing.

The agricultural páramo region in the Cordillera de Mérida strongly depends on water supplies from the high-Andean and periglacial belts. Within the Chama River watershed, this means that most of the 8,450 ha in the agricultural zone, especially potato and garlic cultivation areas, depend on water inflow of approximately 23,700 ha of páramo environments, and especially so from the 150 ha of lagoons and 2,570 ha of swards and marshes (Monasterio et al., 2003). These last two store 70% of the water in all surface water compartments, even though they only represent 7.9% of the total surface area of the watershed. In addition, these environments constitute the most important areas for grazing cattle. Extrapolating the data from the El Banco subwatershed, we estimate that in the long run, these wetland habitats could support between 1,500 and 2,000 head of cattle under moderate to light grazing loads. This equals probably more than 80% of the cattle currently grazing in the high Chama river watershed. A large number of tussock grassland species are only found in areas of low accessibility for grazing, and the native forage species favored by medium to low grazing loads are in an equilibrium state that depends on agricultural dynamics and animal permanence times

in the high-Andean belt. Integrating data obtained at different spatial scales, it is clear that the tendencies for agricultural intensification (changes in the crop types and dynamics) not only produce a large increase in water demands and a consequent increase in pressure on water sources in the páramo but also advance the environmental degradation of wetlands.

On a plot scale, the increase in plots with little or no fallow times (i.e. the fallow strategy has been abandoned) and the introduction of crops with higher water demands generate a change in agricultural dynamics and water use, reflected in new trends toward more private management and less communal administration. At a regional scale, these trends translate into higher pressure on the resources and the need for páramo conservation.

The increased demands on the páramo environmental services of the Cordillera de Mérida, particularly water services, have not been paralleled by effective conservation strategies in the high-Andean and periglacial belts, such as grazing control on the wetlands (Molinillo and Monasterio, 1997b). Neither have control strategies been implemented for more efficient water use at the community and regional levels. One of the main reasons for this is that water-related problems operate at different spatial scales, implying different spheres, stakeholders, and policies (Monasterio and Molinillo, 2002).

Consequently, the subject of water for páramo agriculture and its relationship with conservation in these high-mountain ecosystems will become increasingly important. Hence, there is a need to involve the local population to reach a sustainable use of water resources. This can be achieved through promoting awareness of the relationship between water quality and quantity and the conservation status of páramo ecosystems. In addition, different approaches are needed to allow a better understanding of the role and relationships among the different stakeholders (farmers, grassroots organizations, local authorities, national park services, researchers, and governmental and nongovernmental organizations) involved in water management.

For the purpose of reaching these objectives, the project has supplied basic information, such as the design of cartographic tools (topographic and ecological maps) showing in detail the spatial relation between agricultural and water capture and storage zones in the periglacial belt of each of the watersheds and microwatersheds utilized by each community. The use of these tools, combined with the scale integration approach, is intended to facilitate understanding of the relationships between stakeholders, their roles, and their spheres of influence, stimulating participatory workshops for discussion of the links between conservation of natural ecosystems and agroecosystems (Monasterio et al., 2001).

The identification and quantification of environmental services and their sources of origin in natural areas, in turn, will emphasize the value of natural ecosystems and their role in production systems. Maintaining diversity in the high Andes guarantees the sustainability and functioning of agricultural areas. Moreover, local populations, the institutions responsible for the conservation of natural areas, and researchers will have to cooperate when designing activities and strategies compatible with conservation and enhancement of the environmental functions of protected areas.

All this information, represented through cartographic tools, will be the ecological base to develop and plan local and regional agricultural policies demonstrating conservation alternatives aimed at maintaining and enhancing the role of natural areas for a more socially sustainable and ecological production strategy in páramo agroecosystems.

Within this approach, conservation plans, aimed in general at a regional scale, could not only involve measures to control human impacts on páramo water sources, but also enhance the understanding of agricultural dynamics at a community level and the use of management on a plot scale of the different types of crops. Research projects formulated on a plot scale could provide further detail on the different subjects to include, considering the possible impact of research locally and regionally. Even though these are the obvious consequences of scale integration, other implications at the community level could be essential for conservation and sustainable management at a regional and macroregional level (Monasterio and Molinillo, 2002). One of these is to be able

to take the regional importance of diversity conservation in the high-Andean wetlands to the farmer plots.

CONCLUSIONS

In the páramo of the Cordillera de Mérida, analysis at different scales shows a tight link between agricultural intensification trends and changes in grazing patterns in the high-Andean belt. This situation mainly affects the wetland habitats in the valley bottoms above the agricultural belt, precisely the areas with the best water retention, where the water resources for páramo agriculture are located. The successional dynamics of these vegetation units are mainly controlled by grazing intensity, which could be on the increase because of an increase in the permanence time of animals on each wetland. This is fundamentally due to a decrease in the carrying capacity and supplementary forage supply in the agricultural belt. In terms of plant species composition, this could, in turn, mean the gradual displacement of the dominant swards of native forage species by weeds (natives and exotics), even if species richness is not significantly affected. The spatial scales integration approach could allow producers to identify more clearly the relationship between agricultural management in their plots and high-Andean wetlands. In this context, it is essential to identify and use indicators allowing rapid determination of the status of the hydrological functioning of marshes and swards, so that effective conservation measures can be put in place. Vegetation composition, specifically the ratio between the number of native forage and weed species, is suggested as one of the most significant indicators to be analyzed.

SUMMARY

The relationship between agricultural dynamics and the conservation of páramo ecosystems is analyzed in the central core of the Cordillera de Mérida (Venezuela), particularly in the marshes and swards. These are the origin and storage areas of water for agricultural production in the region. The conceptual and methodological approach is based on the integration of the regional, local, and plot scales through the transversal link of water as an environmental service. On a regional scale, the high-Andean and periglacial belts are identified as the most important providers of water used in the lower belts for agriculture. On a local scale, the cultivated páramo zone and its water sources in the high Chama River watershed is analyzed. On a plot scale, the dynamics of agricultural production and its relationship with grazing impact on species composition of the marshes are discussed. The results indicate that plant community structure and diversity have been influenced by agricultural intensification in the agricultural belt in recent decades. This process, together with a decrease in importance and the disappearance of wheat cultivation, have led to a decrease in carrying capacity for cattle grazing in the agricultural belt, and an increase in the permanence time of cattle in the high-Andean belt. Even though there has been a clear impact on cattle raising, species richness in the swards of El Banco watershed (Chama River affluent) does not show a clear correlation with grazing intensity. This could be due, to a considerable extent, to the fact that a large proportion of the increase in species considered as native weeds and exotics during succession is counterbalanced by a fall in the number of native forage species.

These results seem more evident when the cover of good forage species in comparison to weeds is analyzed. Grazing intensity (number of animals and their residence time in each marsh and sward) and soil water content control the equilibrium between native forage species and weeds. Hence, these variables are suggested as good indicators of the conservation status of marsh and sward environments. Finally, the use of the scale integration approach for developing a better understanding of the key issues for water management and increasing effective participation of the local population in conservation is discussed.

ACKNOWLEDGMENTS

This research is part of the Interdisciplinary Project for Sustainable Agricultural Management in the High Andes of Mérida, Venezuela, financed by the AGENDA PAPA–CDCHT (CV1 PIC-C0201) of the Universidad de los Andes. The SPOT 4 satellite images were provided by CNES and Spot Image within the framework of the Latino Spot project. Páramo communities in the project area (La Toma, Misintá, and El Banco) offered essential support during sampling and mapping. The organization Programa Andes Tropicales (PAT) also offered logistic support during sampling and mapping. We also wish to thank Luis D. Llambí for his help with the translation of the manuscript.

References

Biodiversity Support Program (1995). *A Regional Analysis of Geographic Priorities for Biodiversity Conservation in Latin America and the Caribbean*. Biodiversity Support Program, Washington, D.C., USA.

Drost, H., Mahaney, W., Bezada, M., and Kalm, V. (1999). Measuring the impact of land degradation on agricultural production: a multidisciplinary research approach. *Mountain Research and Development*, 19: 68–70.

Hess, C. (1990). Moving up — moving down: agropastoral land use patterns in the Equatorial Paramos. *Mountain Research and Development*, 10: 333–342.

MAC (Ministerio de Agricultura y Cría) (1958–1997). Anuarios estadísticos agropecuarios. Caracas, Venezuela.

Milchunas, P.G., Sala, O.E., and Lauenroth, W.K. (1988). A generalized model of the effects of grazing by large herbivores on grassland community structure. *American Naturalist*, 132: 87–106.

Molinillo, M. and Monasterio, M. (1997a). Pastoralism in páramo environments: practices, forage, and impact on vegetation in the Cordillera of Mérida, Venezuela. *Mountain Research and Development*, 17: 197–211.

Molinillo, M. and Monasterio, M. (1997b). Pastoreo y conservación en áreas protegidas de la Cordillera de Mérida, Venezuela. In Liberman, M. and Baied, C. (Eds.), *Desarrollo sostenible de ecosistemas de montaña: manejo de areas frágiles en los Andes*. UNU, Instituto de Ecología — UMSA, La Paz, Bolivia.

Molinillo, M. and Monasterio, M. (2002). Patrones de vegetación y pastoreo en ambientes de Páramo. *Ecotropicos*, 15(1): 17–32.

Monasterio, M. (1980a). Las formaciones vegetales de los páramos de Venezuela. In Monasterio, M. (Ed.), *Estudios ecológicos en los páramos Andinos*. Editorial de la Universidad de Los Andes. Mérida, Venezuela, pp. 93–158.

Monasterio, M. (1980b). Poblamiento humano y uso de la tierra en los altos Andes de Venezuela. Las formaciones vegetales de los páramos de Venezuela. In Monasterio, M. (Ed.), *Estudios ecológicos en los páramos Andinos*. Editorial de la Universidad de Los Andes, Mérida, Venezuela, pp. 170–198.

Monasterio, M. and Celecia, M. (1991). El norte de los Andes tropicales. Sistemas naturales y agrarios en la Cordillera de Mérida. *Ambiente*, 68: 2–6.

Monasterio, M., Molinillo, M., Andressen, R., and Gutiérrez, J. (2001). Proyecto Sostenibilidad ecológica y social de la producción agrícola en la cordillera de Mérida: el flujo de los servicios ambientales de los páramos altiandinos para la agricultura papera. Consejo de Desarrollo Científico Humanístico y Tecnológico (CDCHT), Universidad de Los Andes. Mérida, Venezuela, p. 33.

Monasterio, M. and Molinillo, M. (2002). Integrando el desarrollo agrícola y la conservación de áreas frágiles mediante la articulación de escalas espaciales en los Andes de Mérida, Venezuela. In Castroviejo, J. (Ed.), *Uso sostenible, iniciativa privada y reservas de la biosfera (UNESCO) en áreas de montaña: los Andes, la Mata Atlántica y la cordillera Cantábrica*. Editado por el Instituto de Ecología y Mercado, Fundación para el Análisis y los Estudios Sociales FAES, España, p. 16.

Pels, B. and Verweij, P.A. (1992). Burning and grazing in a bunchgrass páramo ecosystem: vegetation dynamics described by a transition model. In Balslev, H. and Luteyn, J.L. (Eds.), *Páramo: An Andean Ecosystem Under Human Influence*. Academic Press, New York, pp. 243–263.

Sarmiento, L. (2000). Water balance and soil loss under long fallow agriculture in the Venezuelan Andes. *Mountain Research and Development*, 20: 246–253.

Sarmiento, L., Monasterio, M., and Montilla, M. (1993). Ecological bases, sustainability, and current trends in traditional agriculture in the Venezuelan high Andes. *Mountain Research and Development*, 13: 167–176.

Sarmiento, L., Llambí, L.D., Escalona, A., and Marquez, N. (2003). Vegetation patterns, regeneration rates and divergence in an old-field succession of the High Tropical Andes. *Plant Ecology*, 166: 63–74.

Velásquez, N. (2001). Dinámica Socio-Ambiental y Modernización Agrícola en los valles Altos Andinos: Mucuchíes y Timotes 1930-1999. Ph.D. thesis, Instituto de Ciencias Ambientales y Ecológicas, Facultad de Ciencias, Universidad de Los Andes, Mérida, Venezuela.

Verweij, P.A. (1995). Spatial and Temporal Modelling of Vegetation Pattern: Burning and Grazing in the Páramo of Los Nevados National Park, Colombia. Ph.D. thesis, ITC Publication no. 30, Enschede, The Netherlands.

Verweij, P.A. and Budde, P.E. (1992). Burning and grazing gradients in páramo vegetation: initial ordination analysis. In Balslev, H. and Luteyn, J.L. (Eds.), *Páramo: An Andean Ecosystem Under Human Influence*. Academic Press, New York, pp. 177–195.

24 Multidimensional (Climatic, Biodiversity, Socioeconomic), Changes in Land Use in the Vilcanota Watershed, Peru

Stephan Halloy, Anton Seimon, Karina Yager, and Alfredo Tupayachi

INTRODUCTION

To investigate the dynamic changes affecting biodiversity across the vertical gradient of the Vilcanota watershed in Peru, we utilize the major vertical profile of the Vilcanota–Urubamba Valley (the Sacred Valley of the Incas at its center). The area combines features of interest for our research, such as a tropical location in a major biodiversity hot spot, which has also been a cultural vortex with thousands of years of occupation and development of resilient sustainable land uses; the point of origin of many indigenous agricultural staples, some of which are now important agricultural crops at a global level; and a unique annually resolved climatic record of more than 500 years in the Quelccaya ice cap to the southeast of the watershed (Thompson et al. 1985). As it descends, the Vilcanota–Urubamba changes its cross section (Figure 24.1), topography, and mesoclimates, traversing an extreme range of climates and environments. These have been described and classified by many researchers (e.g. Brisseau, 1981; Galiano Sánchez, et al., 1995; Gentry, 1993; Sibille, 1997). The watershed starts in the permanent snow and glaciers of the steep peaks above 6300 m (Ausangate), where mean temperatures are below 0°C. We recently recorded (in 2002) the highest vascular plants at 5510 m, close behind the retreating glaciers in this area. High-Andean vegetation develops rapidly down from this level. Around 4900 m, llama and alpaca grazing signal the rising level of human occupation. The highest human occupation found is the house of Pedro Godofredo above Murmurani, at ~5050 m.

The undulating altiplano between 4900 and 4200 m gives way to steep incised valleys as the rivers cut their way down to the Amazon. As in the altiplano, human occupation has developed in these valleys over the centuries, cultivating the valley floors and terracing the steep valley slopes to expand production areas. Apart from the valley topography and gradual increase in temperature, an important environmental factor is the drying of the climate towards the valley floors as a climatic effect of valley wind circulation (Troll, 1968). About 350 km down from its source, the valley finally opens into the foothills of the Andes and the Amazonian lowland forests and savannas, where mean annual temperatures are around 23 to 29°C, and annual rainfall is around 1700 to 2000 mm. Due to the strong orographic gradients, all climate parameters vary in short distances. For example, rainfall slightly to the southeast of the Urubamba at San Gabán and Quince Mil (600 m) reaches 3000 to 6000 mm per year.

Data on species richness will be reviewed, and we will examine information on present impacts affecting the natural and managed biodiversity and the manner in which the latter is distributed. Given the region's rich biodiversity and the reported past levels of prosperity

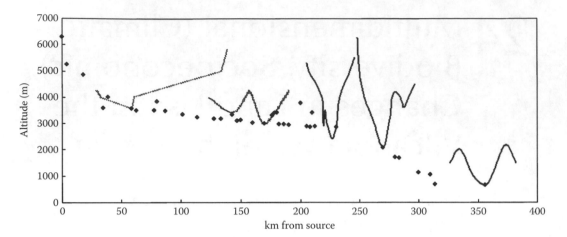

FIGURE 24.1 Topographic profile of the Vilcanota Valley, lengthwise from SSE to NNW with five cross sections approximately W–E to show the changing valley configuration. The Vilcanota is represented by the altitudes of 33 district capitals (dots), some of which are located away from the valley center, hence the higher points. Four additional points complete the profile: village of Santa Barbara (4000 m), outlet of Sibinacocha Lake (4850 m), Rititica summit (5250 m), and the summit of Vizcachani (near the source of the Vilcanota above 6200 m). The five cross sections (full lines) are taken at the level of the capitals (from left to right) Sicuani, Pisac-Cusco, Ollantaytambo, Machu Picchu, and Quellouno.

at a time (>500 years BP) when resource use has been claimed to be more sustainable in the long term, the question that comes to the fore is: Why do human populations now suffer extreme poverty and environments undergo rapid degradation? We examine the temporal dynamics of various components in this three-dimensional space and explore possible drivers in view of human pressures and climate change. Several questions that arise are: Is loss of biodiversity through land use change a consequence of poverty? Is poverty related to a failure to incorporate traditional biodiversity stewardship into modern agricultural systems? Do market pressures tend to decrease the use of traditional agricultural management (e.g. Swinton and Quiroz, 2003; Halloy et al., 2004)?

METHODS

We surveyed, collated, and calculated the information and literature on land use and biodiversity for the Vilcanota–Urubamba watershed. Political (and hence, census) boundaries are not drawn along watershed boundaries, so we selected 33 representative districts along the main axis of the valley. To approach biodiver-

sity at this regional scale, we use proxies (which are more or less relevant and debatable, and provide insights into the system) such as percentages of land use and rates of change (e.g. deforestation, cultivated crops, and grazing), each of which has its own impacts on biodiversity. Cultivated area of each species of crops was collated from all districts, a necessary caveat being that census data are sensitive to human reporting and data-gathering techniques. Many smaller crops and crop areas are not reported, thus biasing the data toward larger areas and crops. However, this is not unlike the bias that occurs in any biodiversity study toward larger, more abundant, and more visible species.

Table 24.1 shows the seven provinces of the Cusco Department, along with some portions in the Vilcanota Valley. Further details on the 33 districts are in Appendix I. Cusco Department has a total area of 71,987 km², slightly larger than the island of Tierra del Fuego. The area of the 33 districts studied here is 29,337 km², or almost half of the department.

Diversity was evaluated as simple species richness, following the Shannon–Weaver information index of diversity ($H = p_i \ln p_i$, where $p_i =$ (abundance of species i)/total abundance;

TABLE 24.1
Provinces of the Cusco Department with districts used in this study, together with their population and area

Province	Capital	Population, Projection 2002	Area (km²)	Density (inhabitants km⁻²)
Total departments	Cusco	1,208,689	71,987	16.8
Acomayo	Acomayo	34,652	948.22	36.5
Calca	Calca	65,330	4,414	14.8
Canchis	Sicuani	107,012	3,999	26.8
Cusco	Cusco	319,422	617	517.7
La Convención	Quillabamba	194,395	30,062	6.5
Quispicanchi	Urcos	89,264	7,565	11.8
Urubamba	Urubamba	56,352	1,439	39.1

Source: From the 1993-1994 Census, Instituto Nacional de Estadística e Informática, Peru (INEI 2003).

[Shannon and Weaver, 1949]), and as frequency distributions (Williams, 1964).

We integrate this study with ongoing research at the regional altitudinal limits of life in the Lake Sibinacocha area. As part of a global network to monitor the effects of global change on biodiversity, we established in 2002 a Global Research Initiative in Alpine Environments (GLORIA) site at 5250 m. This follows a standardized methodology of inventories and temperature measurements for long-term comparisons (Pauli et al., 2002) and is logged as a Global Terrestrial Observation Site (Halloy and Tupayachi, 2004).

VERTICAL DISTRIBUTION OF DIVERSITY

Braun et al. (2002) calculated the number of species of seed plants in an altitudinal profile of Peru from Brako and Zarucchi (1993) (Figure 24.2). They found that the number of species in the Andes above 500 m is more than the total number of Amazonian species in Peru. At the highest levels, over 250 species of seed plants are recorded above 4500 m for the whole of Peru. At the eastern headwaters of the Vilcanota, at the Rititica GLORIA site, we found 24 vascular plants and 28 nonvascular plants (bryophytes and lichens) in a 274-m² sampling area at 5250 m in midwinter 2002. Higher up,

flowering plants were found to 5510 m, right up to the receding ice cliff edge above Rititica.

Gentry (1993) noted that although 43% of Peruvian seed plant species are from lowland Amazonia, 34% grow in lower-Andean forests between 500 and 1500 m, and a remarkable 57% are recorded from Andean cloud forests. The high-Andean region above 3500 m contains approximately 14% of the Peruvian flora.

LAND USE IMPACT

Land-based agriculture contributes 25.4% of the gross domestic product (GDP) and provides 47.5% of employment in the Cusco Department (MAP, 2003). The proportion of total land area that is dedicated to cultivation averages 8% for the whole valley, ranging from less than 1% for Pitumarca and Checacupe districts (limiting ecological conditions near the altitudinal limits of cultivation) to 33% for Quellouno (recent major increase in export crops, principally coffee). Grazing affects almost all lands accessible to stock within the valley. Based on a generous assumption (with present management practices) of one stock unit[1] per hectare, and calculating from all stock censused in the six valley provinces (Sibille, 1997), we obtain that most

---

[1] Stock unit is equivalent to a 45 kg ewe suckling a lamb or a 55 kg pregnant ewe. This amounts to around 0.02 stock units per 1 kg of live weight; 1 stock unit requires 520 kg of dry matter of feed per year.

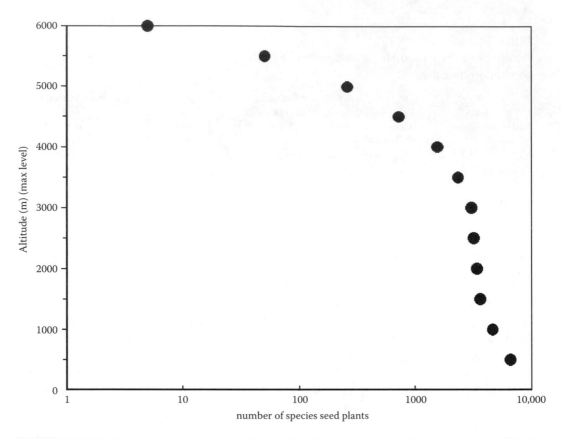

FIGURE 24.2 Number of seed plants at each altitudinal level in Peru, combined from Braun et al. 2002. The GLORIA site and high altitude records.

provinces carry stock requiring 60% (Calca, Quispicanchi) to 150% (Canchis) and 190% (Cusco) of their total land area. Only La Convención requires a minor 3.5% of its land area to feed existing stock. Because only a certain fraction of their total land area is suitable for natural pastures (e.g. 64% for Canchis, 40% for Cusco, and less than 24% for the remaining provinces, INEI in MAP [2003]), the overstocking becomes even more notorious. These are indications of unsustainable levels of overgrazing that exceed the carrying capacity of the land. Fallow and harvested lands also fulfill a role in providing feed for grazing stock, but this is not quantified in censuses.

Although some level of grazing can enhance biodiversity by reducing competition (Fowler 2002), intense overgrazing as suggested by these data leads to depletion of palatable species, reduction of ground cover, and erosion (Duncan et al., 2001). Depending on

management, livestock, as do cultivated plants, will carry with them a variety of commensal/accompanying species including their parasites, as well as transport seed plants that are abundant near their main grazing areas. A 2001 survey around Lake Sibinacocha found that rodent diversity increased around llama and alpaca corrals at an altitude of 4900 m as an effect of anthropogenic enhancement.

The steep terrain of most of the central valley implies high erosion risk: 85% of areas cultivated in the higher areas (310,000 ha) are on steep to moderately steep slopes. They are susceptible to erosion but most are not subject to any soil protection practices at this time (MAP, 2003), unlike ancient mitigation practices of terracing, irrigation, managing soil organic matter, etc.

Deforestation for agricultural land and firewood is claiming large areas of the central valley. For the center of the Valle Sagrado, Galiano

Sánchez et al. (1995) quote deforestation levels of 90% of original forests for valley bottom forests (2700 to 3300 m), 60% for mixed forests of the slopes (3300 to 3700 m), and 20% of the *Polylepis* forests from 3700 to 4800 m. The Ministerio de Agricultura (MAP, 2003) estimated that 50% of the best forests of the department were cut down by 1995, including 15% of the humid lowland forest, more of which is being cut at a rate of 20,000 ha per year. Land use conversion has opened up 630,000 ha in the 22 years from 1972 to 1994, representing an increase of 29.5%.

Introduced species constitute an insufficiently evaluated risk in the area. Weeds of temperate regions are widespread in the middle reaches of the valley, although many weeds in turn have their uses (see subsection titled Species Richness). Irreversible changes are being mediated by exotic species: large areas are reforested with eucalyptus, bringing considerable changes to the landscape and ecosystem, including scenic aspects, soils, erosion, availability of firewood, and capability of native species (including animals and medicinal plants) to survive under their canopy. An other invasive species that has probably had a major impact in this area include trout, widely introduced for subsistence and recreational fishing.

Mining at high altitudes, as well as the impact of large oil deposits found in lowlands (Camisea, Sibille, 1997), provide an incentive and a subsidy to develop roads and infrastructure that then allow penetration into vast new areas, in addition to their direct impacts on devegetation and toxic wastes.

Factors slowing the expansion of land use impacts include difficult access and legislation. Although steepness and lack of roads has provided some protection to more remote parts of the valley, the only formally protected area in the Vilcanota Valley is the Santuario Histórico de Machu Picchu in the Province of Urubamba. With 32,592 ha, it represents almost 23% of the area of that province but only 1% of the area of the 33 districts considered in this study. For comparison, in its land use capability classification, INRENA (2000; in MAP 2003) considers that 66% of departmental lands should be classified as protection land, with only 33% suitable for agriculture (3% arable, 0.4% per-

manent crops, 14% suitable for forestry plantations, and 16% suitable for rangeland management). Yet in the 1994 census of the 33 districts of the Vilcanota, arable and permanent crops alone already cover 8% of the land area, implying that expansion is unsustainable.

RESOURCE DISTRIBUTION IN HUMAN POPULATIONS

The distribution of economic resources can determine the magnitude and type of land use and its effect on biodiversity. Resource distribution is explored from the point of view of land size distribution, distribution of the abundance of crops, and distribution of wealth (social indicators of poverty).

CULTIVATED LAND DISTRIBUTION AND DIVERSITY

The distribution of access to productive land depends on the distribution of cultivated parcel sizes. This overlooks the issue of spatial distribution but is, nevertheless, a large-scale proxy for overall distribution. Plots around a peasant community tend to be of relatively small (typically, much less than 0.5 ha) and even sizes (e.g. for similar cultural landscapes in Peru and Bolivia, see Liberman Cruz, 1987; Pietilä and Jokela, 1988). These areas close to villages produce the mainstay of daily sustenance and hold the highest crop and native plant diversity (Zimmerer, 1997; Ramirez, 2002). In the 17 higher districts (>3000 m, more highly populated) of the Cusco Department, Peru, 93% of properties are less than 5 ha, the mean parcel size is 0.37 ha, and the average cultivated area per person in the overall population is 0.14 ha (INEI, 2003). The distribution of plot sizes controlled by a single family tends to a classic lognormal pattern with occasional large outliers, indicating an imbalance (Halloy et al., 2004). Larger cultivated areas are developed further from houses and are hence tied to the availability of transportation and farm machinery. In the two lower, more market-oriented districts (~650 m), only 22% of properties are less than 5 ha, the mean property size is 1.3 ha, and the cultivated area per person is 0.74 ha.

Larger plot sizes are driven mainly by large-scale cultivation of commercial crops (e.g. coffee and cocoa in lowlands; maize, wheat, *ulluco*, and potatoes in highlands). Much larger cultivated sizes in tropical lowlands are an effect of dynamic colonial expansion into the lowlands and are contrary to ecological expectations (i.e. higher potential yields mean that smaller plots are sufficient for equivalent yields). Older, more established societies tend to produce lognormal distributions of the cultivated areas of crops (e.g. Halloy, 1994; Halloy, 1999), whereas younger colonizing societies have distributions that depart strongly from the lognormal. In the Vilcanota, we can see this, in particular, in the lowering of diversity index (*H*) values in La Convención (below 1.8), despite high species numbers (60 to 75) (Figure 24.3). Many central and highland areas, despite species numbers well below 50, maintain a relatively high diversity (*H* between 1.6 and 2.4), thanks to a more even species distribution. However, some highland areas have very low diversity where crop cultivation becomes ecologically marginal.

WEALTH DISTRIBUTION AND NUTRITION

Despite a wealth of biodiversity and productive land, the 1993 census recorded that 60% of children were chronically malnourished and infant mortality was 91.8 per thousand for the Cusco Department (Table 24.2).

Fecundity (number of children per woman) typically declines with development. The more highly developed Cusco Province shows a rating of 2.8, but poorer and less educated provinces show much higher values (e.g. Quispicanchi 5.8, Urubamba 5.0; Sibille [1997]. In a paradox that is repeated around the world, the areas richest in cultivated plants are the poorest and most malnourished. However, we note that

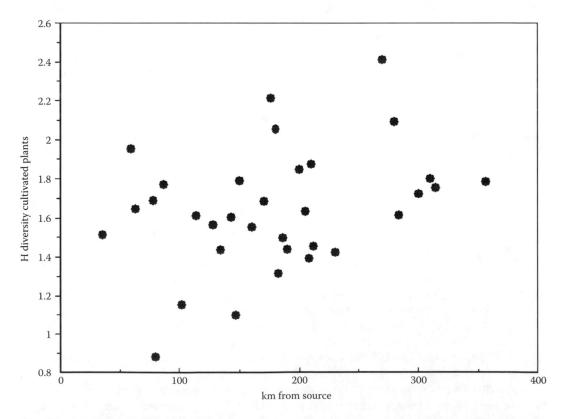

FIGURE 24.3 Shannon–Weaver index of diversity for cultivated plants across 33 districts of the Vilcanota Valley.

TABLE 24.2
Social indicators vs. cultivated plant diversity in some Cusco Provinces, 1993 census

Area	Cusco Department	Canchis Province	Calca Province	Cusco Province
Chronically malnourished children (%)	60.0	59.2	65.5	42.0
Infant mortality rate per 1000	91.8	114.2	86.7	47.7
Number of species of cultivated plants per 1000 inhabitants	3.8	3.6	3.8	0.6

Source: INEI, 2003.

this is not a linear relation, as improved quality of life was found at even higher diversity in traditionally cultivated areas (Halloy et al. 2004).

SPECIES RICHNESS AND DISTRIBUTION OF CULTIVATED SPECIES

A total of 157 categories of cultivated plants were recorded in the 1993 agricultural census. Several census categories represent mixed bags of species in which there may be only one or several species (*Vergel Hortícola Plátano* [vegetable plots planted with bananas], *Vergel Frutícola* [fruit orchards], *Flores* [flowers], etc.; Table 24.3). Hence, estimates of species richness based on the census are underestimates. This species richness is not fixed in time; the actual varieties and species that are grown are continuously changing with a rapid turnover rate (e.g. Halloy, 1999; Ramirez, 2002).

Census data of cultivated crops represents only a fraction of total cultivated plants. For example, for the total area above 3500 m, the INEI 1993 census data records 52 species of cultivated plants in a total of 6679 ha. However, in a small area of 686 ha above 3500 m in Calca Province, Ramirez (2002) recorded 76 species.

In addition, a large number of adventive or "weedy" species accompany cultivation, and additional native species "tolerate" and persist in cultivated areas along road edges, hedges, gullies, etc. Many such species are also used by local populations (Rapoport et al. 1998). For example, Vieyra-Odilon and Vibrans (2001)

report 74 weed species found in maize fields in Mexico that were useful as forage, potherb, medicinal, or ornamental plants. In the high Andes of neighboring Bolivia, Hensen (1992) reports the use of 204 species of plants in the community of Chorojo, Cochabamba, from 3500 to 3800 m, most with forage and medicinal uses. Of these, 24 species were used as food. In every relevé in fallow terrain near La Paz, de Morales (1988) reports that 6 to 12 weedy species are found. Detailed recordings of plant use in the Andes are available in a range of publications (e.g. Brücher, 1989; NRS, 1989; Zimmerer, 1997).

Sibille (1997) (following INEI, 1986) quotes 193 plant products (including 142 arable crops, 37 permanent crops, and 14 grasses) for the whole of Cusco Department, whereas Galiano Sánchez et al. (1995) quote 96 useful species (including this time forestry species) and 685 vascular plant species in a 50 km^2 area of the Sacred Valley, ranging from 2715 to 5300 m. They also recorded 40 nonvascular cryptogams.

The present total of 157 cultivated species in the Vilcanota Valley and 193 for the whole Cusco Department can be compared to 160 species claimed to have been commonly used for food, medicine, and other purposes in precolonial times for Peru (Tapia and Torre, 2003).

It is of some concern for conservation that most of the rarest cultivated plants are natives, whereas many of the common species are exotic.

TABLE 24.3
Most commonly cultivated plants over 33 districts according to 1993 agricultural Census

Census Name	Species/Variety English Name	Scientific Name	Area (ha)	Number of Districts
Café or cafeto	Coffee	*Coffea arabica*	25,511	7
Maiz amiláceo	Starch maize	*Zea mays*	11,375	33
Papa	Potato	*Solanum tuberosum*	8,132	33
Cacao	Cocoa	*Theobroma cacao*	6,581	7
Achiote	*Annatto*	*Bixa orellana*	4,462	6
Coca	Coca	*Erythroxylum coca*	3,705	8
Haba	Broad bean	*Vicia faba*	3,282	30
Yuca	Cassava	*Manihot esculenta*	3,003	8
Cebada grano	Barley grain	*Hordeum vulgare*	2,428	27
Trigo	Wheat	*Triticum aestivum*	2,029	29
Vergel hortícola–plátano	Vegetable garden–banana	Multispecies	1,006	32
Maiz amarillo	Yellow maize	*Zea mays*	1,916	30
Vergel frutícola	Fruit orchard	Multispecies	1,435	29
Arveja (alverjón)	Green pea	*Pisum sativum*	529	29
Olluco	Ulluco	*Ullucus tuberosus*	1,466	29
Oca	Oca, NZ yam	*Oxalis tuberosa*	100	28

Note: The two right columns show total area cultivated (first ten are the species with largest cultivated areas) and number of districts where the crop is recorded (the following six are species with a high number of districts but lower area).

Source: INEI, 2003.

TEMPORAL DYNAMICS

HISTORICAL PERSPECTIVE

It is interesting to compare the present situation with that recorded by the Spaniards in the early 1500s. The area that was then the center of the Inca dominions was praised by chroniclers as a place where "no one ever went hungry" and where "purposely made storage areas were overflowing with vegetables and roots to feed the people and also herbs" (Peró Sancho quoted in Murra, 1975).

Indeed, traditional land use management practices were able to support the livelihoods of households and communities for several millennia and were sufficient for the rise of complex civilizations centuries prior to Spanish occupation. The ample increase in production under the Inca empire may have, in part, depended on its careful environmental husbandry (including tactics of soil conservation, water management and irrigation, management of domesticated plant and animal diversity, and

protection of natural vegetation and fauna) (Halloy et al., 2004).

It is possible that habitat degradation induced by ancient hunter–gatherers and pastoral nomads may have contributed — together with population increase, extended annual occupation, rise of social stratification, and the need to increase production for both social and livelihood needs — to the development of civilizations incorporating the conservation measures in force at the time of arrival of the Spanish (Kessler, 1998).

Despite such measures, it seems likely that considerable destruction of the high-altitude *Polylepis* forests took place long before the arrival of the conquistadores in 1532 (Gade, 1999; Kessler et al., 1998). Before the arrival of the Spanish, the Andean landscape had already experienced significant levels of transformation and degradation. Gade estimates that some 65% of the natural forest had been depleted before the Spanish arrival, shortly after which 90% became depleted (Gade 1999). The

Spanish conquest resulted in increasing deforestation rates as they consumed large amounts of wood for construction and the smelting of ores in mining activities.

Upon arrival, the Spanish implemented agroforestry measures in an attempt to compensate for excessive consumption levels. Unfortunately, such measures did not suffice, and the landscape became mostly depleted of trees. After the decimation of the indigenous population in the 16th century, a majority of the rural landscape was abandoned. Denevan argues that much of the natural landscape was able to recover as a result of the population decline and may have contributed to the early 19th-century misconceptions of the "pristine" landscape (Denevan, 1992). However, the introduction of nonnative species also became commonplace. The Spanish experimented early with the nonnative poplar and capuli trees (Gade 1975), but the most influential species to be introduced in the late 19th century with unprecedented fruition was *Eucalyptus globulus*.

CLIMATE CHANGE

Given the context of intense human and environmental heterogeneity and fluctuations over time, encountering a signal of climate change effects is not a simple matter (e.g. see meta analyses as in Parmesan and Yohe (2003), but such an approach is still to be realized in Peru). However, there are some observations pointing towards vegetation and land use advancing towards higher altitudes in recent decades. Toward the middle of the last century, Troll (1968) observed that in the Central Andes of Peru and Bolivia, maize could be grown up to 3500 m, whereas tuberiferous plants (potato, *oca*, *isaño*, and *ulluco*) and introduced wheat and barley reached their upper limit at 4100 m. Mitchell (1976), followed by Price (1981), also placed the altitudinal limit of cropping at 4100 m. Higher up, the grasslands were grazed by llamas, alpacas, and wild vicuñas. Uninterrupted plant cover ended at around 4700 m, where nightly frosts began. The climatic snowline was indicated at 5300 m.

Interpretation of such reports is problematic, given their nonspecificity in terms of location or dates, as well as issues of time lag between observations and publication. However, these and other authors had extensive experience in geography, and it is unlikely that their observations would be far off the mark. More recently, Tapia and Torre (2003) quote several crops grown up to 4000 m, two species grown up to 4100 m (maca and kañiwa), and one (Papa amarga: *Solanum juzepczukii*) grown up to 4200 m. Potato cultivation today in the Vilcanota headwaters occasionally reaches 4580 m (Chillca; our observations in 2004). Recent attempts to cultivate oats and potato have even been made at 5050 m above Murmurarni, although these were unsuccessful.

Interestingly, archaeological remains show that these and higher areas were cultivated in the more remote past. Archaeological remains of cultivation higher than today have also been noted in the Cordillera Blanca (Cardich, 1985). In 1985, Cardich also observed that since he began making observations, "the limit of cultivation has been moving upward and crops are now grown at higher elevations than during previous decades. Simultaneously, there has been an accelerated recession of glaciers in the high cordilleras, as well as disappearance of snow and consequent opening of passes connecting the Pacific and Atlantic slopes." In summary, 2002 cultivation levels are higher than in the past decades and recent centuries, but still not as high as maximum levels reached at some time in the past, presumably before the Little Ice Age. The first post–Little Ice Age settlers in the Sibinacocha area moved into the valley in 1906 (Pedro Godofredo, personal communication, 2003). Today, there are a number of corrals and settlements.

POPULATION AND LAND USE

An indication of the growing population impact is its increasing concentration in urban centers, from 25% of the department's population in 1940 to 46.5% in the 1993 census (Figure 24.4). For the whole department, infant mortality has declined from 149 per 1000 births in 1979 to 1980 to 101 in 1990 to 1991 (Sibille, 1997). Because of political–economic change, there is also a strong net outmigration, principally towards centers offering employment and natural resources (Lima, Arequipa, and Madre de Dios). Emigration rates are rapidly increasing.

From 8% of the total departmental population in 1961, emigration climbed to 16% in 1972, 18% in 1981, and 21% in 1993 (Sibille, 1997). Emigration from rural areas contributes to important land use changes with mixed impacts on biodiversity: lack of maintenance of terraces and irrigation leading to erosion, lack of control of animals leading to grazing and overgrazing, lack of cultivation leading to weedy successional phases, then back to vegetation that is more diverse, etc.

Sibille (1997) also indicates that agricultural land is decreasing significantly in several areas due to urban encroachment. Thus, Cusco province lost 62% of its arable land in the 10 years from 1985 to 1995, whereas Urubamba lost 25%. At the same time, he reports a loss of some of the more traditional crops to livestock grazing and intensification of farming (e.g. irrigated land has increased 89% in 22 years from 1972). Many irrigation schemes disregard impacts on the overall social and natural web of interactions (Liberman Cruz, 1987), leading to further loss of arable land and native biodiversity.

The advance of the agricultural frontier is particularly evident in the lowland areas, where it is marked by large-scale deforestation. How-ever, although less evident, use pressure is growing in the highlands as well, as manifested by the increasing altitude at which crops and livestock are grown and the increasing intensity and density of cultivation.

THREATENED SPECIES

Although there are no data specific to the Cusco Department, Pulido (2001) reports threatened animal species for Peru have increased from 162 to 222 from 1990 to 1999. Such increases have been recorded around the world in what is often more a matter of increased monitoring and perception than of real change of status in such short times. Amphibian decline noted around the world is also being observed in the Vilcanota region, with local people reporting the apparent reduction or total disappearance of three to four species of frogs in areas close to 4000 m. Although a causal relation has yet to be found, it is of concern that recent sampling above 4400 m found evidence of deadly chytrid fungus infections (believed to be implicated in global decline) in remote populations of the aquatic *Telmatobius marmoratus* (DeVries et al., 2004).

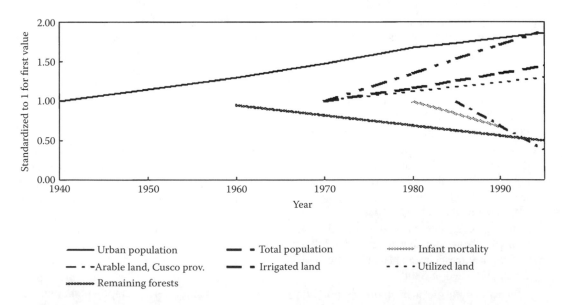

FIGURE 24.4 Combined social and environmental changes in Cusco Department, Peru (data from INEI in MAP, 2003 and Sibille, 1997).

DISCUSSION AND CONCLUSION: MACROECONOMIC DRIVERS

There is considerable understanding of the small-scale effects on biodiversity of land use changes (e.g. other chapters in this book). Although we recognize the rich tapestry of ecology, farm- and people-scale dynamic processes that underpin the large scale, there is also a need to understand the large spatial scale and drivers. The increasing complexity on larger scales creates particular research difficulties, including reliance on secondary information, the importance of historical information, consistency of information across scales and across disciplines, and the translation of methods and language between disciplines.

The remarkable diversity of environments in the Vilcanota Valley arising from the combination of topographic conditions, altitude, and rainfall, together with the mosaic of natural disturbances and dynamic human management strategies, has led to a high-energy system with high biodiversity and high flows of materials among its landscape components.

The main threats to biodiversity in the Vilcanota presently involve land use changes. Mitigating the effects of those changes on biodiversity requires identifying and understanding the drivers of change. This chapter is a contribution to identify the next level of causal interactions between biodiversity, land use changes, and socioeconomic drivers (the macroeconomic system). There is added value in that patterns observed in the Vilcanota are comparable and, hence, can be extrapolated (with careful consideration of differences) to a large range of similar valleys along the Central Andes, and likely provide insights for similar valleys worldwide.

SUMMARY

We explore the multidimensional environment–biodiversity–human–time complex of an important cultural and ecological hub in the Central Andes of Peru: the Vilcanota–Urubamba river catchment (Sacred Valley of the Incas and Cordillera de Vilcanota). The watershed begins at the upper borderline of the biosphere where glaciers are retreating, vegetation and local fauna are rising, and humans are cultivating crops and herding camelids at increasingly higher altitudes. Maximum altitude of potato cropping is now reaching 4580 m, whereas the highest vascular plants were found at 5510 m. The number of cultivated species and varieties censused is up to 34 above 3600 m. The midaltitude Valle Sagrado has been occupied for millennia, but is presently undergoing dramatic changes (deforestation at 60 to 90%, fire, exotic invasions, large eucalyptus plantations, etc) resulting from socioeconomic pressures (poverty, malnutrition in 60% of children, outmigration at >21%, market pressures leading to monocultures, and intensive cropping), causing a restructuring of spatially and temporally integrated land use patterns. The number of cultivated plants can be up to 49 in one district around 3500 m, whereas total number of vascular plants above 3500 m in Peru is >1800. The lower part of the valley reaches the Amazonian lowlands, providing a pathway for intensive exchanges with the higher regions. Up to 75 species of cultivated plants are censused in one lower district, with 6500 species of vascular plants recorded below 500 m. Andean valleys such as the Vilcanota, reaching from glacial peaks to tropical rainforests, provide unique opportunities for understanding the complex interactions between landscape, biodiversity, human cultures, and land use change at a large scale. The complex mix of macroeconomics, culture, and ecology are key causes of land use change and its effect on biodiversity.

References

Brako, L. and Zarucchi, J.L. (1993). *Catalogue of the Flowering Plants and Gymnosperms of Peru*. Missouri Botanical Garden, St. Louis MO, USA.

Braun, G., Mutke, J., Reder, A., and Barthlott, W. (2002). Biotope patterns, phytodiversity and forestline in the Andes, based on GIS and remote sensing data. In Körner, C. and Spehn, E.M. (Eds.), *Mountain Biodiversity: A Global Assessment*. Parthenon Publishing, London, pp. 75–89.

Brisseau, J. (1981). *Le Cuzco dans sa Région: Étude de l'aire d'influence d'une ville andine*. IFEA/CEGET, Bordeaux.

Brücher, H. (1989). *Useful Plants of Neotropical Origin and their Wild Relatives*. Springer-Verlag, Berlin.

Cardich, A. (1985). The fluctuating upper limits of cultivation in the Central Andes and their impact on Peruvian prehistory. In Wendorf, F. and Close, A.E. (Eds.), *Advances in World Archaeology*, Vol. 4. Academic Press, Orlando, pp. 293–333.

de Morales, C.B. (1988). *Manual de Ecología*. Instituto de Ecología, Universidad Mayor de San Andrés (UMSA), La Paz, Bolivia.

Denevan, W. (1992). The Pristine Myth: The Landscape of the Americas in 1492. *Annals of the Association of American Geographers*, 82: 369–385.

DeVries, T.A., Hoernig, G., Sowell, P., Halloy, S.R.P., and Seimon, A. (2005). Identification of Chytridiomycosis in *Telmatobius marmoratus* at 4450m in the Cordillera Vilcanota of Southern Peru. In Lavella, E.O. de la Riva, I. (Eds.). Studies on the Andean Frogs of the Genera Telmatobius and Batrachophrynus. Monografías de Herpetología 7, Valencia.

Duncan, R.P., Webster, R.J., and Jensen, C.A. (2001). Declining plant species richness in the tussock grasslands of Canterbury and Otago, South Island, New Zealand. *New Zealand Journal of Ecology*, 25: 35–47.

Fowler, N.L. (2002). The joint effects of grazing, competition, and topographic position on six savanna grasses. *Ecology*, 83: 2477–2488.

Gade, D. (1999). *Nature and Culture in the Andes*. University of Wisconsin Press, Wisconsin.

Gade, D.W. (1975). *Plants, Man and the Land in the Vilcanota Valley of Peru*. Dr. W. Junk, The Hague.

Galiano Sánchez, W., de Olarte Estrada, J., Tupayachi, A., and Ardiles Jara, A. (1995). Conservación de Recursos Fitogenéticos y Análisis de una Microcuenca Hidrográfica en el Valle Sagrado: Calca-Urubamba. Consejo de Investigación de la Universidad Nacional de San Antonio Abad del Cusco. Informe No 5: 1–51.

Gentry, A.H. (1993). Overview of the Peruvian flora. In Brako, L. and Zarucchi, J.L. (Eds.), *Catalogue of the Flowering Plants and Gymnosperms of Peru*. Missouri Botanical Garden, St. Louis MO, USA, pp. 29–40.

Halloy, S.R.P. (1994). Long term trends in the relative abundance of New Zealand agricultural plants. In Fletcher, D.J. and Manly, B.F.J. (Eds.), *Statistics in Ecology and Environmental Monitoring*, Vol. Otago Conference Series 2. Otago University, Dunedin, New Zealand, pp. 125–142.

Halloy, S.R.P. (1999). The dynamic contribution of new crops to the Agricultural economy: is it predictable? In Janick, J. (Ed.), *Perspectives on New Crops and New Uses*. ASHS Press, Alexandria, Virginia, USA, pp. 53–59.

Halloy, S.R.P. and Tupayachi, A. (2004). Rititica, Cordillera Vilcanota (code: PE-SIB-RIT). In http://www.fao.org/gtos/tems/tsite_show.jsp?TAB=4&TSITE_ID=3249.

Halloy, S.R.P., Ortega Dueñas, R., Yager, K., and Seimon, A. (2005). Traditional Andean Cultivation Systems and Implications for Sustainable Land Use. *Acta Horticulturae*, in 670: 31–55.

Hensen, I. (1992). La flora de la comunidad de Chorojo — Su uso, taxonomía científica y vernacular. *Agruco, Serie Técnica*, 28: 1–37.

INEI (2003). *Censos*. Instituto Nacional de Estadística e Informática (Perú). In http://www.inei.gob.pe/.

Kessler, M. (1998). Forgotten forests of the high Andes. *Plant Talk*, 15: 25–28.

Kessler, M., Bach, K., Helme, N., Beck, S.G., and Gonzales, J. (1998). Floristic diversity of Andean dry forests in Bolivia — an overview. In Breckle, S.W., Schweizer, B., and Arndt, U. (Eds.), *Results of Worldwide Ecological Studies*. Proceedings of the 1st Symposium of the A.F. Schimper-Foundation est. by H. and E. Walter. Günter Heimbach, Stuttgart, Hohenheim, Germany, pp. 219–234.

Liberman Cruz, M. (1987). Uso de la tierra en el altiplano norte de Bolivia como base para la evaluación del impacto ambiental de un proyecto de desarrollo rural. *Rivista di Agricoltura Subtropicale e Tropicale*, Firenze, Anno LXXXI: 207–235.

MAP (2003). Portal Agrario, Ministerio de Agricultura del Perú (MAP). In http://www.portalagrario.gob.pe/polt_cusco22.shtml.

Mitchell, W.P. (1976). Irrigation and community in the central Peruvian highlands. *Am. Anthrop*, 78: 25–44.

Murra, J.V. (1975). *Formación Económica y Política del Mundo Andino*. Instituto de Estudios Peruanos, Lima.

NRS (Ed.) (1989). *Lost Crops of the Incas*. Little-known plants of the Andes with promise for worldwide cultivation. National Research Council, National Academy Press, Washington, D.C., USA.

Parmesan, C. and Yohe, G. (2003). A globally coherent fingerprint of climate change impacts across natural systems. *Nature*, 421: 37–42.

Pauli, H., Gottfried, M., Hohenwallner, D., Reiter, K., and Grabherr, G. (Eds.) (2002). *The GLORIA Field Manual — Multi-Summit Approach*, 4th version (draft), December 2002 edn. Global Observation Research Initiative in Alpine Environments — A contribution to the Global Terrestrial Observing System (GTOS), Vienna.

Pietilä, L. and Jokela, P. (1988). Cultivation of minor tuber crops in Peru and Bolivia. *Journal of Agricultural Science in Finland*, 60: 87–92.

Price, L.W. (1981). *Mountains and Man — A Study of Process and Environment*. University of California Press, Berkeley, California.

Pulido, V. (2001). El Libro Rojo de la fauna silvestre del Perú. Normal Legal: DS 013-99-Agricultura, Lima.

Ramirez, M. (2002). On farm conservation of minor tubers in Peru: the dynamics of oca (*Oxalis tuberosa*) landrace management in a peasant community. *Plant Genetic Resources Newsletter*, 132: 1–9.

Rapoport, E.H., Ladio, A., Raffaele, E., Ghermandi, L., and Sanz, E.H. (1998). Malezas comestibles — hay yuyos y yuyos ... *Ciencia Hoy*, 9: 30–43.

Shannon, C.E. and Weaver, W. (1949). *The Mathematical Theory of Communication*. University of Illinois Press, Urbana, IL.

Sibille, O. (1997). *La Cuenca del Vilcanota y sus Áreas de Influencia*. Coordinación Intercentros de Investigación, Desarrollo y Educación (COINCIDE), Cusco, p. 56.

Swinton, S.M. and Quiroz, R. (2003). Is poverty to blame for soil, pasture and forest degradation in Peru's Altiplano? *World Development*, 31: 1903–1919.

Tapia, M.E. and Torre, ADl (2003). La mujer campesina y las semillas andinas: Género y el manejo de los recursos genéticos. In http://www.fao.org/DOCREP/x0227s/x0227s00.htm#TopOfPage.

Thompson, L.G., Mosley-Thompson, E., Bolzan, J.F., and Koci, B.R. (1985). A 1500 year record of tropical precipitation in ice cores from the Quelccaya Ice Cap, Peru. *Science*, 229: 971–973.

Troll, C. (1968). The Cordilleras of the tropical Americas — aspects of climatic, phytogeographical and agrarian ecology. In Troll, C. (Ed.), *Geoecology of the Mountainous Regions of the Tropical Americas*, Vol. 9. Ferd. Dümmlers Verlag, Bonn, pp. 15–56.

Vieyra-Odilon, L. and Vibrans, H. (2001). Weeds as crops: the value of maize field weeds in the valley of Toluca, Mexico. *Economic Botany*, 55: 426–443.

Williams, C.B. (1964). *Patterns in the Balance of Nature and Related Problems in Quantitative Ecology*. Academic Press, London and New York.

Zimmerer, K. (1997). *Changing Fortunes: Biodiversity and Peasant Livelihood in the Peruvian Andes*. University of California Press, Berkeley.

APPENDIX I

Province	District	Capital	Altitude Capital (m)	Population, Projection 2002	Area (km²)	Density (inhabitants km⁻²)	Education[a]	Housing[b]	n cult. sp+var	Cultivated Area (ha)	ha cultiv. per person	Percentage of Land in Parcels <5 ha	Ratio of Permanent to Annual Crops	Sp.1000 hab⁻¹	Sp.100 ha cultivated⁻¹
La Convención	Quellouno	Quellouno	650	11,900	800	14.9	49.9	79	61	11964	1.01	11.0%	4.1430	5.13	0.51
La Convención	Echarate	Echarate	667	56,776	19,136	3.0	51.7	60	75	26383	0.46	12.2%	5.6588	1.32	0.28
La Convención	Santa Ana	Quillabamba	1047	35,784	359	99.6	73.1	19.5	71	6246	0.17	34.4%	5.6094	1.98	1.14
La Convención	Maranura	Maranura	1120	9,419	150	62.7	58.5	68.3	47	2649	0.28	42.0%	5.4600	4.99	1.77
La Convención	Huayopata	Ipal	1660	9,639	524	18.4	57.8	60.6	38	3726	0.39	47.5%	7.2317	3.94	1.02
La Convención	Santa Teresa	Santa Teresa	1700	10,210	1,340	7.6	53.7	73.5	39	2048	0.2	29.5%	1.0398	3.82	1.90
Urubamba	Machu Picchu	Machu Picchu	2060	3,070	271	11.3	64.3	29.4	28	471.8	0.15	49.6%	0.4497	9.12	5.93
Urubamba	Ollantaytambo	Ollantaytambo	2846	9,188	640	14.4	43.1	65.3	25	2008	0.22	60.5%	0.0020	2.72	1.24
Urubamba	Yucay	Yucay	2857	3,516	71	49.8	71.8	3	24	406.7	0.12	88.6%	0.0564	6.83	5.90
Urubamba	Huayllabamba	Huayllabamba	2866	5,270	102	51.4	63.4	11.4	31	975.1	0.19	78.3%	0.0049	5.88	3.18
Urubamba	Urubamba	Urubamba	2871	17,079	128	133.1	66.9	19.7	38	1291	0.08	88.5%	0.0289	2.22	2.94
Calca	Calca	Calca	2928	16,583	311	53.3	61.5	27.8	41	1717	0.1	73.0%	0.0135	2.47	2.39
Calca	Lamay	Lamay	2941	5,898	94	62.6	30.3	67.3	19	1015	0.17	71.3%	0.0047	3.22	1.87
Calca	Coya	Coya	2951	3,932	71	55.0	42.1	30.1	25	561.2	0.14	84.1%	0.0113	6.36	4.46
Calca	Pisac	Pisac	2972	9,769	148	65.9	40.4	55.1	29	1295	0.13	62.1%	0.0005	2.97	2.24
Calca	San Salvador	San Salvador	2995	5,351	128	41.8	27.1	59.4	22	760.5	0.14	82.0%	0.0031	4.11	2.89
Quispicanchi	Lucre	Lucre	3086	4,256	119	35.8	56.3	19.8	20	829.5	0.19	51.4%	0.0025	4.70	2.41
Quispicanchi	Oropesa	Oropesa	3116	6,482	74	87.1	56.7	16.1	25	675.4	0.1	89.4%	0.0038	3.86	3.70
Quispicanchi	Urcos	Urcos	3150	15,811	135	117.4	43.5	20.9	22	1076	0.07	89.9%	0.0015	1.39	2.05
Quispicanchi	Huaro	Huaro	3157	5,289	106	49.8	55.3	36.7	18	442.4	0.08	93.1%	0.0047	3.40	4.07
Quispicanchi	Quiquijana	Quiquijana	3210	11,017	361	30.5	27.3	50.8	27	1601	0.15	88.8%	0.0024	2.45	1.69
Cusco	San Sebastián	San Sebastián	3299	47,297	89	528.8	79.4	5.3	41	666.6	0.01	85.2%	0.0006	0.87	6.15
Quispicanchi	Cusipata	Cusipata	3310	5,812	248	23.4	38	43	16	670.5	0.12	88.6%	0.0014	2.75	2.39
Paucartambo	Caicay	Caicay	3330	2,603	111	23.5	30.5	41.8	25	622.7	0.24	83.3%	0.0015	9.60	4.02
Urubamba	Maras	Maras	3385	8,063	132	61.2	47.1	39	34	3906	0.48	49.9%	0.0003	4.22	0.87
Cusco	Cusco	Cusco	3399	100,572	116	865.4	85.3	2.7	36	873.6	0.01	54.8%	0.0009	0.36	4.12
Canchis	Checacupe	Checacupe	3446	5,212	962	5.4	37.6	39.5	30	713	0.14	87.1%	0.0017	5.76	4.21
Canchis	Combapata	Combapata	3475	5,967	183	32.7	40.5	32.4	27	843	0.14	65.5%	0.0007	4.52	3.20
Canchis	San Pablo	San Pablo	3486	6,183	524	11.8	39.4	42.9	26	636.5	0.1	88.5%	0.0002	4.21	4.09
Canchis	Sicuani	Sicuani	3554	59,295	646	91.8	60.5	20.7	49	2760	0.05	67.9%	0.0002	0.83	1.78

Canchis	Pitumarca	Pitumarca	3570	7,692	1,118	6.9	24	56.7	20	674.6	0.09	56.2%	0.0000	2.60	2.96
Urubamba	Chinchero	Chinchero	3762	10,166	95	107.5	49.2	21.3	34	3138	0.31	64.7%	0.0001	3.34	1.08
Acomayo	Mosoc Llacta	Mosoc Llacta	3802	1,750	44	40.1	43.3	32.6	6	106.5	0.06	64.7%	0.0000	3.43	5.63

Note: Districts of the Cusco Department with parts in the Vilcanota watershed, ranked by altitude of the district capital. Collated and calculated from 1993–1994 census (INEI, 2003).

[a]Proportion of total population above 15 years with complete or partial primary schooling.
[b]Percentage of housing with no services (water, electricity, and sewage).

Part VI

Synthesis

25 Fire and Grazing — A Synthesis of Human Impacts on Highland Biodiversity

Eva M. Spehn, Maximo Liberman, and Christian Körner

INTRODUCTION

Humans have been influencing highland biota around the world for millennia. Humans depend on *in situ* highland resources. The way they are used, however, also influences the well-being of lowlands, largely because the amount of clean water that can be delivered across long distances depends on catchment value. The functional integrity of highlands depends on stable soils, and these, in turn, depend on a stable plant cover. The long-term functioning and integrity of the mountains' "green coat" depends on a multitude of plant functional types and their interaction with animals and microbes. The richer these biota, the more likely system integrity and functioning will be retained in the event of unprecedented impacts — the "insurance hypothesis" of biodiversity (Yachi and Loreau, 1999; for mountain biodiversity, Körner and Spehn, 2002; Körner, 2004).

The highland biota we see today are the net outcome of the long-term interplay among human activities, regional taxonomic richness, and climatic drivers. This volume brings together observational and experimental evidence of anthropogenic influences on the biological richness of high-elevation ecosystems around the world. Fire and pasturing are the logic focal points of such an assessment, given their dominant role over vast highland areas. All other human activities, which might severely affect ecosystems locally, are less significant on an area basis and on a global scale. Although this volume cannot claim global coverage of this wide theme, it highlights the major trends and processes and offers management guidelines.

Although fire and herbivory are the major agents through which humans transform highland biota, both are natural factors that have driven evolution in nearly all ecosystems around the globe. It is the intensity and frequency (the *dose*) and the timing and mode of impact (the *quality*) through which human action can induce significant departures from the sustainable functioning of highland ecosystems and their biodiversity. In this chapter, we will briefly summarize the main findings presented in this volume and distill a few major lines of evidence, but also suggest major gaps of knowledge that culminated in the Moshi–La Paz research agenda of the Global Mountain Biodiversity Assessment program (GMBA 2003). In this attempt, we will not go by chapter but by themes and overarching issues.

FIRE AND DIVERSITY IN THE HIGHLANDS

Fire is one of the key environmental factors that controls the composition and functioning of biota globally. Fire needs fuel, adequate physical conditions, and ignition to come into action. All these three factors generally tend to reduce the significance of natural fire at high elevation under conditions without human influence. Biomass and productivity tend to decline with elevation, the climate gets cooler, and the precipitation–evaporation ratio increases in most cases. Lightning frequency tends to be lower in mountains, and lightning strikes often hit exposed topography with diminished vegetation cover. Biomass fuel commonly needs to contain less than 15% of moisture to inflame (Lovelock, 1979), but after it starts burning, the heat wave can create favorable situations for the spread of fire in otherwise humid conditions. Once lit, the two important factors determining the rate of spreading and the area extent of fire are wind and topography, both of which, in most highland areas, are not favorable for the spread of fires. In contrast to common belief, mountain ecosystems (except for exposed summits and ridges) are less windy than the forelands and plains (Körner, 2003), and the rough topography and fragmented vegetation, which often occur at high altitudes, restrict the spreading of fire. For all these reasons, natural fires commonly are rare at high elevation (e.g. DeBenedetti and Parsons, 1979), and most natural highland floras are not specifically selected for fire resistance, hence, these are easily transformed if fire frequency is enhanced through human action.

Human intervention may reverse these trends, particularly in the tropics, where the precipitation–evaporation ratio often shows a sharp decline above the montane cloud zone (see the chapters by Fetene et al. on the Bale Mountains and by Hemp on Mt. Kilimanjaro). A major problem in the interpretation of the impact of fire on highland biota is that we mostly lack an unburned reference (Aragon et al., this volume). The current vegetation commonly offers only grades of fire impact, but we do not know how much of the potential flora — and with it, other organisms groups — have already been lost, with only the commonly depauperate, fire-adapted fraction of the original highland flora left after millennia of enhanced burning in an otherwise not particularly fire-prone environment. Several researchers have commented on this issue (Aragon et al. and Wesche, both this volume). We need "control" areas of sufficient extent against which the gradual impact of fire can be rated and ranked. Such reference habitats could be protected areas or topographically isolated mountains that cannot be reached by fire. Given that such refugia will commonly be small and strongly dependent on the surrounding reservoir of taxa, these would always present rather coarse approxima-

tions, and the nature of these habitats would potentially confound the "absence-of-fire effect." This "reference" diversity can assist, however, in estimating the degree of transformation that the vegetation has undergone through the action of fire, naturally occurring ones or lit by man, by calculating a biodiversity intactness index (BII; Scholes and Biggs 2005).

In the first six chapters of this book, a variety of assessments have been presented on the impact of fire in tropical highland ecosystems. The spectrum of effects range from the positive impact of burning in terms of biodiversity to disastrous consequences. The reasons for the broad range of fire effects on diversity are obvious. Frequently burned areas are inhabited by organisms that were selected for coping with fire; hence, regular burning exhibits no or little effect, because this is the very reason for the given biodiversity (e.g. the tropical high-elevation grassland studied by Wesche in Uganda [this volume] or the montane rangeland in Madagascar studied by Rasolonandrasana and Goodman [this volume]). Thus, it would be a misleading conclusion to assume that fire is beneficial for maintaining biodiversity. The question to be asked is whether or not the given vegetation composition fulfills an optimum set of ecosystem services such as land use, ground coverage, soil conservation, biodiversity conservation, and catchment value.

As fire frequency increases, tall woody taxa (first trees, later shrubs) are suppressed, and dwarf shrubs and grassland become dominant. At highest fire frequency, only a few species can cope, and these are commonly poorly palatable tussock grasses and a tiny intertussock flora that is destroyed easily by trampling (Figure 25.1). Intense burning selects for plants with belowground meristems (e.g. grasses), annuals, or geophytes (belowground storage organs such as bulbils). As the latter two categories are commonly rare at high elevation, the pyrophytic mountain flora gets poorer in taxa with altitude, also for this reason. Each of these steps of degradation opens, stepwise, the floor for invasive species, either from the adjacent lower-elevation flora or for exotic ruderal species. In addition, a downslope migration of alpine taxa into burned montane forest areas has been observed (Hemp, this volume). At burned sites in the

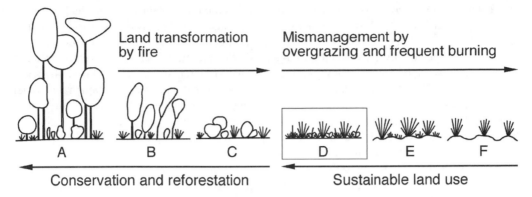

FIGURE 25.1 A schematic representation of land transformation and degradation following fire and misman-agement in the tropics and subtropics. Step D could be a desirable compromise between pasture needs, soil protection, and biodiversity conservation, with a diverse intertussock ground cover becoming key.

highlands of Madagascar, exotic rodents rep-resent 42% of those captured, whereas in unburned areas this is 11% (Rasolonan-drasana and Goodman, this volume). Species diversity may be very low in a given fire-prone community, but the overall diversity in a larger area may suggest no such decline because of a mosaic of differently impacted zones, mosaics that are strongly enhanced by a rich topography (geodiversity). For this rea-son, the judgment of the impact of fire strongly depends on the size of land area con-sidered. Imagine a mosaic of forest remnants interspersed with burned areas: As the latter will contain a very different biocenosis than the first, the overall diversity may actually increase if data for both categories of land cover are pooled, whereas, at the same time, the rare forest flora and fauna may diminish due to the fragmentation. Axmacher et al. (this volume) document such a case for geometrid moths in Africa.

The assessment of the impact of fire should thus address four questions:

1. *Biodiversity:* How far has the result-ant organismic diversity departed from the natural zonal "climax," and what are the biotic losses incurred (loss of rare species, important plant functional types, habitats for certain animals, etc.)? How is the assess-ment affected by pooling diversities across mosaics of habitats, and how

is the individual habitat type affected (scale dependency)?

2. *In situ resources:* To what extent has the functional integrity of the result-ant ground cover been retained, irre-spective of its taxonomic composition? Is the soil well pro-tected year-round? How much is pro-ductivity reduced? How is forage quality affected by the fire-driven changes in species absence, pres-ence, and abundance?

3. *Ex situ resources:* How does the fire-driven transformation affect catch-ment value (water yield), and does it affect landscape attractiveness (tour-ism)?

4. *Socioeconomic factors:* What are the socioeconomic implications of ques-tion 1 to question 3? How are animal production, household fuels, medic-inal plant availability, ownership and land use rights, land use intensity, overall income, safety (erosion, floods, etc.), and population growth affected by any given fire regime?

Based on these assessments and circum-stantial evidence, the general patterns of moun-tain fire regimes in the subtropics and tropics have been determined as the following: Increased fire frequency and intensity has been observed around the globe. Increased human influence is the main cause, but climatic changes have contributed in some areas to this trend (e.g. Mt. Kilimanjaro, Africa; see the following text). Regularly burned areas show little effect on plant species diversity when burning intervals are between one and a few years (e.g. Aragon et al. and Wesche, this volume). Other studies on páramo tussock vegetation showed that fire often leads to degradation (Laegard, 1992; Ramsay and Oxley, 1996) if the regeneration of tussocks takes longer than the burning frequency. Burned areas are commonly poorer in species than unburned areas, particularly when uniform plots are compared and when the unburned control contains forests. Burned mountain areas contain flora and fauna that are almost completely different from unburned areas, and the species spectra in burned areas contain numerous widespread and very common taxa. Woody components of the flora become either completely eliminated or very uniform, as is the case with the *Erica* shrub in the African mountains. Moderate fire frequencies do not necessarily lead to incomplete ground cover and reduction in the bulk number of taxa present. However, in any case, they induce a change in ecosystem functioning that includes facilitation of further fires, reduced water and soil nutrient retention, reduced carbon storage except for black carbon (e.g. charcoal), more uniform packaging of biomass and age structure, and a greater abundance of R-strategy organisms (fast and intense reproduction) vs. K-strategy organisms, which live very long and facilitate niche diversification for smaller taxa. Commonly, plant taxa belonging to the latter type of life strategy produce stronger root systems and protect mountain slopes much better than R-strategists.

The significance of the presence or absence of certain taxa (a functional significance of biodiversity) is best illustrated by the Kilimanjaro case: A recent greater incidence of fires, facilitated by a drier climate in the uppermost montane *Erica* forest belt, had destroyed this ecosystem almost completely and, with it, one of its major functions, trapping cloud water. Hemp (Hemp, this volume) estimated that the impact in terms of the water-yielding to savanna-type forelands of Kilimanjaro by far exceeds the effect of its melting ice cap. Highland fire had eliminated almost completely a key functional plant type that had produced a very significant ecosystem service to the downhill population. Similar dramatic effects of fire-driven land transformation have occurred in the Bale Mountains (which lost almost all their forests), which supply eight major river systems and the Nile (Fetene et al., this volume). In such cases, the loss of a certain group of life-forms (trees) is more significant than the loss of species diversity as such.

From a biodiversity- and ecosystem-functioning point of view, fire is not a desired tool of land management at high elevation. High-mountain biota, the treeless alpine belt in particular, differ in this respect from many lowland ecosystems, the richness and functioning of which depend on recurrent fires. However, once the landscape had been transformed to fire-tolerant highland biota, a moderate use of fire may be sustainable under certain conditions if slopes are not too steep, the follow-up grazing does not lead to soil erosion through trampling, and when the soil (its clay content, in particular) ensures sufficient nutrient and water retention. However a loss of biodiversity, particularly functional diversity, is almost always incurred, but most often we lack the unburned control to quantify the actual losses.

GRAZING AND MOUNTAIN BIODIVERSITY

Animal husbandry represents the major use of highland biota around the globe. Beyond the climatic zone that permits tillage crop farming, grass and herbage must be transformed by animals to provide food to humans. Some old mountain cultures have created man-made high-elevation ecosystems with a very specific flora and fauna, high biodiversity, stable slopes, and high water yield (Körner et al., this volume). However, these traditional land use forms can neither be "exported" to other regions nor do these systems retain their functional integrity if they become either over- or underexploited. In other words, their biodiversity and sustained functioning depends completely on well-dosed human intervention, giving limited leeway to regional population growth or abandonment. The 11 chapters in this book on grazing effects on mountain biodiversity cover a broad range of elevations (and thus mountain climates) from montane temperate forests (about 1600 m) to the high Andes (about 4600 m). The majority of data comes from traditionally managed rangelands. Three chapters deal with montane-forest grazing and forest succession after land clearing, one with firewood collection, and seven present observational and experimental data on the impact of grazing.

Grazing the highlands may be desirable in terms of biodiversity and ecosystem functioning if managed sustainably, and may even increase biodiversity (e.g. Sarmiento et al., this volume). As with fire, grazing and browsing are natural drivers of plant life in all mountains. The issue here is whether the type of replacement of natural ungulate herbivores by domestic ones and the intensity of land use are sustainable and tolerable or destructive to biodiversity and ecosystem functioning. The missing-reference issue is even more problematic with regard to grazing, because all vegetated mountain terrain is naturally grazed to some degree. Quite often, wild-animal grazing has replaced domestic animal grazing; in other cases, wild- and domestic-animal herbivory is additive. Even if only conservation areas with wild animal grazing are taken as a reference, we commonly do not know what a sustainable wild-animal abundance would be, because the top carnivores that controlled herbivore populations have diminished.

We make a distinction here between pasturing as such and the combination of grazing and fire: (1) There are areas that are burned accidentally or for hunting but not grazed by livestock that are still transformed to grassland. (2) There are areas that are transformed by the grazing process alone. (3) There are areas where one facilitates the other. The latter ones are restricted to subtropical and tropical highlands. Grazing can influence fire frequency and intensity, and fire determines what is left or regrown for herbivores, not only in terms of quantity but also in terms of forage quality (Hobbs et al., 1991). The study by Aragon et al. (this volume) in the montane grasslands of northwestern Argentina shows that fire has stronger effects than grazing on biomass and plant cover, favoring more palatable species and

thus also affecting species composition in the long term. It appears that the frequency of fires and grazing events is crucial for biodiversity in these high-elevation grasslands. Disturbances by grazing and fire provide open space for colonization that, in turn, can modify species diversity, promote seedling establishment of certain species, and change the general structure of the community (e.g. Valone and Kelt, 1999). In areas where fires are not easily lit or where burning is not the custom as in most mountain regions in the temperate zone, logging is a frequent precursor of pasturing the mountains. Once more, we deal with millennia-long impacts as illustrated by 7000 years of agropastoralism in the Andes (Browman, 1987), 5000- to 7000-year-old herding traditions in the Alps (Eijgenraam and Anderson, 1991), and similarly old land use practices in the Himalayas.

Many of these traditionally used highlands are extremely rich in plant species. The páramo region from Costa Rica to the north of Peru alone has 5200 plant species of 735 genera and 133 families (Rangel, this volume). Globally, the treeless alpine flora alone includes 4% of the globe's flora but covers only 3% of the inhabitable land surface area. The land area considered here includes the upper-montane forest and the treeline ecotone covering approximately a tenth of the globe's vegetated area (about 10 Mio km^2; Körner et al., 2005) and hosts around 15 to 20% of all plant taxa. Therefore, whatever land use is incurred, it particularly affects rich biota (Körner, 2004).

The Andean páramo is a special case, not only because of the earlier-mentioned species richness but also because of its comparatively low elevation (often as low as 3200 m), which could be forest-covered, particularly in the relatively humid northern part that reaches up to 4000 mm of rainfall annually. Even in the drier parts in the south with only 600 mm of precipitation, there is no climatic reason for the absence of forest. This anomaly has given rise to the assumption that the páramo is a man-made ecosystem (Ellenberg, 1979; Laegard, 1992) and that the restriction of forest patches to scree and boulder slopes, not accessible by fire or grazing animals, is a result of land use. However, many of the typical páramo taxa have been identified as of ancient evolutionary origin associated with today's type of vegetation (Cleef, 1981), and it is now believed that these largest tropical rangeland areas are the result of both natural treelessness and human land use (Luteyn, 1999; Rangel, this volume). Most other high-elevation rangeland below the climatic treeline (about 3900 to 4300 masl in the subtropics and tropics) would be invaded by trees in the absence of fire and grazing. However, for the páramo, this may not be the case, and it is uncertain for the Bale Mountains (see the discussion in Miehe and Miehe, 1994). On Mt. Kilimanjaro and Mt. Kenya, the suppression of fire would definitively induce a succession back to a dense montane forest, with the *Erica* phase becoming stationary possibly only in the uppermost elevations. At lower elevations, other less fire-tolerant taxa would become more abundant (following from the data presented by Hemp, this volume, and Wesche, this volume).

In line with findings for open pastureland, moderate-intensity grazing of temperate montane forests with cattle is increasing rather than decreasing biodiversity. Unlike wild ungulates or goats, cattle mainly feed on grass and profit from minor clearings intentionally opened by farmers by selective logging (Mayer et al., this volume). The complete banning of forest pasturing in higher latitude mountains is thus not desirable, provided stocking rates are low and obligatory browsing livestock (such as goats) are avoided. However, this mode of land use cannot be exported to lower latitudes and to montane forests with their much denser stands. Even in temperate mountains, grazing of montane forests needs a lot of local knowledge and careful management.

Once montane forests are completely clearcut or burned for grazing, it may, however, take very long to recover, as exemplified in northern Argentina by Carilla et al. (this volume). The more rapidly trees invade and grow up, the more diminished the flora becomes. Biodiversity only recovers when the system reaches a late successional stage, in this case (Carilla et al., this volume) with slow-growing *Podocarpus*, which may take several hundred years to obtain a new steady state. Whether, and how fast, such forest recovery may occur will also be strongly deter-

mined by specific climatic conditions (e.g. favorably wet periods) and by external forces such as rural population growth, as was shown for montane *Prosopis* forests in another part of the Argentinean Andes (Morales and Villalba, this volume). Ecologically, montane forest pasture systems with small-size clearings (from timber use) are thus preferable to clear-cutting regimes, also in light of the difficulties to reestablish forests.

High-elevation grassland and open-rangeland grazing in regions that have a long evolutionary history of ungulate presence commonly has little impact on biodiversity as long as full ground cover is retained and stocking rates do not cause the highly palatable species to disappear. In a very detailed analysis, Sarmiento (this volume) shows that such adapted plant communities in the Venezuelan páramo may even lose 30 to 40% of their aboveground biomass without a significant effect on biodiversity. The author demonstrates that grazing can promote plant species diversity by balancing competition among taxa for key resources, but when grazing intensity is enhanced, the already-existing dominants tend to get even more dominant. Therefore, the abundance of the less-palatable dominants vs. that of the highly palatable subdominants is the best measure of appropriate stocking rates (Bustamante et al., Alzerreca et al., this volume). The effects of animal trampling can be more severe than biomass removal, particularly for small shrubs but also on wet ground, as was shown for Andean wetlands by Hernandez et al. (this volume). These authors have demonstrated that plants avoided by cattle may still be essential for the functioning of such systems through their water retention capacity. In this specific case, subterranean necromass (dead leaf sheets) form a sort of "sponge" that is easily destroyed by trampling.

A key question in high-elevation pasturing is that of appropriate animal selection. Molinillo and Monasterio illustrate, by comparing pastures in Bolivia, Argentina, and Venezuela, that "picky" animal types such as cattle, sheep, and alpaca have much more impact on pasture quality and biodiversity than species with a broad food selection, such as the llama. Several studies show that an increase in soil humidity is correlated with grazing intensity and the composition of herds changing to a higher alpaca and sheep proportion (e.g. Molinillo and Monasterio, this volume; Buttolph and Coppock 2004). The more selective animals are, the more restricted is the actual pasture space used, and even low stocking rates may destroy the most valuable areas. This becomes most critical in periodically dry regions, where herds must be sustained on small areas with good ground moisture. Several studies have documented the key role of these moister sites for the Andes (bofedales, etc.; Bustamante et al., Alzerreca et al., Hernandez and Monasterio, this volume) and for the dry inner parts of the Himalayas (Rawat and Adhikari, this volume). Such marsh-type meadows (bofedales, in the Andean altiplano) may represent only 5% of the total land area (as shown for Ladakh by Rawat and Adhikari) but may have to carry, periodically, the full stocking of 100% of the potential grazing land. There is overwhelming evidence that these areas need prime attention in any management plan for sustainable highland land use. The one key message from these and many other works, including the temperate-zone mountains, is that the total area of potential grazing land is an unsuitable reference for the calculation of stocking rates due to the use of microenvironments such as bofedales and marshlands. In addition, transhumance, shepherding, and rotations provide methods of land use that enable recovery of pastures during the growing season (Molinillo and Monasterio, this volume; Preston et al., 2003).

Several authors in this volume provided support for the intermediate disturbance hypothesis for maximum biodiversity in high-altitude grazing land (Sarmiento et al.; Bustamante; Rawat, and Adhikhari, this volume). Moderate grazing increases plant species diversity at local (or patch) scale, as herbivory helps to reduce the height and abundance of the taller, more aggressive species, thereby increasing the competitive ability of other taxa, especially when resources are limited. Disturbance by trampling is especially effective under wet soil moisture conditions, which can vary seasonally as well as spatially. Stocking rates that represent this intermediate disturbance are best assessed by the balanced coexistence of indicator taxa that belong to the trampling-resistant, mechan-

ically important "slope engineer" group and the more vulnerable but highly nutritious group of favorable rangeland species. This mix of robust vs. nutritious species is best represented by the Andean altiplano pastures, which have become dominated by a small group of tussock grasses as tall as 1.5 m and 1 m in diameter (e.g. *Festuca orthophylla, Stipa leptostachya*) and are hardier and less palatable than swards of annual graminoids, which they replace in intensively and selectively grazed areas (Beck et al., 2001). Poorly palatable tussock grasses are found in comparable elevations around the globe and are commonly widely spaced with very little vegetation in between. It is the fate and vigor of this intertussock vegetation that determines regional biodiversity, forage quality, and surface erosion. The intertussock space is key in terms of forage protein content and erosion control. In large parts of the altiplano, intertussock area covers from 80 to 95% of the land area, and it has not been explored in studies separate from tussocks so far. Future research needs to focus on these mosaics of small-stature, often ephemeral taxa, and stocking rates and management plans need to account for this often-overlooked vegetation (Körner et al., this volume).

In one specific chapter for Australia (Green et al., this volume), we are reminded that mountain vegetation adjusted to grazing and trampling is nonexistent in Australia, New Zealand, and the tropic alpine grasslands of New Guinea, the flora of which evolved without ungulates. The major grazing animals in the alpine zone are insects. Early settlers have nearly destroyed the Australian alpine vegetation by livestock grazing, and it has been calculated that rehabilitation and revegetation of the eroded landscape has cost twice the financial benefits of the 100 years of pasturing, not counting the losses in terms of clean water provision and hydroelectric energy.

A case of unsustainable high-elevation land use (the "Teresken syndrome") in the eastern Pamir is presented in two chapters. Akhmadov et al. (this volume) report on the pasture and soil degradation and desertification in Tajikistan that led to a massive productivity decline (down to 10 to 20% of its original productivity) and an increase in poisonous and unpalatable species. Breckle and

Wucherer (this volume) show the consequences of the lack of external energy sources
(coal supply by the former Soviet Union) since
the independence of the state of Tajikistan.
Large high-elevation land areas either have
been cleared from forests or are too dry for
tree growth, as is the case in eastern Pamir.
Shortage of firewood led to shrub and brush
harvesting, also a widespread practice in the
páramos and Andean altiplano. In the case of
the Pamir, the single-most prominent dwarf
shrub (teresken or *Ceratoides papposa*) in the
alpine desert plateau is excavated for its rootstock for household fuel; this shrub taps deep
moisture and represents a prime food source
for goats, sheep, and camels and stabilizes
erosion. The shortage of fuel and the poverty
of the region lead to actions that diminish
diversity, create erosion, remove fodder, and,
in the end, exhaust this energy supply. A similar case is the excavation of *Azorella com-
pacta* in the Bolivian altiplano to supply firewood for drying borax, a mineral excavated in
the region for industrial use. These last two
cases illustrate best the links between land
care, biodiversity, and poverty, which are
addressed in the following section.

SOCIOECONOMIC ASPECTS OF MOUNTAIN BIODIVERSITY

The previous chapters made it quite clear that
land care is the result of a decision process that
is rooted in human expectations and needs.
Exemplified by the situation in the transboundary mountain rangelands between Lesotho and
South Africa, it is made obvious that land care
needs to create incentives for local stakeholders; otherwise, it will not come into action
(Everson and Morris, this volume). By shifting
the fire regime from random burning (mostly
annual) to well-timed biannual burning, biodiversity, vegetation cover, and productivity
increased, but the critical step toward such a
fire regime was the initiation of jobs for conservation programs. These links between local
benefits and sustainable land management have
been widely explored around the globe. There
is a wealth of evidence from other regions, as
for instance reviewed for the Himalayas in

Nepal (Basnet, this volume) and for the European Alps. There are encouraging examples that
natural resource degradation can be limited by
diffusing knowledge about natural resource
stewardship using manageable practices. Participatory approaches involving herders in the
assessment of and management decisions on
livestock husbandry and sustainable resource
use provide a sound basis for negotiation among
stakeholders with different interests (Inam-ur-
Rahim and Maselli, 2004). Active participation
of the local population is key and the bottom-
line message from all mountain land-care programs.

Monasterio and Molinillo (this volume)
point out that land care needs focal areas both
in terms of conservation and pastoral
resources. Given the key function of high-
Andean marshlands, despite their small fraction with regard to land area, they illustrate
both the sensitivity of these wetlands and the
value of indicator plant species to assess management success. They make the point that no
other part of the Andean ecosystem is as
strongly connected to low-elevation well being
as these wetlands, because they determine
regional water availability. Their connection to
the lowlands is perhaps one of the strongest
arguments for sustainable highland management. Gravity works in one direction, and
whatever happens upstream affects downstream life conditions. Halloy et al. (this volume) make a plea for acknowledging the far-
ranging consequences of highland land care for
the complex mosaic of interdependencies
along a valley catena. They showed that biodiversity research, both in the wild and domestic
realm, needs to account for such larger-scale
processes and interdependencies.

CONCLUSION

There is no question that humanity has become
a major player in the shaping of landscapes and
the biodiversity that they contain throughout
the majority of the world's mountain areas. The
transformations that occurred in the distant
past were imposed on these high-elevation
biota by a society that was, in large part, self-
supporting and fully dependent on the sustained services of their mountain ecosystems.

The arrival of modern times with easier accessibility of mountain areas, e.g. for tourism or for mountain dwellers to seek markets and job opportunities, provides new ways of ensuring livelihood in the mountains. In many tropical mountain regions, there is an increase in population and basic life support needs and an increased demand for resources per capita. Whatever measures one applies to conserve the functional integrity of highland ecosystems and their biotic richness, there is no way to succeed without integrating the local people and their needs. It is, however, an illusion that this is enough. The highlands commonly do not offer the extractable resources that permit coverage of the population's growing demands. Hence, there is hardly any way out of the vicious cycle of poverty and land destruction in many mountain regions without external resources. The key, then, is in the way these are provided. One very limited avenue is employment as part of conservation programs; another is the creation of and access to markets for special products. A third approach is tourism, which has its own problems. However, the most significant remedy by far has not been explored yet — the services highland farmers and pastoralists can provide by careful catchment management.

Many billions of dollars are extracted from mountain ecosystems worldwide in the form of clean water and hydroelectric energy. It has been estimated that nearly half of mankind depends on mountain water resources (Liniger et al., 1998; Messerli, 2004). There is no question that the amount and quality of water yielded by mountain catchments is driven by land management. Lowland societies have not yet paid for this service and take it for granted. It has been estimated that land care in mountain watersheds can increase water yield per hectare of managed land by 10% (Körner, 2004, Körner et al., this volume). Well-maintained pastures with good ground cover and soil structure evaporate less than ungrazed rangeland, they store water temporarily and hence improve dose yielding. All these characteristics prevent erosion, thus preventing filling dams with sediments.

There is an urgent need for these services through sustainable land use in the highlands to be acknowledged, quantified, made public, and funded. Without such a lowland–highland contract, the long-term fate of the steep slopes in overpopulated mountain watersheds is not very promising, and with this, biological richness will continue to decline. Although an extreme case, because of the lack of wild mammalian grazing, the protection of the Snowy Mountains in Australia from livestock grazing only became a reality once it was realized that the financial benefits of land care are a multiple of those of pastoralism (Green, et al. ; Körner et al., both this volume). However, for most other mountain regions with ungulates, there is consensus that land use, both in the form of fire management and grazing, is not necessarily negative for mountain forests and open mountain rangelands if land use quality and intensity are under control. There are many examples in which sustainable land use, in fact, has created new, stable, and attractive mountain ecosystems.

The integrity and biological richness of mountain biota will continue to depend on human land care. This volume illustrates many facets of the links between land use and biodiversity, with the latter representing the most sensitive indicator of the degree of sustainability. The absence or presence and the abundance of certain plant species, plant life-forms, and plant functional types are very sensitive indicators of the quality of land management, as shown in many contributions in this book. These organisms integrate mismanagement or sustainability over long periods. However, we often do not know how far historical land use has already transformed biota to judge the current conditions. Perhaps it is a dream to see that the quality of highland management will be assessed (and paid for by lowlanders) based on such biological indicators, but it would ultimately benefit the local population and those who profit from catchment value and conservation. The link between water and biodiversity should become the core of any highland management plan.

References

Beck, S., Paniagua, N., Yevara, M., and Libermann, M. (2001). La vegetacion y uso de la tierra del altiplano y de los valles en el oeste de Tarija, Bolivia. In Beck, S., Paniagua, N., Preston, D. (Eds.). *Historia, Ambiente y Sociedad en Tarija, Bolivia*. La Paz, Bolivia, Instituto de Ecologia. pp. 47–93.

Browman, D.L. (1987). Pastoralism in highland Peru and Bolivia. In Browman, D.L. (Ed.). *Arid Land Use Strategies and Risk Management in Highland Peru and Bolivia*. Westview, Boulder, CO. pp. 121–151.

Buttolph, L.P., Coppock, D.L. (2004). Influence of deferred grazing on vegetation dynamics and livestock productivity in an Andean pastoral system. *Journal of Applied Ecology*. 41. pp. 664–674.

Cleef, A.M. (1981). The vegetation of the páramos of the Colombian Cordillera Oriental. *Dissertationes Botanicae*. 61, 321 pp. Vaduz J. Cramer. Also published in El Cuaternario de Colombia 9 (T. Van der Hammen, Ed.). Amsterdam.

DeBenedetti, S.H., Parsons, D.J. (1979). Natural fire in subalpine meadows: a case description from the Sierra Nevada. *Journal of Forestry*. pp. 477–479.

Eijgenraam, F., Anderson, A. (1991). A window on life in the Bronze Age. *Science* 254, pp. 187–188.

Ellenberg, H. (1979). Man's influence on tropical mountian ecosystems in South America. *Journal of Ecology*. 67. pp. 401–416.

GMBA (2003). Moshi-La Paz Research Agenda on land use effects on subtropical and tropical mountain biodiversity. DIVERSITAS Newsletter 5, pp. 12–14.

Hemp, A. (2005). Climate change driven forest fires marginalize the impact of ice cap wasting on Kilimanjaro. *Global Change Biology* 11. (7), pp. 1013–1023.

Hobbs, T., Schimel, D., Owensby, C., Ojima, D. (1991). Fire and grazing in the tallgrass prairie: contingent effects on nitrogen budgets. *Ecology* 72, pp. 1374–1382.

Hofstede, R.G.M., Mondragon Castillo, M.X., Rocha Osorio C.M., (1995a). Biomass of grazed, burnt, and undisturbed Páramo grasslands, Colombia. I. Aboveground vegetation. *Arctic and Alpine Research* 27, pp. 1–12.

Hofstede, R.G.M., Chilito, E.J., Sandoval, E.M. (1995 b). Vegetative structure, microclimate, and leaf growth of a páramo tussock grass species, in undisturbed, burnt and grazed conditions. *Vegetatio* 119, pp. 53–65.

Inam-ur-Rahim, Maselli, D. (2004). Improving sustainable grazing management in mountain rangelands of the Hindu Kush-Himalaya: an innovative participatory assessment method in Northern Pakistan. *Mount Res Dev* 24:124–133.

Körner, Ch., Spehn, E.M. (Eds.). (2002). *Mountain Biodiversity. A Global Assessment*. Parthenon Publishing, Boca Raton.

Körner, Ch. (2003). *Alpine plant life*. Second Ed. Springer Verlag, Berlin.

Körner, Ch. (2004). Mountain Biodiversity, Its Causes and Function. *Ambio*. 7. Special Report. 13. pp. 11–17.

Körner, Ch., Ohsawa, M. et al. (2005). Mountain systems. In *Condition and Trends Assessment/Millennium Ecosystem Assessment*. Island Press, Washington, D.C., chap. 24.

Laegard, S. (1992). Influence of fire in the grass páramo vegetation of Ecuador. In Balslev, H., Luteyn, J.L. (Eds.). *Páramo: An Andean ecosystem under human influence*. Academic Press, London, pp. 151–170.

Liniger, H.P., Weingartner, R., Grosjean, M., Kull, C., MacMillan, L., Messerli, B., Bisaz, A., Lutz, U. (1998). Mountains of the World, Water Towers for the 21st Century — A Contribution to Global Freshwater Management. *Mountain Agenda*. Paul Haupt, Bern. 28 pp.

Lovelock, J.E. (1979). *Gaia: A New Look at Life on Earth*. Oxford University Press, Oxford.

Luteyn, J.L. (1999). Páramo: a checklist of plant diversity, geographical distribution, and botanical literature. *Memoirs of the New York Botanical Garden* 84. New York.

Messerli, B. (2004). Mountains of the World — Vulnerable Water Towers for the 21st Century. *Ambio* 7, Special Report 13, pp. 29–34.

Miehe, S., Miehe, G. (1994). Ericaceous forests and heathlands in Bale Mountains of South Ethiopia. *Ecology and Man's Impact*. Traute Warnke, Verlag, Hamburg, Germany.

Preston, D., Fairbairn, J., Paniagua, N., Maas, G., Yevara, M., Beck, S. (2003). Grazing and environmental change on the Tarija altiplano, Bolivia. *Mountain Research and Development* 23, pp. 141–148.

Ramsay, P.M., Oxley, E.R.B. (1996). Fire temperatures and postfire plant community dynamics in Ecuadorian grass páramo. *Vegetatio* 124, pp. 129–144.

Scholes, R.J. Biggs, R. (2005). a biodiversity intactness index. *Nature* 434:45–49.

Swinton, S.M., Quiroz, R. (2003). Is poverty to blame for soil, pasture and forest degradation in Peru's Altiplano? *World Development* 31, pp. 1903–1919.

Valone, T., Kelt, D. (1999). Fire and grazing in shrub-invaded arid grassland community: independent or interactive ecological effects? *Journal of Arid Environments* 42, pp. 15–28.

Yachi, S., Loreau, M. (1999). Biodiversity and ecosystem productivity in a fluctuating environment: the insurance hypothesis. *Proceedings of the National Academy of Science USA* 96, pp. 1463–1468.

26 The Moshi–La Paz Research Agenda on "Land Use Effects on Tropical and Subtropical Mountain Biodiversity"

The following list of priority scientific questions emerged from discussions at the two GMBA workshops (Moshi, Tanzania, 2002; La Paz, Bolivia, 2003), along with the need to address them in a concerted approach across the world's highland biota, with particular focus on the tropics and subtropics. This list exemplifies the required research, but is not exhaustive.

THE USE OF HIGHLAND VEGETATION AND HUSBANDRY SYSTEMS (FOOD)

What is the sustainable annual biomass production of herbaceous (including grass) and woody vegetation in tropical and subtropical highlands?

What is the differential contribution of dominant (tussock grasses and shrubs) and subdominant species to overall yield and biodiversity?

What is the role of nitrogen-fixing plants and their phosphate limitation for overall yield?

What is the impact of grazing and trampling on pasture species diversity (including wildlife and other trophic levels)?

How can these impacts be modified by shepherding or the use of fences?

What is the significance of animal type, stocking rate, and animal weight for pasture integrity?

How does plant diversity affect the capacity to recover after disturbance (resilience and ecosystem integrity)?

How do wild and domestic animals interact on common pastureland?

FIRE ECOLOGY

What is the optimal fire frequency for sustainable yield and biodiversity?

What is the differential effect of fire on dominant (tussock grasses and shrubs) vs. subdominant species?

Are there alternatives to burning for improving forage quality while at the same time maintaining or increasing biodiversity?

How does burning affect runoff and the export of solid organic and inorganic matter and dissolved nutrients?

What is the relative significance of climatic vs. anthropogenic controls of fire?

HIGHLAND CROPPING, HUNTING AND GATHERING, AND MEDICINAL PLANTS

Establishing wild-medical-plant inventories of mountain regions.

What is the effect of wildlife management on richness of mountain biota?

Upper-montane crops and medicinal plant gardens: genetic treasures — how to treat and protect them?

What are the sustainable cropland rotation systems on high-elevation slopes?

REGENERATION AND REVEGETATION OUTSIDE FORESTS

What is the effect of animal grazing and fire on regeneration and revegetation?

What are the techniques and their levels of success of land surface stabilization on steep slopes with local plant species?

What is the significance of persistent soil seed banks and natural seed migration into disturbed mountain areas for biodiversity and revegetation success?

How to cope with scrub invasion in grassland? Leave it, destroy it, or use it?

UPPER-MONTANE FOREST AND THE HIGH-ELEVATION TREELINE (FIBER AND FUEL)

Livestock and mountain forests — what is the optimal balance between food and fiber production at high biodiversity?

How would the high elevation treelines look if they are natural? How do climate and disturbances control them?

Regeneration mechanisms and rates of forest regeneration in former high-elevation pasturelands (native vs. exotic species).

Remnants of the former climatic treeline: a source for reforestation and biodiversity?

Firewood in the upper-montane forest — how to use it sustainably?

What are the effects of non-tree biomass harvesting (e.g. shrubs and cushion plants) on biodiversity?

What are the differential effects of high-elevation forests and non-forest vegetation on habitat quality and animal biodiversity?

CROSSCUTTING RESEARCH ISSUES

Although significant for biodiversity, all of these themes have links to watershed management. Hence, the evaluation of hydrological consequences and erosion risk (which affect biodiversity) are central to all forms of land use.

In each of these fields, there is indigenous knowledge and experience that needs to be implemented and documented.

For each theme, simple and common methods as well as common protocols need to be adopted, and training is needed (field courses).

Interaction of land use with climate change needs to be considered in view of the most rapid climatic change happening in tropical mountains.

Results need to be communicated in ways that convince local stakeholders. Most promising is the practical involvement of local people in the research work and field demonstrations.

RESEARCH IMPLEMENTATION

This research agenda is designed to catalyze new research activities with a comparative emphasis. The GMBA network activities aim to create consortia for joint research and can offer coordination and platform support to facilitate multiple-site approaches, using common research protocols. GMBA can assist funding agencies with lists of qualified international peers. Global Mountain Biodiversity Assessment encourages researchers in mountain ecology to adopt research plans that account for these pending themes and hopes that national research agencies will consider these aspects in their funding priorities.

Subject Index

A

aerial biomass distribution, 94
Afroalpine flora classification, 189
Afromontane archipelagos, 25
Afromontane vegetation, 41
age structure, 279
agricultural calendar, 142
agricultural intensification, 312
agroforestry, 329
alfalfa hay, 179
alpaca, 139, 157
alpine, definition, 3
alpine herbfield, 215, 218
alpine land area, global fraction, 4
altitudinal gradient, 72
Andean wetlands, 189
Andes, Peru, 155
Andringitra National Park Madagascar, 80
animal husbandry, 342
anthropogenic highland biota, 3
architectonic models, 189
Argentina, Jujuy, 277
    Los Toldos, 92
    Northwest, 91, 265
    Quebrada de Humahuaca, 277
    Salta, 92
Argentinean yungas, 266
Australia, 213, 345
avalanche protection, 256

B

Bale Mountains, Ethiopia, 25, 342
bare soil, 125
barley cultivation, 27
Berger-Parker index, 172
biomass, 8, 14
    allocation, 99
    distribution, 91
    sequestration, 273
bofedales, 141, 144, 161, 169, 345
bogs, 218
Bolivia, 169

C

cattle, 8, 92, 157, 251
Chama River, 314
Changpa, 201
Changthang Plateau, 201
Chingaza Nature Park, 110
chuscales, 105
climate change, 53, 329
climatic zones, compression of, 5

cloud water, trapping, 68, 342
Colombia, 105, 112
competitive displacement, 165
Cordillera Central, 113
Cordillera de Mérida, Venezuela, 142, 189, 311, 317
Cordillera Oriental, 113
corrals, 142
cryptogams, 15
cultivated species, 327
Cumbres Calchaquies, 141
cushion, 189

D

deforestation, 229, 324
degradation, 242
dendroecology, 266, 278
desert steppe, 205
desertification, 227, 235
desirable species, 160
diet, 209
Dischma Valley, 251
diversity, Afroalpine plants, 41
    bofedales, 144
    elevational gradient, 62
    functional , 15, 189
    habitat , 5
    highland biota, 4
    hot spots, 287
    of plants, 29, 155
    small mammals, 81
    vertical gradient, 323
Drakensberg Park, 287
dynamic equilibrium model, 165

E

economic resources distribution, 325
ecosystem functioning, 3
Ecuador, 112
El Banco watershed, 314
El Nino-ENSO, 76, 178
electricity, 218
elevational gradient, 62
emigration from rural areas, 330
endangered species, 112
endemism, 112
ephemeroids, 246
equativity, 93
equativity of life-forms, 99
ericaceous bushes, 25
ericoid bush, 80
erosion, 160, 219, 229, 324
Ethiopia, 25, 41

351

Plant Species Index

Animal Species Index